TECHNOLOGY GUIDE

TO ACCOMPANY

UNDERSTANDABLE STATISTICS
EIGHTH EDITION
BRASE/BRASE

Yixun Shi
Bloomsburg University of Pennsylvania

Charles Henry Brase
Regis University

Corrinne Pellillo Brase
Arapahoe Community College

Laurel Tech Integrated Publishing Services

HOUGHTON MIFFLIN COMPANY   BOSTON   NEW YORK

Sponsoring Editor: Lauren Schultz
Development Editor: Catherine Mergen
Editorial Associate: Kasey McGarrigle
Editorial Assistant: Eric Moore
Senior Manufacturing Coordinator: Karen Banks
Senior Marketing Manager: Ben Rivera

Printed in the U.S.A.

ISBN: 0-618-50156-8

4 5 6 7 8 9-CRS-09 08 07

Contents

Part II: EXCEL Guide

Part III: MINITAB Guide

APPENDIX

xi

Preface

The use of computing technology can greatly enhance a student's learning experience in statistics. This *Technology Guide* provides basic instruction, examples, and lab activities for four different tools:

TI-83 Plus and TI-84 Plus
Microsoft® Excel with Analysis ToolPak for Windows®
MINITAB Version 14
SPSS Version 13

The TI-83 Plus and TI-84 Plus are versatile, widely-available graphing calculators made by Texas Instruments. This guide shows how to use their statistical functions, including plotting capabilities.

Excel is an all-purpose spreadsheet software package. This guide shows how to use Excel's built-in statistical functions and how to produce some useful graphs. Excel is not designed to be a complete statistical software package. In many cases, macros can be created to produce special graphs, such as box-and-whisker plots. However, this guide only shows how to use the existing, built-in features. In most cases, the operations omitted from Excel are easily carried out on an ordinary calculator. The Analysis ToolPak is part of Excel and can be installed from the same source as the basic Excel program (normally, a CD-ROM) as an option on the installer program's list of Add-Ins. Details for getting started with the Analysis ToolPak are in Chapter 1 of this guide. No additional software is required to use the Excel functions described.

MINITAB is a statistics software package suitable for solving problems. It can be packaged with the text. Contact Houghton Mifflin for details regarding price and platform options.

SPSS is a powerful tool that can perform many statistical procedures. This guide shows how the manage data and perform various statistical procedures using this software.

The lab activities that follow accompany the text *Understandable Statistics,* 8th edition by Brase and Brase. On the following page is a table to coordinate this guide with *Understanding Basic Statistics,* 3nd edition by Brase and Brase. Both texts are published by Houghton Mifflin.

In addition, over one hundred data files from referenced sources are described in the Appendix. These data files are available on the HM StatPass CD-ROM or via download from the Houghton Mifflin Web site

http://math.college.hmco.com/students

Understandable Statistics,
Eighth Edition
Charles Brase and Corrinne Pellillo Brase

Understanding the differences between Understandable Statistics and Understanding Basic Statistics

Understandable Statistics is the full, two-semester introductory statistics textbook, which is now in its Eighth Edition.

Understanding Basic Statistics is the brief, one-semester version of the larger book. It is currently in its Third Edition and is being revised for copyright year 2007.

Unlike other brief texts, Understanding Basic Statistics is not just the first six or seven chapters of the full text. Rather, topic coverage has been shortened in many cases and rearranged, so that the essential statistics concepts can be taught in one semester.

The major difference between the two tables of contents is that Regression and Correlation are covered much earlier in the brief book. In the full text, these topics are covered in Chapter 10. In the brief text, they are covered in Chapter 4.

Analysis of a Variance (ANOVA) is not covered in the brief text.

Each book has 12 chapters. The full text is a hardcover book, while the brief is softcover.

The same pedagogical elements are used throughout both texts.

The same supplements package is shared by both texts.

Following are the two Tables of Contents, side-by-side:

	Understandable Statistics (full)	Understanding Basic Statistics (brief)
Chapter 1	Getting Started	Getting Started
Chapter 2	Organizing Data	Organizing Data
Chapter 3	Averages and Variation	Averages and Variation
Chapter 4	Elementary Probability Theory	Correlation and Regression
Chapter 5	The Binomial Probability Distribution and Related Topics	Elementary Probability Theory
Chapter 6	Normal Distributions	The Binomial Probability Distribution and Related Topics
Chapter 7	Introduction to Sampling Distributions	Normal Distributions
Chapter 8	Estimation	Introduction to Sampling Distributions
Chapter 9	Hypothesis Testing	Estimation
Chapter 10	Correlation and Regression	Hypothesis Testing
Chapter 11	Chi-Square and F Distributions	Inferences About Differences
Chapter 12	Nonparametric Statistics	Additional Topics Using Inference

PART I

TI-83 PLUS and TI-84 PLUS GRAPHING CALCULATORS GUIDE

FOR

UNDERSTANDABLE STATISTICS

EIGHTH EDITION

CHAPTER 1 GETTING STARTED

ABOUT THE TI-83 PLUS AND TI-84 PLUS GRAPHING CALCULATORS

Calculators with built-in statistical support provide tremendous aid in performing the calculations required in statistical analysis. The Texas Instruments TI-83 Plus and TI-84 Plus graphing calculators have many features that are particularly useful in an introductory statistics course. Among the features are

(a) Data entry in a spreadsheet-like format

The TI-83 Plus and TI-84 Plus have six columns (called lists L_1, L_2, L_3, L_4, L_5, and L_6) in which data can be entered. The data in a list can be edited, and new lists can be created by doing arithmetic using existing lists.

Sample Data Screen

L1	L2	L3	1
11.2	8.1	1	
8.7	6.2	2	
6.8	3.6	-1	
3.2	.9	2	
4.5	1	1	
------	------	1	
		5	

L1(1)=11.2

(b) Single-variable statistics: mean, standard deviation, median, maximum, minimum, quartiles 1 and 3, sums

(c) Graphs for single-variable statistics: histograms, box-and-whisker plots

(d) Estimation: confidence intervals using the normal distribution, or Student's t distribution, for a single mean and for a difference of means; intervals for a single proportion and for the difference of two proportions

(e) Hypothesis testing: single mean (z or t); difference of means (z or t); proportions, difference of proportions, chi-square test of independence; two variances; linear regression, one-way ANOVA

(h) Two-variable statistics: linear regression

(i) Graphs for two-variable statistics: scatter diagrams, graph of the least-squares line

In this *Guide* we will show how to use many features on the TI-83 Plus and TI-84 Plus graphing calculators to aid you as you study particular concepts in statistics. **Lab Activities** coordinated to the text *Understandable Statistics* are also included.

USING THE TI-83 PLUS AND TI-84 PLUS

The TI-83 Plus and TI-84 Plus have several functions associated with each key, and they both use the same keystroke sequences to perform those functions. With TI-83 Plus, all the yellow items on the keypad are accessed by first pressing the yellow 2nd key. These keys are in blue color with TI-84 Plus. For both calculators, the items in green are accessed by first pressing the green ALPHA key. The four arrow keys enable you to move through a menu screen or along a graph. Below is the image of a TI-83 Plus calculator. The keypad of TI-84 Plus is exactly the same as that of TI-83 Plus.

The calculator uses screen menus to access additional operations. For instance, to access the statistics menu, press [STAT]. Your screen should look like this:

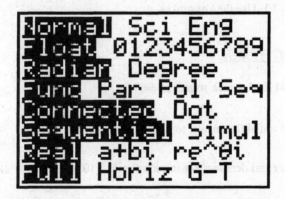

Use the arrow keys to highlight your selection. Pressing [ENTER] selects the highlighted item.

To leave a screen, press either [2nd] **[QUIT]** or [CLEAR], or select another menu from the keypad.

Now press [MODE]. When you use your calculator for statistics, the most convenient settings are as shown.

You can enter the settings by using the arrow keys to highlight selections and then pressing [ENTER].

COMPUTATIONS ON THE TI-83 PLUS AND TI-84 PLUS

In statistics, you will be evaluating a variety of expressions. The following examples demonstrate some basic keystroke patterns.

Example Evaluate $-2(3) + 7$.

Use the following keystrokes: [(-)][2][×][3][+][7][ENTER] The result is 1.

To enter a negative number, be sure to use the key $\boxed{(-)}$ rather than the subtract key. Notice that the expression **-2*3 + 7** appears on the screen. When you press $\boxed{\text{ENTER}}$, the result **1** is shown on the far right side of the screen.

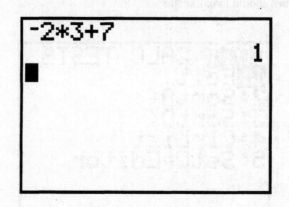

Example

(a) Evaluate $\dfrac{10-7}{3.1}$ and round the answer to three places after the decimal.

A reliable approach to evaluating fractions is to enclose the numerator in parentheses, and if the denominator contains more than a single number, enclose it in parentheses as well.

$\dfrac{10-7}{3.1} = (10 - 7) \div 3.1$ Use the keystrokes

$\boxed{(}\boxed{(}\boxed{1}\boxed{0}\boxed{-}\boxed{7}\boxed{)}\boxed{\div}\boxed{3}\boxed{.}\boxed{1}\boxed{\text{ENTER}}$.

The result is .9677419355, which rounds to .968.

(b) Evaluate $\dfrac{10-7}{\frac{3.1}{2}}$ and round the answer to three places after the decimal.

Place both numerator and denominator in parentheses: $(10 - 7) \div (3.1 \div 2)$.
Use the keystrokes

$\boxed{(}\boxed{(}\boxed{1}\boxed{0}\boxed{-}\boxed{7}\boxed{)}\boxed{\div}\boxed{(}\boxed{3}\boxed{.}\boxed{1}\boxed{\div}\boxed{2}\boxed{)}\boxed{\text{ENTER}}$.

The result is 1.935483871 or 1.935 rounded to three places after the decimal.

Example

Several formulas in statistics require that we take the square root of a value. Note that a left parenthesis, (, is automatically placed next to the square root symbol when $\boxed{\text{2nd}}\boxed{\sqrt{\ }}$ is pressed. Be sure to close the parentheses after typing in the radicand.

(a) Evaluate $\sqrt{10}$ and round the result to three places after the decimal.

$\boxed{\text{2nd}}\boxed{\sqrt{\ }}\boxed{1}\boxed{0}\boxed{)}\boxed{\text{ENTER}}$

The result is 3.16227766 and rounds to 3.162.

(b) Evaluate $\frac{\sqrt{10}}{3}$ and round the result to three places after the decimal.

[2nd][√][1][0][)][÷][3][ENTER]

The result rounds to 1.054.

Be careful to close the parentheses. If you do not close the parentheses, you will get the result of $\sqrt{\frac{10}{3}} \approx 1.826$.

Example Some expressions require us to use powers.

(a) Evaluate 3.2^2.

In this case, we can use the [x²] key.

[3][.][2][x²][ENTER]

The result is 10.24.

(b) Evaluate 0.4^3.

In this case we use the [^] key.

[.][4][^][3][ENTER]

The result is 0.064.

ENTERING DATA

To use the statistical processes built into the TI-83 Plus and TI-84 Plus, we first enter data into lists. Press the [STAT] key. Next we will clear any existing data lists. With EDIT highlighted, select **4:ClrList** using the arrow keys, and press [ENTER].

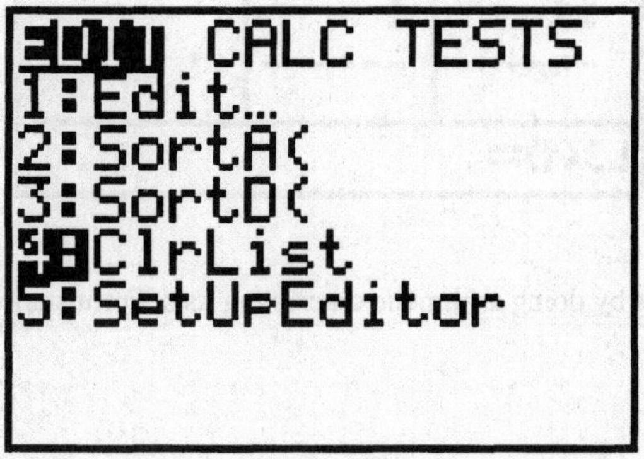

```
EDIT  CALC TESTS
1:Edit…
2:SortA(
3:SortD(
4:ClrList
5:SetUpEditor
```

Then type in the six lists separated by commas as shown. Press ENTER.

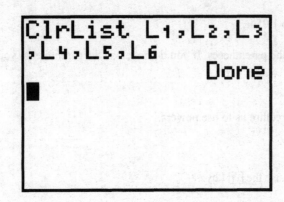

Next press STAT again and select item **1:Edit**. You will see the data screen and are now ready to enter data.

Let's enter the numbers

$$2 \quad 5 \quad 7 \quad 9 \quad 11 \quad \text{in list } L_1$$

$$3 \quad 6 \quad 9 \quad 11 \quad 1 \quad \text{in list } L_2$$

Press ENTER after each number, and use the arrow keys to move between L_1 and L_2.

To correct a data entry error, highlight the entry that is wrong and enter the correct data value.

```
L1        L2        L3      3
2         3         ███████
5         6
7         9
9         11
11        1
------    ------    ------

L3(1)=
```

We can also create new lists by doing arithmetic on existing lists. For instance, let's create L_3 by using the formula $L_3 = 2L_1 + L_2$.

Highlight L_3. Then type in $2L_1 + L_2$ and press ENTER.

L1	L2	▓	3
2	3	------	
5	6		
7	9		
9	11		
11	1		
------	------		

L3 =2L1+L2▓

The final result is shown.

L1	L2	L3	3
2	3	▓	
5	6	16	
7	9	23	
9	11	29	
11	1	23	
------	------	------	

L3(1)=7

To leave the data screen, press 2nd [QUIT] or press another menu key, such as STAT.

COMMENTS ON ENTERING AND CORRECTING DATA

(a) To create a new list from old lists, the lists must all have the same number of entries.

(b) To delete a data entry, highlight the data value you wish to correct and press the DEL key.

(c) To insert a new entry into a list, highlight the position directly below the place you wish to insert the new value. Press 2nd [INS] and then enter the data value.

LAB ACTIVITIES TO GET STARTED USING THE TI-83 PLUS AND TI-84 PLUS

1. Practice doing some calculations using the TI-83 Plus and TI-84 Plus: Be sure to use parentheses around a numerator or denominator that contains more than a single number. Round all answers to three places after the decimal.

(a) $\dfrac{5-2.3}{1.3}$ Ans. <u>2.077</u>

(b) $\dfrac{8-3.3}{\frac{3}{2}}$ Ans. <u>3.133</u>

(c) $-2(3.4)+5.8$ Ans. <u>-1</u>

(d) $-4(-1.7)-2.1$ Ans. <u>4.7</u>

(e) $\sqrt{5.3}$ Ans. <u>2.302</u>

(f) $\sqrt{6+3(2)}$ Ans. <u>3.464</u>

(g) $\sqrt{\dfrac{8-2.7}{5-1}}$ Ans. <u>1.151</u>

(h) 1.5^2 Ans. <u>2.25</u>

(i) $(5-7.2)^2$ Ans. <u>4.84</u>

(j) $(0.7)^3(0.3)^2$ Ans. <u>0.031</u>

2. Enter the following data into the designated list. Be sure to clear all lists first.

L_1: 3   7   9.2   12   −4

L_2: −2   9   4.3   16   10

L_3: $L_1 - 2$

L_4: $-2L_2 - L_1$

L_1: Change the data value in the second position to 6.

RANDOM SAMPLES (SECTION 1.2 OF *UNDERSTANDABLE STATISTICS*)

The TI-83 Plus and TI-84 Plus graphics calculators have a random number generator, which can be used in place of a random number table. Press [MATH]. Use the arrow keys to highlight **PRB**. Notice that **1:rand** is selected.

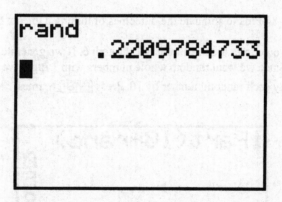

When you press [ENTER], rand appears on the screen. Press [ENTER] again and a random number between 0 and 1 appears.

```
rand
             .2209784733
```

To generate a random whole number with up to 3 digits, multiply the output of **rand** by 1000 and take the integer part. The integer-part command is found in the [MATH] menu under **NUM**, selection **3:iPart.** Press [ENTER] to display the command on the main screen.

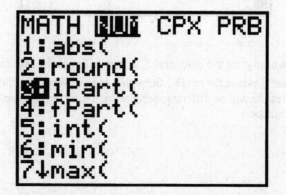

Now let's put all commands together to generate random whole numbers with up to 3 digits. First display **iPart.** Then enter **1000*rand** and close the parentheses. Now, each time you press ENTER, a new random number will appear. Notice that the numbers might be repeated.

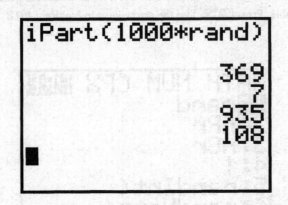

```
iPart(1000*rand)
                369
                  7
                935
                108
```

To generate random whole numbers with up to 2 digits, multiply **rand** by 100 instead of 1000. For random numbers with 1 digit, multiply **rand** by 10.

Simulating experiments in which outcomes are equally likely is an important use of random numbers.

Example

Use the TI-83 Plus or TI-84 Plus to simulate the outcomes of tossing a die six times. Record the results.

In this case, the possible outcomes are the numbers 1 through 6. If we generate a random number outside of this range we will ignore it. Since we want random whole numbers with 1 digit, we will follow the method prescribed earlier and multiply each random number by 10. Press ENTER 6 times.

```
iPart(10*rand)
                 0
                 5
                 8
                 9
                 2
                 2
```

Of the first six values listed, we only use the outcomes 5, 2, and 2. To get the remaining outcomes, keep pressing ENTER until three more digits in the range 1 through 6 appear. When we did this, the next three such digits were 1, 4, and 3. Your results will be different, because each time the random number generator is used, it gives a different sequence of numbers.

LAB ACTIVITIES FOR RANDOM SAMPLES

1. Out of a population of 800 eligible county residents, select a random sample of 20 for prospective jury duty. By what value should you multiply **rand** to generate random whole numbers with 3 digits? List the first 20 numbers corresponding to people for prospective jury duty.

2. We can stimulate dealing bridge hands by numbering the cards in a bridge deck from 1 to 52. Then we draw a random sample of 13 numbers without replacement from the population of 52 numbers. A bridge deck has 4 suits: hearts, diamonds, clubs, and spades. Each suit contains 13 cards: those numbered 2 through 10, a jack, a queen, a king, and an ace. Decide how to assign the numbers 1 through 52 to the cards in the deck. Use the random number generator on the TI-83 Plus or TI-84 Plus to get the numbers of the 13 cards in one hand. Translate the numbers to specific cards and tell what cards are in the hand. For a second game, the cards would be collected and reshuffled. Using random numbers from the TI-83 Plus or TI-84 Plus, determine the hand you might get in a second game.

CHAPTER 2 ORGANIZING DATA

HISTOGRAMS (SECTION 2.2 OF *UNDERSTANDABLE STATISTICS*)

The TI-83 Plus and TI-84 Plus graphing calculators draw histograms for data entered in the data lists. The calculators follow the convention that if a data value falls on a class limit, then it is tallied in the frequency of the bar to the right. However, if you specify the lower class boundary of the first class and the class width (see *Understandable Statistics* for procedures to find these values), then no data will fall on a class boundary. The following example shows you how to draw histograms.

Example

Throughout the day from 8 A.M. to 11 P.M., Tiffany counted the number of ads occurring every hour on one commercial T.V. station. The 15 data values are

$$10 \quad 12 \quad 8 \quad 7 \quad 15 \quad 6 \quad 5 \quad 8$$
$$8 \quad 10 \quad 11 \quad 13 \quad 15 \quad 8 \quad 9$$

To make a histogram with 4 classes:

First enter the data in L_1.

Next we will set the graphing window. However, we need to know both the lower class boundary for the first class and the class width. Use the techniques in Section 2.2 of *Understandable Statistics* to find these values.

The smallest data value is 5, so the lower class boundary of the first class is 4.5.

The class width is

$$\frac{\text{largest data value} - \text{smallest data value}}{\text{Number of classes}} \text{ } increased \text{ } to \text{ } the \text{ } next \text{ } integer$$

$$\frac{15 - 5}{4} = 2.5 \text{ } increased \text{ } to \text{ } 3$$

Now we set the graphing window. Press the [WINDOW] key.

Select **Xmin** and enter the *lower class boundary* of the first class. For this example, **Xmin = 4.5.**

Select **Xscl** and enter the *class width*. For this example, **Xscl = 3.**

Setting **Xmax = 17** ensures that the histogram fits on the screen, since our highest data value is 15.

Setting **Ymin = -5** and **Ymax = 15** makes sure the histogram fits comfortably on the screen.

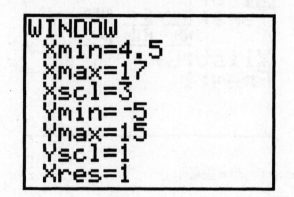

The next step is to select histogram style for the plot. Press [2nd] **[STAT PLOT].**

Highlight **1:Plot1** and press [ENTER].

Then on the next screen select **On** and the histogram shape.

In this example we stored the data in L_1. Set Xlist to L1 by pressing [2nd] **[L1].** Each data value is listed once for each time it occurs, so **1** is entered for **Freq.**

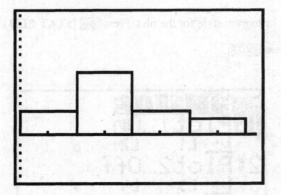

We are ready to graph the histogram. Press [GRAPH].

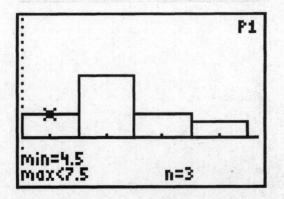

Press [TRACE] and notice that a blinking cursor appears over the first bar. The class boundaries of the bar are given as **min** and **max.** The frequency is given by **n.** Use the right and left arrow keys to move from bar to bar.

After you have graphed the histogram, you can use [TRACE] and the arrow keys to find the class boundaries and frequency of each class. The calculator has done the work of sorting the data and tallying the data for each class.

LAB ACTIVITIES FOR HISTOGRAMS

1. A random sample of 50 professional football players produced the following data.

Weights of Pro Football Players (data file Svls02.txt)

The following data represents weights in pounds of 50 randomly selected pro football linebackers.

Source: The Sports Encyclopedia Pro Football 1960–1992

225	230	235	238	232	227	244	222
250	226	242	253	251	225	229	247
239	223	233	222	243	237	230	240
255	230	245	240	235	252	245	231
235	234	248	242	238	240	240	240
235	244	247	250	236	246	243	255
241	245						

Enter this data in L_1. Scan the data for the low and high values. Compute the class width for 5 classes and find the lower class boundary for the first class. Use the TI-83 Plus or TI-84 Plus to make a histogram with 5 classes. Repeat the process for 9 classes. Is the data distribution skewed or symmetric? Is the distribution shape more pronounced with 5 classes or with 9 classes?

2. Explore the other data files found in the Appendix of this *Guide*, such as

       Disney Stock Volume (data file Svls01.txt)

       Heights of Pro Basketball Players (data file Svls03.txt)

       Miles per Gallon Gasoline Consumption (data file Svls04.txt)

       Fasting Glucose Blood Tests (data file Svls05.txt)

       Number of Children in Rural Families (data file Svls06.txt)

Select one of these files and make a histogram with 7 classes. Comment on the histogram shape.

3. (a) Consider the data

1	3	7	8	10
6	5	4	2	1
9	3	4	5	2

Place the data in L_1. Use the TI-83 Plus or TI-84 Plus to make a histogram with $n = 3$ classes. Jot down the class boundaries and frequencies so that you can compare them to part (b).

(b) Now add 20 to each data value of part (a). The results are

$$
\begin{array}{ccccc}
21 & 23 & 27 & 28 & 30 \\
26 & 25 & 24 & 22 & 21 \\
29 & 23 & 24 & 25 & 22
\end{array}
$$

By using arithmetic in the data list screen, you can create L_2 and let $L_2 = L_1 + 20$.

Make a histogram with 3 classes. Compare the class boundaries and frequencies with those obtained in part (a). Are each of the boundary values 20 more than those of part (a)? How do the frequencies compare?

(c) Use your discoveries from part (b) to predict the class boundaries and class frequency with 3 classes for the data values below. What do you suppose the histogram will look like?

$$
\begin{array}{ccccc}
1001 & 1003 & 1007 & 1008 & 1010 \\
1006 & 1005 & 1004 & 1002 & 1001 \\
1009 & 1003 & 1004 & 1005 & 1002
\end{array}
$$

Would it be safe to say that we simply shift the histogram of part (a) 1000 units to the right?

(d) What if we multiply each of the values of part (a) by 10? Will we effectively multiply the entries for using class boundaries by 10? To explore this question create a new list L_3 using the formula

$L_3 = 10L_1$. The entries will be

$$
\begin{array}{ccccc}
10 & 30 & 70 & 80 & 100 \\
60 & 50 & 40 & 20 & 10 \\
90 & 30 & 40 & 50 & 20
\end{array}
$$

Compare the histogram with three classes to the one from part (a). You will see that there does not seem to be an exact correspondence. To see why, look at the class width and compare it to the class width of part (a). The class width is always increased to the next integer value no matter how large the integer data values are. Consequently, the class width for the data in part (d) was increased to 31 rather than 40.

4. Histograms are not effective displays for some data. Consider the data

$$
\begin{array}{cccccc}
1 & 2 & 3 & 6 & 5 & 7 \\
9 & 8 & 4 & 12 & 11 & 15 \\
14 & 12 & 6 & 2 & 1 & 206
\end{array}
$$

Use the TI-83 Plus or TI-84 Plus to make a histogram with 2 classes. Then change to 3 classes, on up to 10 classes. Notice that all the histograms lump the first 17 data values into the first class, and the one data value, 206, in the last class. What feature of the data causes this phenomenon? Recall that

$$
\text{Class width} = \frac{\text{largest value} - \text{smallest value}}{\text{number of class}} \; \textit{increased to the next integer.}
$$

How many classes would you need before you began to see the first 17 data values distributed among several classes? What would happen if you simply did not include the extreme value 206 in your histogram?

CHAPTER 3 AVERAGES AND VARIATION

ONE-VARIABLE STATISTICS (SECTIONS 3.1 AND 3.2 OF *UNDERSTANDABLE STATISTICS*)

The TI-83 Plus and TI-84 Plus graphing calculators support many of the common descriptive measures for a data set. The measured supported are

Mean $\bar{x}$

Sample standard deviations S_x

Population standard deviation σ_x

Number of data values n

Median

Minimum data value

Maximum data value

Quartile 1

Quartile 3

$\sum x$

$\sum x^2$

Although the mode is not provided directly, the menu choice **SortA** sorts the data in ascending order. By scanning the sorted list, you can find the mode fairly quickly.

Example

At Lazy River College, 15 students were selected at random from a group registering on the last day of registration. The times (in hours) necessary for these students to complete registration follow:

1.7   2.1   0.8   3.5   1.5   2.6   2.1   2.8
3.1   2.1   1.3   0.5   2.1   1.5   1.9

Use the TI-83 Plus or TI-84 Plus to find the mean, sample standard deviation, and median. Use the **SortA** command to sort the data and scan for the mode, if it exists.

First enter the data into list L_1.

Press [STAT] again, and highlight **CALC** with item **1:1-Var Stats.**

```
EDIT CALC TESTS
1:1-Var Stats
2:2-Var Stats
3:Med-Med
4:LinReg(ax+b)
5:QuadReg
6:CubicReg
7↓QuartReg
```

Press [ENTER]. The command **1-Var Stats** will appear on the screen. Then tell the calculator to use the data in list L_1 by pressing [2nd] [L_1].

```
1-Var Stats L₁
```

Press [ENTER]. The next two screens contain the statistics for the data in List L_1.

Note the arrow ↓ on the last line. This is an indication that you may use the down-arrow key to display more results.

```
1-Var Stats
x̄=1.97333333
Σx=29.6
Σx²=67.68
Sx=.8136923486
σx=.7861014919
↓n=15
```

Scroll down. The second full screen shows the rest of the statistics for the data in list L_1.

```
1-Var Stats
↑n=15
 minX=.5
 Q₁=1.5
 Med=2.1
 Q₃=2.6
 maxX=3.5
■
```

To sort the data, first press [STAT] to return to the edit menu. Then highlight **EDIT** and item **2:SortA(.**

```
EDIT CALC TESTS
1:Edit…
2:SortA(
3:SortD(
4:ClrList
5:SetUpEditor
```

Press [ENTER]. The command **SortA** appears on the screen. Type in L_1.

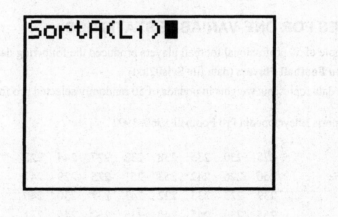

```
SortA(L₁)■
```

When you press [ENTER], the comment **Done** appears on the screen.

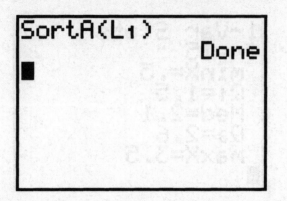

Press [STAT] and return to the data entry screen. Notice that the data in list L_1 is now sorted in ascending order.

A scan of the data shows that the mode is 2.1, since this data value occurs more than any other.

LAB ACTIVITIES FOR ONE-VARIABLE STATISTICS

1. A random sample of 50 professional football players produced the following data.

Weights of Pro Football Players (data file Svls02.txt)

The following data represents weights in pounds of 50 randomly selected pro football linebackers.

Source: The Sports Encyclopedia Pro Football 1960–1992

225	230	235	238	232	227	244	222
250	226	242	253	251	225	229	247
239	223	233	222	243	237	230	240
255	230	245	240	235	252	245	231
235	234	248	242	238	240	240	240
235	244	247	250	236	246	243	255
241	245						

Enter this data in L_1. Use **1-Var Stats** to find the mean, median, and standard deviation for the weights. Use **SortA** to sort the data and scan for a mode if one exists.

2. Explore some of the other data sets found in the Appendix of this *Guide*, such as

Disney Stock Volume (data file Svls01.txt)

Heights of Pro Basketball Players (data file Svls03.txt)

Miles per Gallon Gasoline Consumption (data file Svls04.txt)

Fasting Glucose Blood Tests (data file Svls05.txt)

Number of Children in Rural Families (data file Svls06.txt)

Use the TI-83 Plus or TI-84 Plus to find the mean, median, standard deviation, and mode of the data.

3. In this problem we will explore the effects of changing data values by multiplying each data value by a constant, or by adding the same constant to each data value.

(a) Consider the data

1	8	3	5	7
2	10	9	4	6
3	5	2	9	1

Enter the data into list L_1 and use the TI-83 Plus or TI-84 Plus to find the mode (if it exits), mean, sample standard deviation, range, and median. Make a note of these values, since you will compare them to those obtained in parts (b) and (c).

(b) Now multiply each data value of part (a) by 10 to obtain the data

10	80	30	50	70
20	100	90	40	60
30	50	20	90	10

Remember, you can enter these data in L_2 by using the command $L_2 = 10L_1$.

Again, use the TI-83 Plus or TI-84 Plus to find the mode (if it exists), mean, sample standard deviation, range, and median. Compare these results to the corresponding ones of part (a). Which values changed? Did those that changed change by a factor of 10? Did the range or standard deviation change? Referring to the formulas for these measures (see Section 3.2 of *Understandable Statistics*), can you explain why the values behaved the way they did? Will these results generalize to the situation of multiplying each data entry by 12 instead of by 10? What about multiplying each by 0.5? Predict the corresponding values that would occur if we multiplied the data set of part (a) by 1000.

(c) Now suppose we add 30 to each data value of part (a)

31	38	33	35	37
32	40	39	34	36
33	35	32	39	31

To enter this data, create a new list L_3 by using the command $L_3 = L_1 + 30$.

Again use the TI-83 Plus or TI-84 Plus to find the mode (if it exists), mean, sample standard deviation, range, and median. Compare these results to the corresponding ones of part (a). Which values changed? Of those that are different, did each change by being 30 more than the corresponding value of part (a)? Again look at the formulas for range and standard deviation. Can you predict the observed behavior from the formulas? Can you generalize these results? What if we added 50 to each data value of part (a)? Predict the values for the mode, mean, sample standard deviation, range, and median.

GROUPED DATA (SECTION 3.3 OF *UNDERSTANDABLE STATISTICS*)

The TI-83 Plus and TI-84 Plus support grouped data, because it allows you to specify the frequency with which each data value occurs in a separate list. The frequencies must be *whole* numbers. This means that numbers with decimal parts are not permitted as frequencies.

Example

A random sample of 44 automobiles registered in Dallas, Texas show the ages of the cars to be

Age (in years)	Midpoint	Frequency
1 – 3	2	9
4 – 6	5	12
7 – 9	8	20
10–12	11	3

Estimate the mean and standard deviation of the ages of the cars.

Put the midpoints in list L_1 and the corresponding frequencies in list L_2.

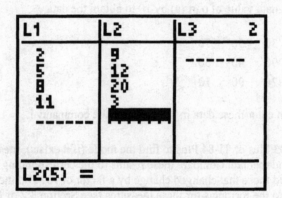

Press STAT and, with **CALC** highlighted, select item **1:1-Var Stats**.

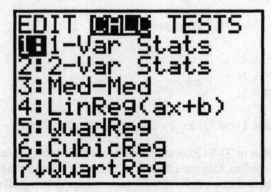

Designate the data lists, with the first list holding the data and the second list holding the frequencies.

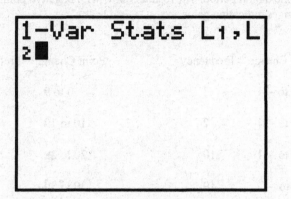

When you press ENTER, the statistics for the grouped data will appear. Note that $n = 44$ is the correct value for the number of data values.

1-Var Stats
$\bar{x}$=6.159090909
Σx=271
Σx^2=1979
Sx=2.684522233
σx=2.653840971
↓n=44

From the last screen, we see that the estimate for the mean is 6.16 and the estimate for the sample standard deviation is 2.68.

LAB ACTIVITIES FOR GROUPED DATA

1. A teacher rating form uses the scale values –2, –1, 0, 1, 2 with –2 meaning "strongly disagree" and 2 meaning "strongly agree." The statement "The professor grades fairly" was answered as follows for all courses in the sociology department.

Score:	–2	–1	0	1	2
Number of responses:	30	125	200	180	72

Use the TI-83 Plus or TI-84 Plus to calculate the mean response and the standard deviation for this statement.

2. A stock market analyst looked at 80 key stocks and recorded the total gains and losses (in points) per share for each stock over a one month period. The results follow with negative point changes indicating losses and positive point changes indicating gains.

Point Change	Frequency	Point Change	Frequency
−40 to −31	3	0 to 9	20
−30 to −21	2	10 to 19	14
−20 to −11	10	20 to 29	7
−10 to −1	19	30 to 39	5

First find the class midpoints. Then use the TI-83 Plus or TI-84 Plus to estimate the mean point change and standard deviation for this *population* of stocks.

BOX-AND-WHISKER PLOTS (SECTION 3.4 OF *UNDERSTANDABLE STATISTICS*)

The box-and-whisker plot is based on the five-number summary values found under **1-Var Stats:**

Lowest value

Quartile 1, or Q_1

Median

Quartile 3, or Q_3

Highest value

Example

Let's make a box-and-whisker plot using the data about the time it takes to register for Lazy River College students who waited till the last day to register. The times (in hours) are

1.7	2.1	0.8	3.5	1.5	2.6	2.1	2.8
3.1	2.1	1.3	0.5	2.1	1.5	1.9	

In a previous example we found the **1-Var Stats** for this data. First enter the data into list L_1.

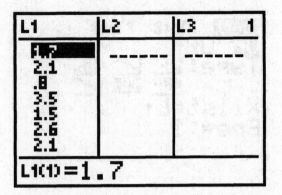

Press WINDOW to set the graphing window. Use Xmin = 0.5, since that is the smallest value. Use Xmax = 3.5, since that is the largest value. Use Xscl = 1. Use Ymin = −5 and Ymax = 5 to position the axis in the window.

```
WINDOW
 Xmin=.5
 Xmax=3.5
 Xscl=1
 Ymin=-5
 Ymax=5
 Yscl=1
 Xres=1
```

Press 2nd [STAT PLOT] and highlight **1:Plot.** Press ENTER.

```
STAT PLOTS
1:Plot1...On
    L1      1
2:Plot2...Off
    L3   L4    +
3:Plot3...Off
    L1   L2    □
4↓PlotsOff
```

Highlight **On,** and select the box plot. The Xlist should be L_1 and Freq should be **1.**

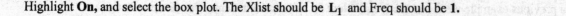

Press GRAPH .

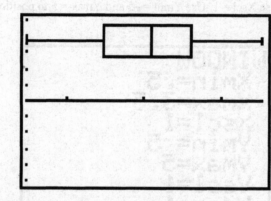

Press TRACE and use the arrow keys to display the five-number summary.

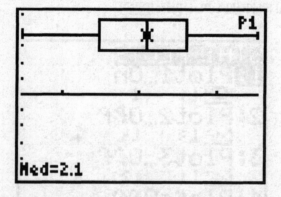

You now have many descriptive tools available: histograms, box-and-whisker plots, averages, and measures of variation such as the standard deviation. When you use all of these tools, you get a lot of information about the data.

LAB ACTIVITIES FOR BOX-AND-WHISKER PLOTS

1. One of the data sets included in the Appendix gives the miles per gallon gasoline consumption for a random sample of 55 makes and models of passenger cars.

   Miles per Gallon Gasoline Consumption (data file Svls04.txt)

   The following data represents miles per gallon gasoline consumption (highway) for a random sample of 55 makes and models of passenger cars.

   Source: Environmental Protection Agency

30	27	22	25	24	25	24	15
35	35	33	52	49	10	27	18
20	23	24	25	30	24	24	24
18	20	25	27	24	32	29	27
24	27	26	25	24	28	33	30
13	13	21	28	37	35	32	33
29	31	28	28	25	29	31	

   Enter these data into list L_1, and then make a histogram and a box-and-whisker plot, compute the mean, median, and sample standard deviation. Based on the information you obtain, respond to the following questions:

   (a) Is the distribution skewed or symmetric? How is this shown in both the histogram and the box-and-whisker plot?

   (b) Look at the box-and-whisker plot. Are the data more spread out above the median or below the median?

   (c) Look at the histogram and estimate the location of the mean on the horizontal axis. Are the data more spread out above the mean or below the mean?

   (d) Do there seem to be any data values that are unusually high or unusually low? If so, how do these show up on a histogram or on a box-and-whisker plot?

   (e) Pretend that you are writing a brief article for a newspaper. Describe the information about the data in non-technical terms. Be sure to make some comments about the "average" of the data values and some comments about the spread of the data.

2. (a) Consider the test scores of 30 students in a political science class.

85	73	43	86	73	59	73	84	62	100
75	87	70	84	97	62	76	89	90	83
70	65	77	90	84	80	68	91	67	79

   For this population of test scores find the mode, median, mean, range, variance, standard deviation, CV, the five-number summary, and make a box-and-whisker plot. Be sure to record all of these values so you can compare them to the results of part (b).

(b) Suppose Greg was in the political science class of part (a). Suppose he missed a number of classes because of illness, but took the exam anyway and made a score of 30 instead of 85 as listed as the first entry of the data in part (a). Again, use the calculator to find the mode, median, mean, range, variance, standard deviation, *CV*, the five-number summary, and make a box-and-whisker plot using the new data set. Compare these results to the corresponding results of part (a). Which average was most affected: mode, median, or mean? What about the range, standard deviation, and coefficient of variation? How do the box-and-whisker plots compare?

(c) Write a brief essay in which you use the results of parts (a) and (b) to predict how an extreme data value affects a data distribution. What do you predict for the results if Greg's test score had been 80 instead of 30 or 85?

CHAPTER 4 ELEMENTARY PROBABILITY THEORY

There are no specific TI-83 Plus or TI-84 Plus activities for the basic rules of probability. However, notice that the TI-83 Plus and TI-84 Plus have menu items for factorial notation, combinations $C_{n,r}$ and permutations $P_{n,r}$. To find these functions, press [MATH] and then highlight **PRB**.

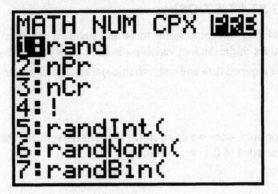

To compute $C_{12,6}$, first clear the screen and type 12. Then use the [MATH] key to select **PRB** and item **3:nCr**. Press [ENTER]. Type the number 6 and then press [ENTER].

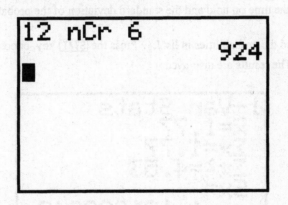

The menu option **nPr** works in a similar fashion and gives you the value of $P_{n,r}$.

CHAPTER 5 THE BINOMIAL PROBABILITY DISTRIBUTION AND RELATED TOPICS

DISCRETE PROBABILITY DISTRIBUTIONS (SECTION 5.1 OF *UNDERSTANDABLE STATISTICS*)

The TI-83 Plus and TI-84 Plus will compute the mean and standard deviation of a discrete probability distribution. Just enter the values of the random variable x in list L_1 and the corresponding probabilities in list L_2. Use the techniques of grouped data and one-variable statistics to compute the mean and standard deviation.

Example

How long do we spend on hold when we call a store? One study produced the following probability distribution, with the times recorded to the nearest minute.

Time on hold, x:	0	1	2	3	4	5
$P(x)$:	0.15	0.25	0.40	0.10	0.08	0.02

Find the expected value of the time on hold and the standard deviation of the probability distribution.

Enter the times in L_1 and the probabilities in list L_2. Press the $\boxed{\text{STAT}}$ key, choose **CALC** and use **1-Var Stats** with lists L_1 and L_2. The results are displayed.

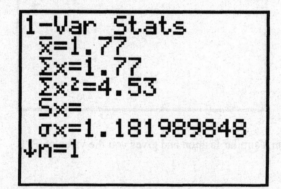

The expected value is $\bar{x} = 1.77$ minutes with standard deviation $\sigma = 1.182$.

LAB ACTIVITIES FOR DISCRETE PROBABILITY DISTRIBUTIONS

1. The probability distribution for scores on a mechanical aptitude test is

Score, x:	0	10	20	30	40	50
$P(x)$:	0.130	0.200	0.300	0.170	0.120	0.080

Use the TI-83 Plus or TI-84 Plus to find the expected value and standard deviation of the probability distribution.

2. Hold Insurance has calculated the following probabilities for claims on a new $20,000 vehicle for one year if the vehicle is driven by a single male under 25.

$ Claims, x:	0	1000	5000	100,000	200,000
P(x):	0.63	0.24	0.10	0.02	0.01

What is the expected value of the claim in one year? What is the standard deviation of the claim distribution? What should the annual premium be to include $400 in overhead and profit as well as the expected claim?

BINOMIAL PROBABILITIES (SECTION 5.2 OF *UNDERSTANDABLE STATISTICS*)

For a binomial distribution with n trials, r successes, and the probability of success on a single trial p, the formula for the probability of r successes out of n trials is

$$P(r) = C_{n,r} p^r (1-p)^{n-r}$$

To compute probabilities for specific values of n, r, and p, we can use the TI-83 Plus or TI-84 Plus graphing calculator.

Example

Consider a binomial experiment with 10 trials and probability of success on a single trial $p = 0.72$. Compute the probability of 7 successes out of n trials.

On the TI-83 Plus or TI-84 Plus, the combinations function **nCr** is found in the MATH menu under **PRB**.

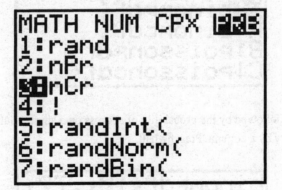

To use the function, we first type the value of n, then **nCr**, and finally the value of r. Here, $n = 10$, $r = 7$, $p = 0.72$ and $(1 - p) = 0.28$. By the formula, we raise 0.72 to the power 7 and 0.28 to the power 3. The probability of 7 successes is 0.264.

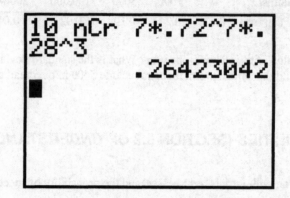

Using the Probability Distributions on the TI-83 Plus and TI-84 Plus

The TI-83 Plus and TI-84 Plus fully support the binomial distribution and has it built in as a function. To access the menu that contains the probability functions, press [2nd] [DISTR]. Then scroll to item **0:binompdf(** and press [ENTER].)

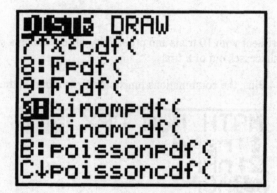

Type in the number of trials, followed by the probability of success on a single trial, followed by the number of successes. Separate each entry by a comma. Press [ENTER].

LAB ACTIVITIES FOR BINOMIAL PROBABILITIES

1. Consider a binomial distribution with $n = 8$ and $p = 0.43$. Use the formula to find the probability of r successes for r from 0 through 8.

2. Consider a binomial distribution with $n = 8$ and $p = 0.43$. Use the built-in binomial probability distribution function to find the probability of r successes for r from 0 through 8.

CHAPTER 6 NORMAL DISTRIBUTIONS

CONTROL CHARTS (SECTION 6.1 OF *UNDERSTANDABLE STATISTICS*)

Although the TI-83 Plus or TI-84 Plus does not have a control chart option built into **STAT PLOT,** we can use one of the features in **STAT PLOT** combined with the regular graphing options to create a control chart.

Example

Consider the data from the data files in the Appendix regarding yield of wheat at Rothamsted Experiment Station over a period of thirty years. Use the TI-83 Plus or TI-84 Plus to make a control chart for this data using the target mean and standard deviation values.

Yield of Wheat at Rothamsted Expreiment Station, England (data file Tscc01.txt)

The following data represent annual yield of wheat in tonnes (one ton = 1.016 tonne)
for an experimental plot of land at Rothamsted Experiment Station U.K. over a period
of thirty consecutive years.

Source: Rothamsted Experiment Station U.K.

We will use the following target production values:
target mu = 2.6 tonnes
target sigma = 0.40 tonnes

1.73	1.66	1.36	1.19	2.66	2.14	2.25	2.25	2.36	2.82
2.61	2.51	2.61	2.75	3.49	3.22	2.37	2.52	3.43	3.47
3.20	2.72	3.02	3.03	2.36	2.83	2.76	2.07	1.63	3.02

First we enter the data by row. In list L_1 put the year numbers 1 through 30. In list L_2 put the corresponding annual yield. Again, read the data by row.

L1	L2	L3	1
1	1.73	------	
2	1.66		
3	1.36		
4	1.19		
5	2.66		
6	2.14		
7	2.25		

L1(1)=1

Then press [2nd] [STAT PLOT] and select **1:Plot1**.

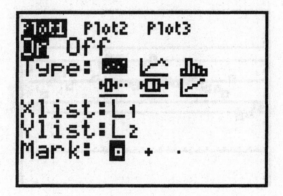

Highlight **On**. Select scatter diagram for type, L_1 for **Xlist**, and L_2 for **Ylist**. Select the symbol you like for **Mark**.

Next Press [Y=].

For Y_1 enter the target mean. In this case, enter 2.6.

For Y_2 enter the mean + 2 standard deviations. In this case, enter 2.6 + 2(.4).

For Y_3 enter the mean – 2 standard deviations. In this case, enter 2.6 – 2(.4).

For Y_4 enter the mean + 3 standard deviations. In this case, enter 2.6 + 3(.4).

For Y_5 enter the mean – 3 standard deviations. In this case, enter 2.6 – 3(.4).

The last step is to set the graphing window. Press WINDOW. Since there were 30 years, set **Xmin** to 1 and **Xmax** to 30. To position the graph in the window, set **Ymin** to –1 and **Ymax** to 5.

```
WINDOW
 Xmin=1
 Xmax=30
 Xscl=1
 Ymin=-1
 Ymax=5
 Yscl=1
 Xres=1
```

When you press GRAPH, the control chart appears.

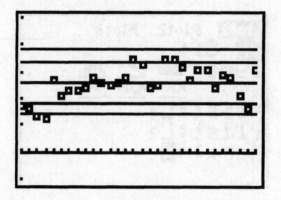

LAB ACTIVITIES FOR CONTROL CHARTS

1. Look in the Appendix and find the data file for **Futures Quotes for the Price of Coffee Beans** (Tscc04.txt). Make a control chart using the data and the target mean and target standard deviation given. Read the data by row. Are there any out-of-control signals? Explain.

2. Look in the Appendix and find the data file for **Incidence of Melanoma Tumors** (Tscc05.txt). Make a control chart using the data and the target mean and target standard deviation given. Read the data by row. Are there any out-of-control signals? Explain.

THE AREA UNDER ANY NORMAL CURVE (SECTION 6.3 OF *UNDERSTANDABLE STATISTICS*)

The TI-83 Plus and TI-84 Plus give the area under any normal distribution and shades the specified area. Press [2nd] **[DISTR]** to access the probability distributions. Select **2:normalcdf(** and press [ENTER].)

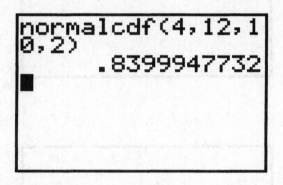

Then type in the lower bound, upper bound, μ, and σ in that order separated by commas. To find the area under the normal curve with $\mu = 10$ and $\sigma = 2$ between 4 and 12, select **normalcdf(** and enter the values as shown. Then press [ENTER].)

```
normalcdf(4,12,1
0,2)
          .8399947732
■
```

Drawing the normal distribution

To draw the graph, first set the window to accommodate the graph. Press the [WINDOW] button and enter values as shown.

```
WINDOW
 Xmin=2
 Xmax=18
 Xscl=3
 Ymin=-.1
 Ymax=.3
 Yscl=1
 Xres=1
```

Press [2nd] [DISTR] again and highlight **DRAW.** Select **1:ShadeNorm(.**

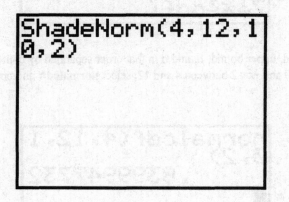

Again enter the lower limit, upper limit, μ, and σ separated by commas.

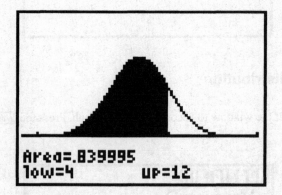

Finally press [ENTER].

LAB ACTIVITIES FOR THE AREA UNDER ANY NORMAL CURVE

1. Find the area under a normal curve with mean 10 and standard deviation 2, between 7 and 9. Show the shaded region.

2. To find the area in the right tail of a normal distribution, select the value of 5σ for the upper limit of the region. Find the area under a normal curve with mean 10 and standard deviation 2 that lies to the right of 12.

3. Consider a random variable x that follows a normal distribution with mean 100 and standard deviation 15. Shade the regions corresponding to the probabilities and find

   (a) Shade the region corresponding to $P(x < 90)$ and find the probability.

   (b) Shade the region corresponding to $P(70 < x < 100)$ and find the probability.

   (c) Shade the region corresponding to $P(x > 115)$ and find the probability.

   (d) If the random variable were larger than 145, would that be an unusual event? Explain by computing $P(x > 145)$ and commenting on the meaning of the result.

CHAPTER 7
INTRODUCTION TO SAMPLING DISTRIBUTIONS

In this chapter, use the TI-83 Plus or TI-84 Plus graphing calculator to do computations. For example, to compute the z score corresponding to a raw score from an $\bar{x}$ distribution, we use the formula

$$z = \frac{\bar{x} - \mu}{\frac{\sigma}{\sqrt{n}}}$$

To evaluate z, be sure to use parentheses as necessary.

Example

If a random sample of size 40 is taken from a distribution with mean $\mu = 10$ and standard deviation $\sigma = 2$, find the z score corresponding to $x = 9$.
We use the formula

$$z = \frac{9 - 10}{\frac{2}{\sqrt{40}}}$$

Key in the expression using parentheses: $(9 - 10) \div (2 \div \sqrt{\ } (40)$ [ENTER].

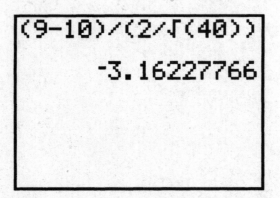

The result rounds to $z = -3.16$.

CHAPTER 8 ESTIMATION

CONFIDENCE INTERVALS FOR A POPULATION MEAN (SECTION 8.1 AND 8.2 OF *UNDERSTANDABLE STATISTICS*)

The TI-83 Plus and TI-84 Plus fully support confidence intervals. To access the confidence interval choices, press [STAT] and select **TESTS**. The confidence interval choices are found in items 7 through B:

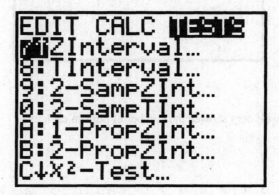

Example (σ is known with a Large Sample)

Suppose a random sample of 250 credit card bills showed an average balance of $1200. Also assume that the population standard deviation is $350. Find a 95% confidence interval for the population mean credit card balance.

Since σ is known and we have a large sample, we use the normal distribution. Select **7:ZInterval.**

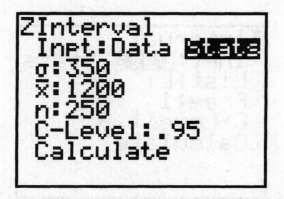

In this example, we have summary statistics, so we will select the STATS option for input. We enter the value of σ, the value of $\overline{x}$ and the sample size n. Use 0.95 for the C-Level.

Highlight **Calculate** and press ENTER to get the results. Notice that the interval is given using standard mathematical notation for an interval. The interval for μ goes from \$1156.6 to \$1232.4.

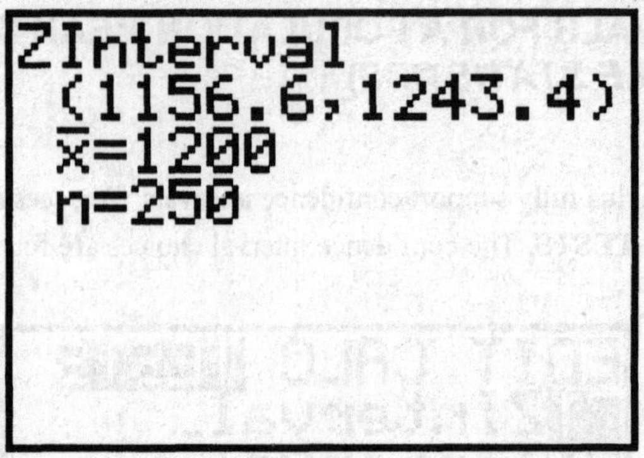

```
ZInterval
 (1156.6,1243.4)
x̄=1200
n=250
```

Example (σ is unknown)

A random sample of 16 wolf dens showed the number of pups in each to be

> 5 8 7 5 3 4 3 9
>
> 5 8 5 6 5 6 4 7

Find a 90% confidence interval for the population mean number of pups in such dens.

In this case we have raw data, so enter the data in list L_1 using the **EDIT** option of the STAT key. Since σ is unknown, we use the t distribution. Under **Tests** from the STAT menu, select item **8:TInterval.** Since we have raw data, select the DATA option for Input. The data is in list L_1, and occurs with frequency 1. Enter 0.90 for the C-Level.

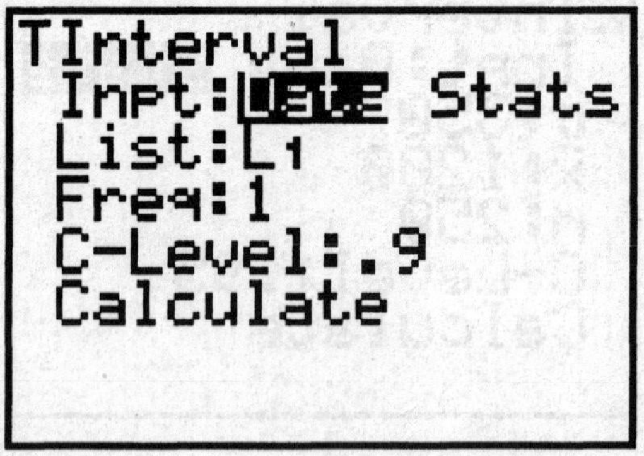

```
TInterval
Inpt:DATA Stats
List:L₁
Freq:1
C-Level:.9
Calculate
```

Highlight Calculate and press [ENTER]. The result is the interval from 4.84 pups to 6.41 pups.

```
TInterval
 (4.8431,6.4069)
x̄=5.625
Sx=1.784189825
n=16
```

LAB ACTIVITIES FOR CONFIDENCE INTERVALS FOR A POPULATION MEAN

1. Markey Survey was hired to do a study for a new soft drink, Refresh. 20 people were given a can of Refresh and asked to rate it for taste on a scale of 1 to 10 (with 10 being the highest rating). The ratings were

5	8	3	7	5	9	10	6	6	2
9	2	1	8	10	2	5	1	4	7

   Find an 85% confidence interval for the population mean rating of Refresh.

2. Suppose a random sample of 50 basketball players showed the average height to be 78 inches. Also assume that the population standard deviation is 1.5 inches.

   (a) Find a 99% confidence interval for the population mean height.
   (b) Find a 95% confidence interval for the population mean height.
   (c) Find a 90% confidence interval for the population mean height.
   (d) Find an 85% confidence interval for the population mean height.
   (e) What do you notice about the length of the confidence interval as the confidence level goes down? If you used a confidence level of 80%, would you expect the confidence interval to be longer or shorter than that of 85%? Run the program again to verify your answer.

CONFIDENCE INTERVALS FOR THE PROBABILITY OF SUCCESS p IN A BINOMIAL DISTRIBUTION (SECTION 8.3 OF *UNDERSTANDABLE STATISTICS*)

To find a confidence interval for a proportion, press the [STAT] key and use option **A:1-PropZInt** under **TESTS.** Notice that the normal distribution will be used.

Example:

The public television station BPBS wants to find the percent of its viewing population who give donations to the station. 300 randomly selected viewers were surveyed, and it was found that 123 made contributions to the station. Find a 95% confidence interval for the probability that a viewer of BPBS selected at random contributes to the station.

The letter x is used to count the number of successes (the letter r is used in the text). Enter 123 for x and 300 for n. Use 0.95 for the C-level.

```
1-PropZInt
 x:123
 n:300
 C-Level:.95
 Calculate
```

Highlight **Calculate** and press ENTER. The result is the interval from 0.35 to 0.47.

```
1-PropZInt
 (.35434,.46566)
 p̂=.41
 n=300
```

LAB ACTIVITIES FOR CONFIDENCE INTERVALS FOR THE PROBABILITY OF SUCCESS p IN A BINOMIAL DISTRIBUTION

1. Many types of error will cause a computer program to terminate or give incorrect results. One type of error is punctuation. For instance, if a comma is inserted in the wrong place, the program might not run. A study of programs written by students in an introductory programming course showed that 75 out of 300 errors selected at random were punctuation errors. Find a 99% confidence interval for the proportion of errors made by beginning programming students that are punctuation errors. Next, find a 90% confidence interval. Is this interval longer or shorter?

2. Sam decided to do a statistics project to determine a 90% confidence interval for the probability that a student at West Plains College eats lunch in the school cafeteria. He surveyed a random sample of 12 students and found that 9 ate lunch in the cafeteria. Can Sam use the program to find a confidence interval for the population proportion of students eating in the cafeteria? Why or why not? Try **1-PropZInt** with $n = 12$ and $r = 9$. What happens? What should Sam do to complete his project?

CONFIDENCE INTERVALS FOR $\mu_1 - \mu_2$ (INDEPENDENT SAMPLES)
CONFIDENCE INTERVALS FOR $p_1 - p_2$ (LARGE SAMPLES) (SECTION 8.5 OF *UNDERSTANDABLE STATISTICS*)

For tests of difference of means, press the $\boxed{\text{STAT}}$ key, highlight **TESTS** and use the option **9:2-SampZInt** for confidence intervals for the difference of means when σ_1 and σ_2 are known (if the populations are not normal distributions, large samples are required). Use option **0:2-SampTInt** for confidence intervals for the difference of means when σ_1 and σ_2 are unknown. Option **B:2-PropZInt** gives confidence intervals for the difference of two proportions.

For the difference of means, you have the option of using either summary statistics or raw data in lists as input. **Data** allows you to use raw data from a list, while **Stats** lets you use summary statistics.

When finding confidence intervals for differences of means when σ_1 and σ_2 are unknown, use sample estimates s_1 and s_2 for corresponding population standard deviations σ_1 and σ_2. For confidence intervals for difference of means using small samples, be sure to choose **YES** for Pooled.

Example (Difference of means, σ_1 and σ_2 are known, large samples)

A random sample of 45 pro football players produced a sample mean height of 6.18 feet with population standard deviation 0.37 feet. A random sample of 40 pro basketball players produced a mean height of 6.45 feet with population standard deviation 0.31. Find a 90% confidence interval for the difference of population heights.

We select option **9:2-SampZInt**, since σ_1 and σ_2 are known and we have large samples, we use the normal distribution. Since we have summary statistics, select **Inpt: Stats.** Then enter the specified values. The data entry requires two screens.

```
2-SampZInt            2-SampZInt
 Inpt:Data █Stats█     ↑σ2:.31
 σ1:.37                 x̄1:6.18
 σ2:.31                 n1:45
 x̄1:6.18                x̄2:6.45
 n1:45                  n2:40
 x̄2:6.45                C-Level:.9
↓n2:40                  Calculate
```

Highlight **Calculate** and press $\boxed{\text{ENTER}}$.

```
2-SampZInt
 (-.3914, -.1486)
 x̄1=6.18
 x̄2=6.45
 n1=45
 n2=40
```

The interval is from –0.39 to –0.14. Since zero lies outside the interval, we conclude that at the 90% level, the mean height of basketball players is different from that of football players.

LAB ACTIVITIES FOR CONFIDENCE INTERVALS FOR $\mu_1 - \mu_2$ OR FOR $p_1 - p_2$

1. The following data is based on random samples of red foxes in two regions of Germany. The number of cases of rabies was counted for a random sample of 16 areas in each of two regions.

Region 1:	10	2	2	5	3	4	3	3	4	0	2	6	4	8	7	4
Region 2:	1	1	2	1	3	9	2	2	3	4	3	2	2	0	0	2

   (a) Use a confidence level $c = 90\%$. Does the interval indicate that the population mean number of rabies cases in Region 1 is greater than the population mean number of rabies cases in Region 2 at the 90% level? Why or why not?

   (b) Use a confidence level $c = 95\%$. Does the interval indicate that the population mean number of rabies cases in Region 1 is greater than the population mean number of rabies cases in Region 2 at the 95% level? Why or why not?

   (c) Determine whether the population mean number of rabies cases in Region 1 is greater than the population mean number of rabies cases in Region 2 at the 99% level. Explain your answer. Verify your answer by running **2-SampTInt** with a 99% confidence level.

2. A random sample of 30 police officers working the night shift showed that 23 used at least 5 sick leave days per year. Another random sample of 45 police officers working the day shift showed that 26 used at least 5 sick leave days per year. Find the 90% confidence interval for the difference of population proportions of police officers working the two shifts using at least 5 sick leave days per year. At the 90% level, does it seem that the proportions are different? Does there seem to be a difference in proportions at the 99% level? Why or why not?

CHAPTER 9 HYPOTHESIS TESTING

The TI-83 Plus and TI-84 Plus fully support hypothesis testing. Use the $\boxed{\text{STAT}}$ key, then highlight **TESTS.** The options used in Chapter 9 are given on the two screens.

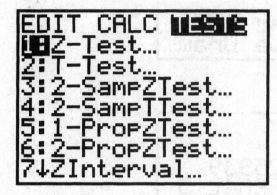

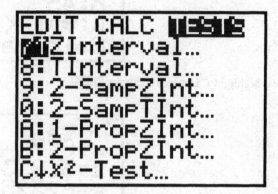

TESTING A SINGLE POPULATION MEAN (SECTION 9.1–9.2 OF *UNDERSTANDABLE STATISTICS*)

When the value of σ is known, z-test based on normal distribution is used to test a population mean, provided that the population has a normal distribution or the sample size is large (at least 30). If the value of σ is unknown, then t-test will be used if either the population has approximately a normal distribution or the sample size is large. Select option **1:Z-Test** for z-test, or **2:T-Test** for t-test. As with confidence intervals, we have a choice of entering raw data into a list using the **Data** input option or using summary statistics with the **Stats** option. The null hypothesis is $H_0: \mu = \mu_0$. Enter the value of μ_0 in the spot indicated. To select the alternate hypothesis, use one of the three μ options $\neq \mu_0, < \mu_0,$ or $> \mu_0$. Output consists of the value of the sample mean $\bar{x}$ and its corresponding z value. The P value of the sample statistic is also given.

Example (Testing a mean, when σ is known)

Ten years ago, State College did a study regarding the number of hours full-time students worked each week. The mean number of hours was 8.7. A recent study involving a random sample of 45 full time students showed that the average number of hours worked per week was 10.3. Also assume that the population standard deviation was 2.8. Use a 5% level of confidence to test if the mean number of hours worked per week by full-time students has increased.

Because we have a large sample, we will use the normal distribution for our sample test statistic. Select option **1:Z-Test.** We have summary statistics rather than raw data, so use the **Stats** input option. Enter the appropriate values on the screen.

Next highlight **Calculate** and press ENTER.

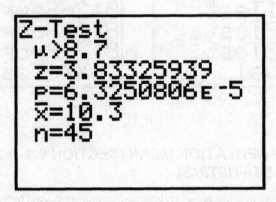

We see that the z value corresponding to the sample test statistic $\bar{x} = 10.3$ is $z = 3.83$. The critical value for a 5% level of significance and a right-tailed test is $z_0 = 1.645$. Clearly the sample z value is to the right of z_0, and we reject H_0. Notice that the P value = 6.325E-5. This means that we move the decimal 5 places to the left, giving a P value of 0.000063. Since the P value is less than 0.05, we reject H_0.

Graph

The TI-83 Plus and TI-84 Plus give an option to show the sample test statistic on the normal distribution. Highlight the **Draw** option on the **Z-Test** screen. Because the sample z is so far to the right, it does not appear on this window. However, its value does, and a rounded P value shows as well.

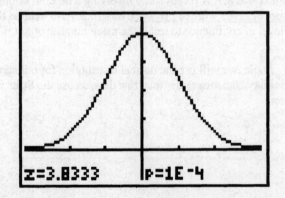

To do hypothesis testing of the mean when σ is unknown, we use Student's t distribution. Select option **2:T-Test.** The entry screens are similar to those for the Z-Test.

LAB ACTIVITIES FOR TESTING A SINGLE POPULATION MEAN

1. A random sample of 65 pro basketball players showed their heights (in feet) to be

6.50	6.25	6.33	6.50	6.42	6.67	6.83	6.82
6.17	7.00	5.67	6.50	6.75	6.54	6.42	6.58
6.00	6.75	7.00	6.58	6.29	7.00	6.92	6.42
5.92	6.08	7.00	6.17	6.92	7.00	5.92	6.42
6.00	6.25	6.75	6.17	6.75	6.58	6.58	6.46
5.92	6.58	6.13	6.50	6.58	6.63	6.75	6.25
6.67	6.17	6.17	6.25	6.00	6.75	6.17	6.83
6.00	6.42	6.92	6.50	6.33	6.92	6.67	6.33
6.08							

   (a) Enter the data into list L_1. Use **1-Var Stats** to determine the sample standard deviation.

   (b) Use the **T-Test** option to test the hypothesis that the average height of the players is greater than 6.2 feet, at the 1% level of significance.

2. In this problem, we will see how the test conclusion is possibly affected by a change in the level of significance.

   Teachers for Excellence is interested in the attention span of students in grades 1 and 2, now as compared to 20 years ago. They believe it has decreased. Studies done 20 years ago indicate that the attention span of children in grades 1 and 2 was 15 minutes. A study sponsored by Teachers for Excellence involved a random sample of 20 students in grades 1 and 2. The average attention span of these students was (in minutes) $\bar{x} = 14.2$ with standard deviation $s = 1.5$.

   (a) Conduct the hypothesis test using $\alpha = 0.05$ and a left-tailed test. What is the test conclusion? What is the P value?

   (b) Conduct the hypothesis test using $\alpha = 0.01$ and a left-tailed test. What is the test conclusion? How could you have predicted this result by looking at the P value from part (a)? Is the P value for this part the same as it was for part (a)?

3. In this problem, let's explore the effect that sample size has on the process of testing a mean. Run **Z-Test** with the hypotheses $H_0 : \mu = 200$, $H_1 : \mu > 200$, $\alpha = 0.05$, $\bar{x} = 210$ and $\sigma = 40$.

   (a) Use the sample size $n = 30$. Note the P value and z score of the sample test statistic and test conclusion.

   (b) Use the sample size $n = 50$. Note the P value and z score of the sample test statistic and test conclusion.

   (c) Use the sample size $n = 100$. Note the P value and z score of the sample test statistic and test conclusion.

   (d) In general, if your sample statistic is close to the proposed population mean specified in H_0, and you want to reject H_0, would you use a smaller or a larger sample size?

TESTS INVOLVING A SINGLE PROPORTION (SECTION 9.3 OF *UNDERSTANDABLE STATISTICS*)

To conduct a hypotheses test of a single proportion, select option **5:1-PropZTest.** The null hypothesis is H_0: $p = p_0$. Enter the value of p_0. The number of successes is designated by the value x. Enter the value. The sample size or number of trials is n. The alternate hypothesis will be prop $\neq p_0$, $< p_0$, or $> p_0$. Highlight the appropriate choice. Finally highlight **Calculate** and press ENTER. Notice that the **Draw** option is available to show the results on the standard normal distribution.

TEST INVOLVING PAIRED DIFFERENCES (DEPENDENT SAMPLES) (SECTION 9.4 OF *UNDERSTANDABLE STATISTICS*)

To perform a paired difference test, we put our paired data into two columns, and then put the differences between corresponding pairs of values in a third column. For example, put the "before" data in list L_1, the "after" data in L_2. Create $L_3 = L_1 - L_2$.

Example

Promoters of a state lottery decided to advertise the lottery heavily on television for one week during the middle of one of the lottery games. To see if the advertising improved ticket sales, the promoters surveyed a random sample of 8 ticket outlets and recorded weekly sales for one week before the television campaign and for one week after the campaign. The results follow (in ticket sales) where B stands for "before" and A for "after" the advertising campaign.

B:	3201	4529	1425	1272	1784	1733	2563	3129
A:	3762	4851	1202	1131	2172	1802	2492	3151

Test the claim that the television campaign increased lottery ticket sales at the 0.05 level of significance.

We want to test to see if $D = B - A$ is less than zero, since we are testing the claim that the lottery ticket sales are greater after the television campaign. We will put the "before" data in L_1, the "after" data in L_2, and put the differences in list L_3 by highlighting the header L£ entering $L_1 - L_2$, and pressing ENTER.

```
 L1      L2       L3        3
3201    3762     -561
4529    4851     -322
1425    1202      223
1272    1131      141
1784    2172     -388
1733    1802      -69
2563    2492       71

L3 =L1-L2
```

Next we will conduct a *t*-test on the differences in list L_1. Select option **2:T-Test** from the **TESTS** menu. Select **Data** for **Inpt**. Since the null hypotheses is that $d = 0$, we use the null hypothesis $H_0: d = \mu_0$ with $\mu_0 = 0$. Since the data is in L_3, enter that as the List. Since the test is a left-tailed test, select $\mu :< \mu_0$.

```
T-Test
 Inpt:DATA Stats
 μ0:0
 List:L₃
 Freq:1
 μ:≠μ0  <μ0  >μ0
 Calculate Draw
```

Highlight **Calculate** and press ENTER.

```
T-Test
 μ<0
 t=-1.178455137
 P=.138559057
 x̄=-115.875
 Sx=278.1132542
 n=8
```

We see that the sample *t* value is −1.17 with a corresponding *P* value of 0.138. Since the *P* value is greater than 0.05, we do not reject H_0.

LAB ACTIVITIES USING TESTS INVOLVING PAIRED DIFFERENCES (DEPENDENT SAMPLES)

1. The data are pairs of values where the first entry represents average salary (in thousands of dollars/year) for male faculty members at an institution and the second entry represents the average salary for female faculty members (in thousands of dollars/year) at the same institution. A random sample of 22 U.S. colleges and universities was used (source: Academe, Bulletin of the American Association of University Professors).

(34.5, 33.9)	(30.5, 31.2)	(35.1, 35.0)	(35.7, 34.2)	(31.5, 32.4)
(34.4, 34.1)	(32.1, 32.7)	(30.7, 29.9)	(33.7, 31.2)	(35.3, 35.5)
(30.7, 30.2)	(34.2, 34.8)	(39.6, 38.7)	(30.5, 30.0)	(33.8, 33.8)
(31.7, 32.4)	(32.8, 31.7)	(38.5, 38.9)	(40.5, 41.2)	(25.3, 25.5)
(28.6, 28.0)	(35.8, 35.1)			

(a) Put the first entries in L_1, the second in L_2, and create L_3 to be the difference $L_1 - L_2$.

(b) Use the **T-Test** option to test the hypothesis that there is a difference in salary. What is the P value of the sample test statistic? Do we reject or fail to reject the null hypothesis at the 5% level of significance? What about at the 1% level of significance?

(c) Use the **T-Test** option to test the hypothesis that female faculty members have a lower average salary than male faculty members. What is the test conclusion at the 5% level of significance? At the 1% level of significance.

2. An audiologist is conducting a study on noise and stress. Twelve subjects selected at random were given a stress test in a room that was quiet. Then the same subjects were given another stress test, this time in a room with high-pitched background noise. The results of the stress tests were scores 1 through 20, with 20 indicating the greatest stress. The results, where B represents the score of the test administered in the quiet room and A represents the scores of the test administered in the room with the high-pitched background noise, are shown below.

Subject	1	2	4	5	6	7	8	9	10	11	12
B	13	12	16	19	7	13	9	15	17	6	14
A	18	15	14	18	10	12	11	14	17	8	16

Test the hypothesis that the stress level was greater during exposure to noise. Look at the P value. Should you reject the null hypotheses at the 1% level of significance? At the 5% level?

TESTS OF DIFFERENCE OF MEANS (INDEPENDENT SAMPLES) (SECTION 9.5 OF *UNDERSTANDABLE STATISTICS*)

We consider the $\bar{x}_1 - \bar{x}_2$ distribution. The null hypothesis is that there is no difference between means so $H_0: \mu_1 = \mu_2$, or $H_0: \mu_1 - \mu_2 = 0$.

Example

Sellers of microwave French fry cookers claim that their process saves cooking time. McDougle Fast Food Chain is considering the purchase of these new cookers, but wants to test the claim. Six batches of French fries cooked in the traditional way. These times (in minutes) are

15   17   14   15   16   13

Six batches of French fries of the same weight were cooked using the new microwave cooker. These cooking times (in minutes) are

11   14   12   10   11   15

Test the claim that the microwave process takes less time. Use $\alpha = 0.05$.

Put the data for traditional cooking in list L_1 and the data for the new method in List L_2. Since we have small samples, we want to use the Student's t distribution. Select option **4:2-SampTTest**. Select **Data** for **Inpt,** use the alternate hypothesis $\mu_1 > \mu_2$, select **Yes** for **Pooled:**

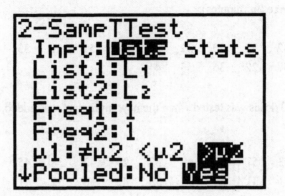

The output is on two screens.

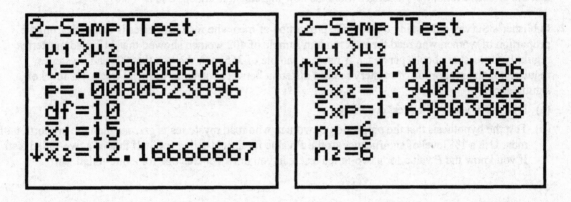

Since the P value is 0.008, which is less than 0.05, we reject H_0 and conclude that the new method cooks food faster.

TESTING A DIFFERENCE OF PROPORTIONS (SECTION 9.5 OF *UNDERSTANDABLE STATISTICS*)

To conduct a hypothesis test for a difference of proportions, select option **6:2-PropZTest** and enter appropriate values.

LAB ACTIVITIES FOR TESTING DIFFERENCE OF MEANS (INDEPENDENT SAMPLES) OR PROPORTIONS

1. Calm Cough Medicine is testing a new ingredient to see if its addition will lengthen the effective cough relief time of a single dose. A random sample of 15 doses of the standard medicine was tested, and the effective relief times were (in minutes):

42	35	40	32	30	26	51	39	33	28
37	22	36	33	41					

   A random sample of 20 doses was tested when the new ingredient was added. The effective relief times were (in minutes):

43	51	35	49	32	29	42	38	45	74
31	31	46	36	33	45	30	32	41	25

   Assume that the standard deviations of the relief times are equal for the two populations. Test the claim that the effective relief time is longer when the new ingredient is added. Use $\alpha = 0.01$.

2. Publisher's Survey did a study to see if the proportion of men who read mysteries is different from the proportion of women who read them. A random sample of 402 women showed that 112 read mysteries regularly (at least six books per year). A random sample of 365 men showed that 92 read mysteries regularly. Is the proportion of mystery readers different between men and women? Use a 1% level of significance.

   (a) Find the P value of the test conclusion.

   (b) Test the hypothesis that the proportion of women who read mysteries is *greater* than the proportion of men. Use a 1% level of significance. Is the P value for a right-tailed test half that of a two-tailed test? If you know the P value for a two-tailed test, can you draw conclusions for a one-tailed test?

CHAPTER 10 REGRESSION AND CORRELATION

LINEAR REGRESSION (SECTION 10.1–10.3 OF *UNDERSTANDABLE STATISTICS*)

Important Note: Before beginning this chapter, press 2nd [CATALOG] (above 0) and scroll down to the entry DiagnosticOn. Press ENTER twice. After doing this, the regression correlation coefficient r will appear as output with the linear regression line.

The TI-83 Plus and TI-84 Plus graphing calculators have automatic regression functions built-in as well as summary statistics for two variables. To locate the regression menu, press STAT, and then select **CALC.** The choice
8:LinReg(a + bx) performs linear regression on the variables in L_1 and L_2. If the data are in other lists, then specify those lists after the **LinReg(a + bx)** command.

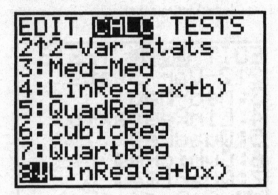

Let's look at a specific example to see how to do linear regression on the TI-83 Plus or TI-84 Plus.

Example

Merchandise loss due to shoplifting, damage, and other causes is called shrinkage. Shrinkage is a major concern to retailers. The managers of H.R. Merchandise believe there is a relationship between shrinkage and number of clerks on duty. To explore this relationship, a random sample of 7 weeks was selected. During each week the staffing level of sales clerks was kept constant and the dollar value of the shrinkage was recorded.

Number of Sales Clerks:	12	11	15	9	13	8
Shrinkage:	15	20	9	25	12	31

(a) Find the equation of the least-squares line, with number of sales clerks as the explanatory variable.

First put the numbers of sales clerks in L_1 and the corresponding amount of shrinkage in L_2.

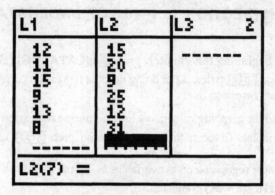

Use the **CALC** menu and select **8:LinReg(a+bx)**.

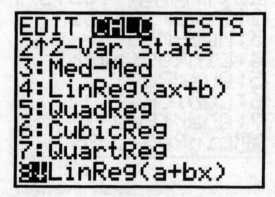

Designate the lists containing the data. Then press ENTER.

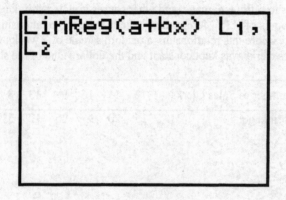

```
LinReg
 y=a+bx
 a=54.48
 b=-3.16
 r²=.9638610039
 r=-.9817642303
```

Note that the linear regression equation is $y = 54.48x - 3.16x$.

We also have the value of the Pearson correlation coefficient r and r^2.

(b) Make a scatter diagram and show the least-squares line on the plot.

To graph the least squares line, press $\boxed{Y=}$. Clear all existing equations from the display. To enter the least-squares equation directly, press $\boxed{VARS}$. Select item **5:Statistics.**

```
VARS  Y-VARS
1:Window…
2:Zoom…
3:GDB…
4:Picture…
5:Statistics…
6:Table…
7:String…
```

Press $\boxed{ENTER}$ and then select **EQ** and then **1:RegEQ.**

```
XY  Σ  EQ  TEST PTS
1:RegEQ
2:a
3:b
4:c
5:d
6:e
7↓r
```

You will see the least-squares regression equation appear automatically in the equations list.

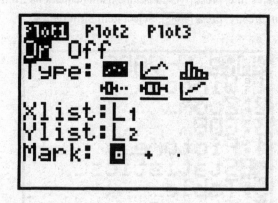

We want to graph the least-squares line with the scatter diagram. Press [2nd] **[STAT PLOT]**. Select **1:Plot1.** Then select **On,** scatter diagram picture, and enter L_1 for Xlist and L_2 for Ylist.

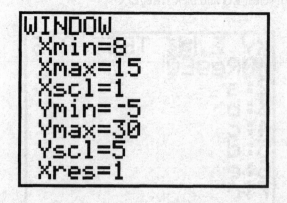

The last step before we graph is to set the graphing window. Press [WINDOW]. Set Xmin to the lowest x value, Xmax to the highest x value, Ymin at -5 and Ymax at above the highest y value.

```
WINDOW
  Xmin=8
  Xmax=15
  Xscl=1
  Ymin=-5
  Ymax=30
  Yscl=5
  Xres=1
```

Now press GRAPH.

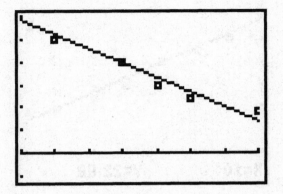

(c) Predict the shrinkage when 10 clerks are on duty.

Press 2nd **[CALC].** This is the Calculate function for graphs and is different from the CALC you highlight from the STAT menu. Highlight **1:Value.**

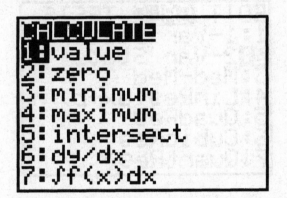

Press ENTER. The graph will appear. Enter 10 next to **X=.**

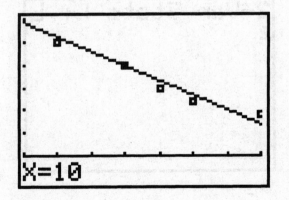

Press [ENTER]. The predicted value for y appears.

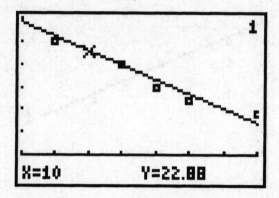

To display the summary statistics for x and y, press [STAT], highlight **CALC**. Select **2:2-Var Stats**. Press [ENTER].

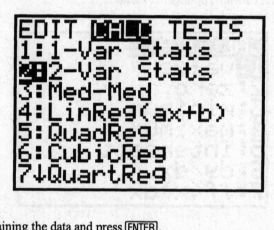

Enter the lists containing the data and press [ENTER].

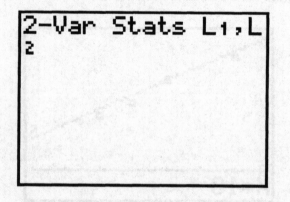

Scroll to the statistics of interest.

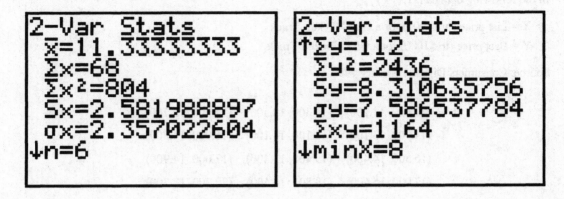

LAB ACTIVITIES FOR LINEAR REGRESSION

For each of the following data sets, do the following.

(a) Enter the data, putting x values in L_1 and y values in L_2.

(b) Find the equation of the least-squares line.

(c) Find the value of the correlation coefficient r.

(d) Draw a scatter diagram and show the least-squares line on the scatter diagram.

(e) Use **2-Var Stats** to find all the sums necessary for computing the standard error of estimate S_e.

1. **Cricket Chirps Versus Temperature** (data files Slr02L1.txt and Slr02L2.txt)

In the following data pairs (X, Y)

   X = Chirps/s for the striped ground cricket

   Y = Temperature in degrees Fahrenheit

Source: *The Song of Insects* by Dr. G.W. Pierce, Harvard College Press

(20.0, 88.6)	(16.0, 71.6)	(19.8, 93.3)
(18.4, 84.3)	(17.1, 80.6)	(15.5, 75.2)
(14.7), 69.7)	(17.1, 82.0)	(15.4, 69.4)
(16.2, 83.3)	(15.0, 79.6)	(17.2, 82.6)
(16.0, 80.6)	(17.0, 83.5)	(14.4, 76.3)

2. List Price Versus Best Price for a New GMC Pickup Truck (data files Slr01L1.txt and Slr01L2.txt)

In the following data pairs (X, Y)

   X = List price (in $1000) for a GMC Pickup Truck

   Y = Best price (in $1000) for a GMC Pickup Truck

Source: Consumers Digest, February 1994

(12.400, 11.200)	(14.300, 12.500)	(14.500, 12.700)
(14.900, 13.100)	(16.100, 14.100)	(16.900, 14.800)
(16.500, 14.400)	(15.400, 13.400)	(17.000, 14.900)
(17.900, 15.600)	(18.800, 16.400)	(20.300, 17.700)
(22.400, 19.600)	(19.400, 16.900)	(15.500, 14.000)
(16.700, 14.600)	(17.300, 15.100)	(18.400, 16.100)
(19.200, 16.800)	(17.400, 15.200)	(19.500, 17.000)
(19.700, 17.200)	(21.200, 18.600)	

3. Diameter of Sand Granules Versus Slope on a Natural Occurring Ocean Beach (data files Slr03L1.txt and Slr03L2.txt)

In the following data pairs (X,Y)

   X = Median diameter (mm) of granules of sand

   Y = Gradient of beach slope in degrees

The data is for naturally occurring ocean beaches.

Source: *Physical Geography* by A.M. King, Oxford Press, England

(0.170, 0.630)	(0.190, 0.700)	(0.220, 0.820)
(0.235, 0.880)	(0.235, 1.150)	(0.300, 1.500)
(0.350, 4.400)	(0.420, 7.300)	(0.850, 11.300)

4. National Unemployment Rate Male Versus Female (data files Slr04L1.txt and Slr04L2.txt)

In the following data pairs (X, Y)

   X = National unemployment rate for adult males

   Y = National unemployment rate for adult females

Source: Statistical Abstract of the United States

(2.9, 4.0)	(6.7, 7.4)	(4.9, 5.0)
(7.9, 7.2)	(9.8, 7.9)	(6.9, 6.1)
(6.1, 6.0)	(6.2, 5.8)	(6.0, 5.2)
(5.1, 4.2)	(4.7, 4.0)	(4.4, 4.4)
(5.8, 5.2)		

CHAPTER 11 CHI-SQUARE AND *F* DISTRIBUTIONS

CHI-SQUARE TEST OF INDEPENDENCE (SECTION 11.1 OF *UNDERSTANDABLE STATISTICS*)

The TI-83 Plus and TI-84 Plus calculators support tests for independence. Press the $\boxed{\text{STAT}}$ key, select **TESTS,** and option **C: χ^2 –Test.** The original observed values need to be entered in a matrix.

Example

A computer programming aptitude test has been developed for high school seniors. The test designers claim that scores on the test are independent of the type of school the student attends: rural, suburban, urban. A study involving a random sample of students from each of these types of institutions yielded the following information, where aptitude scores range from 200 to 500 with 500 indicating the greatest aptitude and 200 the least. The entry in each cell is the observed number of students achieving the indicated score on the test.

School Type

Score	Rural	Suburban	Urban
200–299	33	65	82
300–399	45	79	95
400–500	21	47	63

Using the option **C: χ^2 -Test…,** test the claim that the aptitude test scores are independent of the type of school attended at the 0.05 level of significance.

We need to use a matrix to enter the data. Press $\boxed{\text{2nd}}$ **[MATRX].** Highlight **EDIT,** and select **1:[A].** Press $\boxed{\text{ENTER}}$.

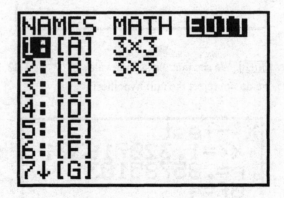

Since the contingency table has 3 rows and 3 columns, type 3 (for number of rows), press [ENTER], type 3 (for number of columns), press [ENTER]. Then type in the observed values. Press [ENTER] after each entry. Notice that you enter the table by rows. When the table is entered completely, press [2nd] [MATRX] again.

```
MATRIX[A] 3 ×3
[ 33      65      82  ]
[ 45      79      95  ]
[ 21      47      63  ]




3,3=63
```

This time, select **2:[B]**. Set the dimensions to be the same as for matrix [A]. For this example, [B] should be 3 rows × 3 columns.

Now press the [STAT] key, highlight TESTS, and use option **C: χ^2–Test.** This tells you that the observed values of the table are in matrix [A]. The expected values will be placed in matrix [B].

```
X²-Test
 Observed: [A]
 Expected: [B]
 Calculate Draw
```

Highlight **Calculate** and press [ENTER]. We see that the sample value of χ^2 is 1.32. The P value is 0.858. Since the P value is larger than 0.05, we do not reject the null hypothesis.

```
X²-Test
 X²=1.320716301
 P=.857851031
 df=4
```

If you want to see the expected values in matrix [B], type [2nd] [MATRX] and select [B] under **NAMES.**

Graph

Notice that one of the options of the χ^2 test is to graph the χ^2 distribution and show the sample test statistic on the graph. Highlight **Draw** and press ENTER. Before you do this, be sure that all STAT PLOTS are set to Off, and that you have cleared all the entries in the Y= menu.

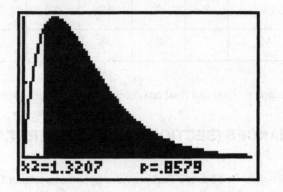

LAB ACTIVITIES FOR CHI-SQUARE TEST OF INDEPENDENCE

1. We Care Auto Insurance had its staff of actuaries conduct a study to see if vehicle type and loss claim are independent. A random sample of auto claims over six months gives the information in the contingency table.

Total Loss Claims per Year per Vehicle

Type of vehicle	$0–999	$1000–2999	$3000–5999	$6000+
Sports car	20	10	16	18
Truck	16	25	33	9
Family Sedan	40	68	17	7
Compact	52	73	48	12

Test the claim that car type and loss claim are independent. Use $\alpha = 0.05$.

2. An educational specialist is interested in comparing three methods of instruction:

   SL – standard lecture with discussion
   TV – video taped lectures with no discussion
   IM – individualized method with reading assignments and tutoring, but no lectures.

   The specialist conducted a study of these three methods to see if they are independent. A course was taught using each of the three methods and a standard final exam was given at the end. Students were put into the different method sections at random. The course type and test results are shown in the next contingency table.

Final Exam Score

Course Type	< 60	60–69	70–79	80–89	90–100
SL	10	4	70	31	25
TV	8	3	62	27	23
IM	7	2	58	25	22

Test the claim that the instruction method and final test scores are independent, using $\alpha = 0.01$.

TESTING TWO VARIANCES (SECTION 11.5 OF *UNDERSTANDABLE STATISTICS*)

Under the **TESTS** menu option **D:2-SampFTest** provides hypothesis testing for two variances. Again there is a choice of data entry styles: raw data in lists, or summary statistics for which you provide the sample standard deviations and sample sizes.

Example

Two paint manufacturing processes are under study. A random sample of 35 applications of the paint produced under the first method shows that the paint's life has a standard deviation of 1.5 years. For the second method, a random sample of 40 applications shows a standard deviation of 1.3 years. Use a 5% level of significance to test if the variances are equal.

Use the **D:2-StampFTest** option. Select **Stats** and enter the data as requested. Note that we follow the convention of always entering the larger standard deviation as S_{x1}.

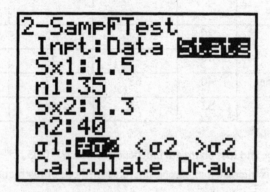

Highlight **Calculate** and press [ENTER].

```
2-SampFTest
 σ1≠σ2
 F=1.331360947
 P=.3867273941
 Sx1=1.5
 Sx2=1.3
↓n1=35
```

The sample F statistic is 1.33 with P value 0.387. We do not reject H_0.

ONE-WAY ANOVA (SECTION 11.5 OF *UNDERSTANDABLE STATISTICS*)

The TI-83 Plus and TI-84 Plus support one-way ANOVA tests. Press the [STAT] key, and under **TESTS** select option **F:ANOVA(**. The format of the command for ANOVA is **ANOVA (L_1, L_2, ...)** where all the treatment lists are given.

Example

A psychologist has developed a series of tests to measure a person's level of depression. The composite scores range from 50 to 100, with 100 representing the most severe depression level. A random sample of 12 patients with approximately the same depression level (as measured by the tests) was divided into 3 different treatment groups. Then, one month after treatment was completed, the depression level of each patient was again evaluated. The after-treatment depression levels are given.

Treatment 1	70	65	82	83	71
Treatment 2	75	62	81		
Treatment 3	77	60	80	75	

Use the option **F: ANOVA** to test the claim that the population means are all the same, at the 5% level of significance.

Enter the treatment data in lists L_1, L_2, and L_3 respectively.

Select the **F:ANOVA (** option from the **TESTS** menu. Then type the names of the 3 lists containing the data. Separate the list names by commas.

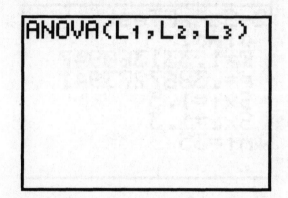

Press ENTER. The output is on two screens.

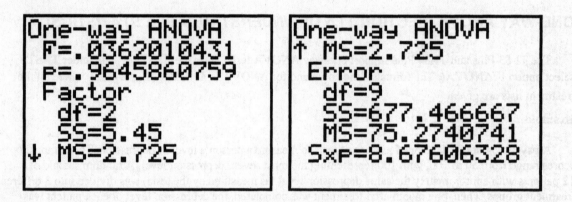

We see that the sample F is $F = 0.036$ with P value 0.9646. We do not reject H_0 and conclude that there is no evidence that some of the means are different.

LAB ACTIVITIES FOR ANALYSIS OF VARIANCE

1. A random sample of 20 overweight adults was randomly divided into 4 groups, and each group was given a different diet plan. The weight loss for each individual (in pounds) after 3 months follows.

Plan 1	18	10	20	25	17
Plan 2	28	12	22	17	16
Plan 3	16	20	24	8	17
Plan 4	14	17	18	5	16

Test the claim that the population mean weight loss is the same for the four diet plans, at the 5% level of significance.

2. A psychologist is studying the time it takes rats to respond to stimuli after being given doses of different tranquilizing drugs. A random sample of 18 rats was divided into 3 groups. Each group was given a different drug. The response time to stimuli was measured (in seconds). The results follow.

Drug A	3.1	2.5	2.2	1.5	0.7	2.4
Drug B	4.2	2.5	1.7	3.5	1.2	3.1
Drug C	3.3	2.6	1.7	3.9	2.8	3.5

Test the claim that the population mean response times for the three drugs are the same, at the 5% level of significance.

3. A research group is testing various chemical combinations designed to neutralize and buffer the effects of acid rain on lakes. Eighteen lakes of similar size in the same region have all been affected in the same way by acid rain. The lakes are divided into four groups and each group of lakes is sprayed with a different chemical combination. An acidity index is then taken after treatment. The index ranges form 60 to 100 with 100 indicating the greatest acid rain pollution. The results follow.

Combination I	63	55	72	81	75
Combination II	78	56	75	73	82
Combination III	59	72	77	60	
Combination IV	72	81	66	71	

Test the claim that the population mean acidity index after each of the four treatments is the same, at the 0.01 level of significance.

PART II

EXCEL® GUIDE

FOR

UNDERSTANDABLE STATISTICS

EIGHTH EDITION

CHAPTER 1 GETTING STARTED

GETTING STARTED WITH EXCEL

Microsoft® Excel is an all-purpose spreadsheet application with many functions. We will be using Excel 97. This guide is not a general Excel manual, but it will show you how to use many of Excel's built-in statistical functions. You may need to install the Analysis ToolPak from the original Excel software if your computer does not have it. To determine if your installation of Excel includes the Analysis ToolPak, open Excel, click on **Tools** in the main menu, and then see if the ToolPak is listed in the **Add-Ins** dialog box. If it is, place a check by Analysis ToolPak. If you do not see a listing for Analysis ToolPak, then you will need to install it from the original Excel installation source.

If you are familiar with Windows-based programs, you will find that many of the editing, formatting, and file handling procedures of Excel are similar to those you have used before. You use the mouse to select, drag, click and double-click as you would in any other Windows program.

If you have any questions about Excel not answered in this guide, consult the Excel manual or select **Help** on the menu bar.

The Excel Window

When you have opened Excel, you should see a window like this:

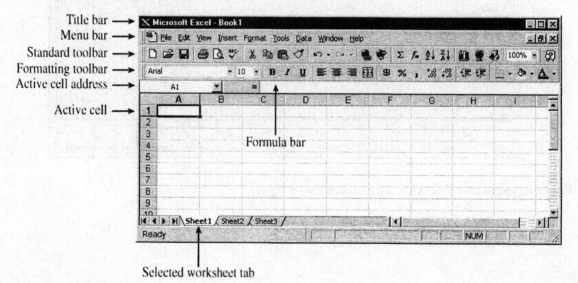

The Excel Workbook

An Excel file is called a Workbook. Notice that in the display shown above, the title bar shows Microsoft Excel - Book1. This means that we are working in Book 1.

Each workbook consists of one or more worksheets. In the worksheet above, the tabs near the bottom of the screen show that we are working with Sheet 1. To change worksheets, click on the appropriate tab. Alternatively, you can right-click the arrows just to the left of the worksheet tabs to get a list of all the worksheets in the projects, and then select a worksheet.

The Cells in the Worksheet

When you look at a worksheet, you should notice horizontal and vertical grid lines. If they are missing, you will need to activate that feature. To do so,

1. Select **Tools** from the menu, and then select **Options** from the drop-down menu.

2. Click on the tab showing **View.** Be sure that the **Gridlines** option is checked.

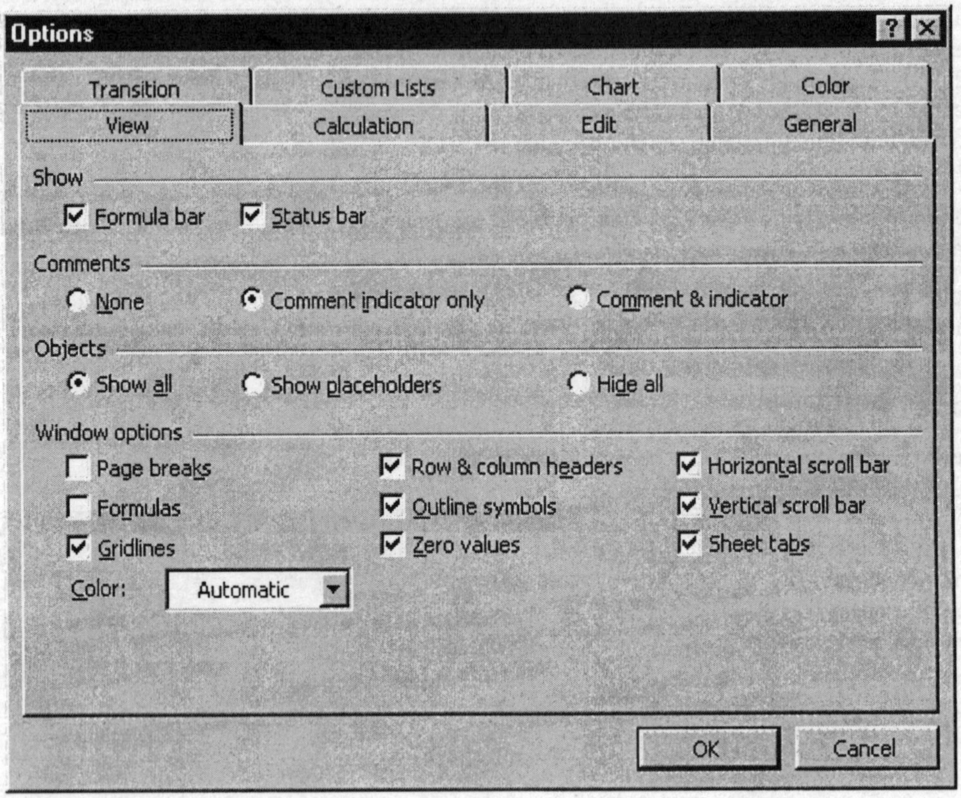

It is convenient to have checks by all the options shown above.

Cell Addresses

The cells are formed by intersecting rows and columns. A cell's address consists of a column letter followed by a row number. For example, the address B3 designates the cell that lies in Column B, Row 3. When Cell B3 is highlighted, it is the active cell. This means we can enter text or numbers into Cell B3.

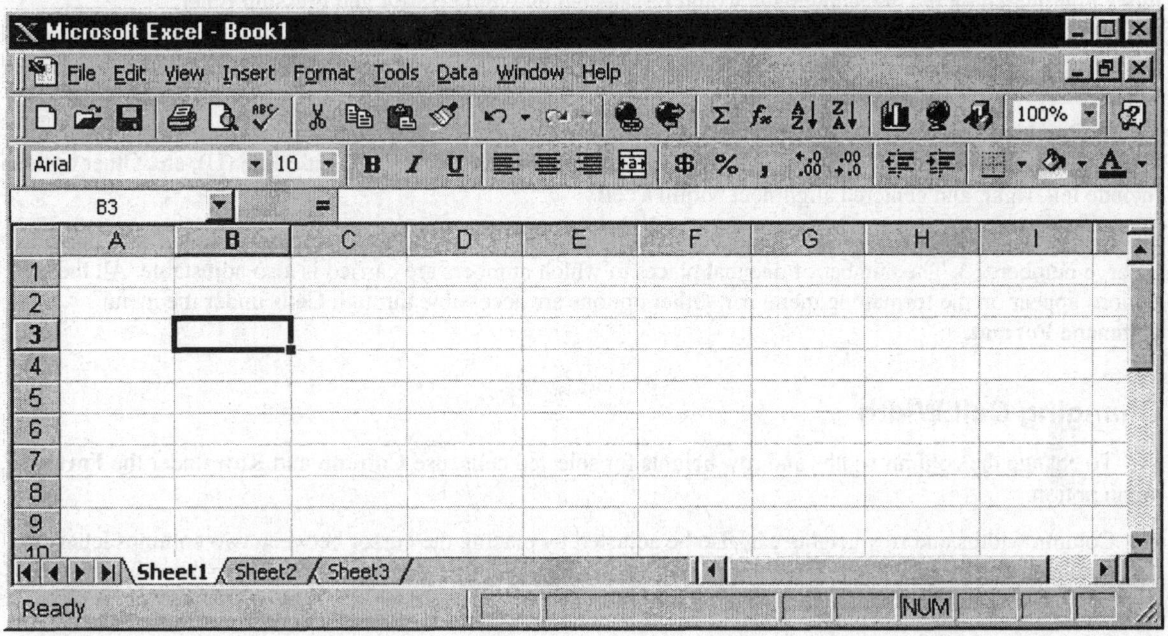

Selecting Cells

To select a cell, position the cursor in the cell and click the left mouse button.

Sometimes you will want to select several cells at once, in order to format them (as described next). To select a rectangular block of cells, position the cursor in a corner cell of the block, hold down the left mouse button, drag the cursor to the cell in the block's opposite corner, and release the button. The selected cells will be highlighted, as shown below.

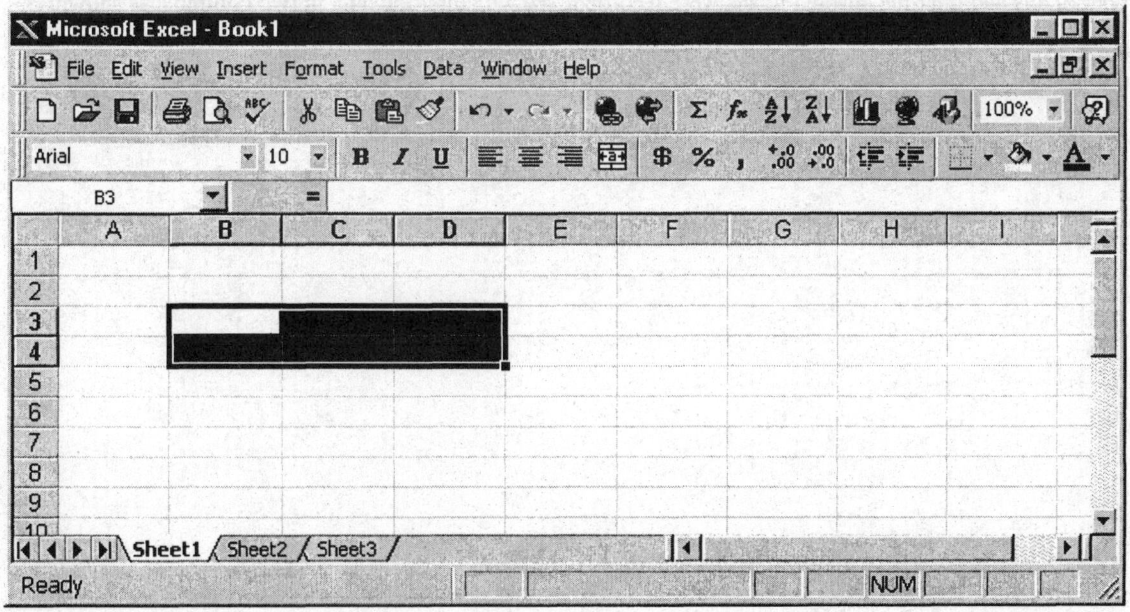

To select an entire column, click on the letter above it; to select an entire row, click on the number to its left. To select every cell in the worksheet, click on the gray blank rectangle in the upper left corner of the worksheet (above row header 1 and left of column header A).

You can also select a block of cells by typing the two corner cells into the active cell address window. The block highlighted on the preceding page would be selected by typing B3:E4 and pressing Enter.

Formatting Cell Contents

In Excel, you may place text or numbers in a cell. As in other Windows applications, you can format the text or numbers by using the formatting toolbar buttons for bold (**B**), italics (*I*), underline (U), etc. Other options include left, right, and centered alignment within a cell.

Numbers can be formatted to represent dollar amounts ($) or percents (%) and can be shown with commas in large numbers (,). The number of decimal places to which numbers are carried is also adjustable. All these options appear on the formatting menu bar. Other options are accessible through **Cells** under the menu command **Format.**

Changing Cell Width

To change the column widths and row heights for selected cells, use **Column** and **Row** under the **Format** menu option.

Column widths and row heights can also be adjusted by placing the cursor between two columns letters or row numbers. When the cursor changes appearance, hold down the left mouse button, move the column or row boundary, and release.

All these instructions may seem a little mysterious. Once you try them, however, you will find that they are fairly easy to remember.

ENTERING DATA

In Excel we enter data and labels in the cells. It is common to select a column for the data and place a label as the first entry in the column.

Let's enter some data on television advertising. For each of twenty hours of prime-time viewing, both the number of ads and the time devoted to ads were recorded. We will enter the data in two columns, as shown:

```
X Microsoft Excel - Book1                                          _ □ X
  File  Edit  View  Insert  Format  Tools  Data  Window  Help        _  5 X
```

	A	B	C	D	E	F	G	H	I
1	Ad Count	Min/Hour							
2	25	11.5							
3	23	10.7							
4	28	10.2							
5	15	9.3							
6	13	11.3							
7	24	11							
8	27	15							
9	22	12							
10	17	10							
11	19	10.5							
12	20	14.3							
13	22	11.7							
14	18	14.9							
15	19	10.7							
16	23	12.3							
17	13	10.1							
18	23	11.2							
19	21	10.8							
20	22	10.3							
21	25	15.7							
22									

```
Sheet1 / Sheet2 / Sheet3 /
Ready                                                   NUM
```

Entering and Correcting Data

To select a cell for content entry, move the mouse pointer to the cell and click. Then type the label or data and press Enter. Excel automatically moves to select the next cell in the same column. If you want to enter information in a different cell, just click on it.

Errors are easily fixed. If you notice a mistake before you press Enter, simply back-arrow to the mistake and correct it. If you notice the error after you have pressed Enter, select the affected cell and then click on the formula bar to add a typing cursor to the cell contents displayed. Use standard keyboard editing techniques to make corrections, then press Enter.

If you want to erase the contents of a cell or range of cells but keep the formatting, select the cells and use ➤ **Edit** ➤ **Clear** ➤ **Contents** from the main menu (or just press Delete). Other options under **Edit** ➤ **Clear** have slightly different effects. The ➤ **Clear** ➤ **Format** option keeps the content but clears the format. The ➤ **Clear** ➤ **All** option clears both content and format. ➤ **Clear** ➤ **Format** is especially useful for changing percent data back into decimal format.

Arithmetic Options on the Standard Toolbar

Summing Data in a Column

On the standard toolbar, the $\boxed{\Sigma}$ button automatically sums the values in the selected cells. When we sum the contents of an entire column, Excel places the sum under the selected cells. It is a good idea to type the label *total* next to the cell where the total appears. Below, we selected Column A, pressed the $\boxed{\Sigma}$ button, and then typed the word *total* in the corresponding row of Column C. We see that the total of Column A is 419.

	A	B	C	D	E	F	G	H	I
	Ad Count	Min/Hour							
1	Ad Count	Min/Hour							
2	25	11.5							
3	23	10.7							
4	28	10.2							
5	15	9.3							
6	13	11.3							
7	24	11							
8	27	15							
9	22	12							
10	17	10							
11	19	10.5							
12	20	14.3							
13	22	11.7							
14	18	14.9							
15	19	10.7							
16	23	12.3							
17	13	10.1							
18	23	11.2							
19	21	10.8							
20	22	10.3							
21	25	15.7							
22	419		Total						
23									

Sheet1 / Sheet2 / Sheet3 /

Ready   Sum=838   NUM

Sorting Data

The two buttons ▲↓ and ▼↓ sort the data in ascending and descending order, respectively. To sort just one column, highlight that column and press one of the buttons. To sort two or more columns by ascending or descending order of the data in the first column, highlight all the columns and click the appropriate button. In general, we will simply sort one column of data at a time, as shown.

X Microsoft Excel - Book1ads									_ □ ×
File Edit View Insert Format Tools Data Window Help									_ 8 ×

	A	B	C	D	E	F	G	H	I
1	Ad Count	Min/Hour							
2	13	11.5							
3	13	10.7							
4	15	10.2							
5	17	9.3							
6	18	11.3							
7	19	11							
8	19	15							
9	20	12							
10	21	10							
11	22	10.5							
12	22	14.3							
13	22	11.7							
14	23	14.9							
15	23	10.7							
16	23	12.3							
17	24	10.1							
18	25	11.2							
19	25	10.8							
20	27	10.3							
21	28	15.7							
22									
23									

Cell A2 = 13

Sheet1 / Sheet2 / Sheet3 /

Ready   Sum=419   NUM

Notice that the data in the first column is now in ascending order. The data in the second column has not moved.

If you decide that it was a mistake to sort the data this way, and you have not made any other changes since you did the sort, you can use ➤ **Edit** ➤ **Undo** from the main menu. The data will appear in their original order.

Copying Cells

To copy one cell or a block of cells to another location on the worksheet,

(a) Select the cells you wish to copy

(b) From the main menu, select ➤ **Edit** ➤ **Copy.** (The shortcut for this process is **Ctrl-C.**) Notice that the range of cells being copied now has a blinking border around it.

(c) Select the upper-left cell of the block that will receive the copy.

(d) Press Enter. When you press Enter, the copy process is complete and the blinking border around the original cells disappears.
Note: Even if you use ➤ **Edit** ➤ **Paste** or the shortcut Ctrl-V to paste, you must still press Enter to remove the blinking border around the original cells.

To copy one cell or a range of cells to another worksheet or workbook, follow steps (a) and (b) above. For step (c), be sure you are in the destination worksheet or workbook and that the worksheet or workbook is activated. Then proceed with step (d).

USING FORMULAS

A formula is an expression that generates a numerical value in a cell, usually based on values in other cells. Formulas usually involve standard arithmetic operations. Excel uses + for addition, - (hyphen) for subtraction, * for multiplication, / for division, and ^ (carat) for exponentiation (raising to a power).

For instance, if we want to divide the contents of Cell A2 by the contents of Cell B2 and place the results in Cell C2, we do the following:

(a) Make Cell C2 the active cell.

(b) Click in the formula bar and type =A2/B2.

(c) Press Enter

The value in Cell C2 will be the quotient of the values in Cells A2 and B2.

If, for a whole series of rows, we wanted to divide the entry in Column A by the entry in Column B and put the results in Column C, we could repeat the above process over and over. However, the typing would be tedious. We can accomplish the same thing more easily by copying and pasting:

(a) Enter =A2/B2 in Cell C2 as described above

(b) Move the cursor to the lower right corner of Cell C2. The cursor should change shape to small black cross (+). Now hold down the left mouse button and drag the + down until all the cells in Column C in which you want the calculation done are highlighted.

(c) Release the mouse button and press Enter. The cell entries in Column C should equal the quotients of the same-row entries in Columns A and B.

Now, if you click on one of the lower cells in Column C, you will notice that the row number in the cell addresses is not 2 but rather the number of the new cell's row. In general, when a formula is copied from one cell to another, the cell addresses in the formula are automatically adjusted. If the formula =D3+C7 is copied to a new cell three columns right and two rows up from the old one, the pasted formula comes out as =G1+F5. (Three columns right from D is G, two rows up from 3 is 1, and so on.)

Sometimes you will want to prevent the automatic address adjustment. To do this put a dollar sign before any row or column number you want to keep from changing. When the formula =D$3+$C7 is copied to a new cell three columns right and two rows up from the old one, the pasted formula comes out as =G$3+$C5. We will call an address with two $ signs in an *absolute* address, because it always refers to the same cell, no matter where the formula is copy/pasted to. A cell with only on $ sign in it, or none at all, we will call a *relative* address, because the cell referred to can change as the formula is pasted from one location to another.

SAVING WORKBOOKS

After you have entered data into an Excel spreadsheet, it is a good idea to save it. On the main menu, click ➤ **File** ➤ **Save As.** A dialog box will appear, similar to the one at the top of the next page.

Save As dialog box

```
Save in:  3½ Floppy (A:)

Book1ads

                                              Save
                                              Cancel
                                              Options...

File name:      Book1ads
Save as type:   Microsoft Excel Workbook
```

If you are in a college computer lab, you might save your files to a floppy disk. We named the workbook on TV ads Book1ads.

It is a good idea to save your workbook periodically as you are working on it. After you have saved the workbook for the first time, you can save updates during your working session by using the save button on the standard toolbar. This is the third button from the left; it looks like a diskette.

To retrieve an Excel Workbook, go to the main menu and click ➤ **File** ➤ **Open.** Select the drive containing your workbook and then the desired workbook file.

Printing Your Worksheets

Clicking the printer button on the standard toolbar (the icon that looks like a little printer) will open a printer dialog box.

Print dialog box

```
Printer
Name:    HP LaserJet 8000 Series PS          Properties
Status:  Idle
Type:    HP LaserJet 8000 Series PS
Where:   \\HP_Network_Printers\192.168.1.17
Comment:                                     ☐ Print to file

Print range              Copies
● All                    Number of copies:  1
○ Page(s)  From:   To:
                                            ☑ Collate
Print what
○ Selection   ○ Entire workbook
● Active sheet(s)

Preview              OK        Cancel
```

If you select a range of cells on the worksheet before you print, you may print the selected material. Notice that you can tell the printer what to print by clicking next to Selection, Active sheet(s) or Entire Workbook. Clicking the Preview button lets you see what you will print before the page is printed.

LAB ACTIVITIES FOR GETTING STARTED

1. Go to your computer lab (or use your own computer) and open Excel. Check to see if you have the Analysis ToolPak add-in. If so, be sure it is activated.

2. If you have not already done so, enter the TV Ad Count and Min/Hour data into the workbook. Use Column A for the Ad Count and Column B for the Min/Hour data.

3. Save the workbook as Book1ads.

4. Select the cells containing the labels and data and print.

5. (a) In Column C, place the quotients A/B for the Rows 2 through 21. Use the formula bar in Cell C2, and drag with the little + symbol to complete the quotients for all rows. Note that if the calculation mode (see Chapter 2) has been set to Manual, you may need to use the key combination **Shift-F9** after the cells are highlighted.

   (b) Use the sum button ⬛Σ on the toolbar to total up the Ad Count column and to total up the Min/Hour column. Place a label in Column D adjacent to the totals.

   (c) Select the data in Columns A, B, C, and D and print it.

6. In this problem we will copy a column of data and sort the copy.

   (a) Select Column A (Ad Count) and copy it to Column D.

   (b) Select Column D and sort it in ascending order (use only the original data, not the sum).

   (c) Select both Column A (Ad Count) and Column B (Min/Hour). Sort these columns by Column A in descending order. Are the data entries of 13 in Column A still next to the data entries 11.3 and 10.1 in Column B? Are the data in Column B sorted?

RANDOM SAMPLES (SECTION 1.2 OF *UNDERSTANDABLE STATISTICS*)

Excel has several random number generators. The one we will find most convenient is the function **RANDBETWEEN(bottom,top).** This function generates a random integer between (inclusively) whatever integer is put in for "bottom" and whatever integer is put in for "top."

To use RANDBETWEEN, select a cell in the active worksheet. Click in the formula bar, type =, and on the standard toolbar click the Paste Function button

Pressing this button calls up a two-column menu, similar to the one at the top of the next page. Select All in the Function category, and then scroll down on the Function name side until you reach RANDBETWEEN. Note that this command is present only if you have the Analysis ToolPak checked under ➤ **Tools** ➤ **Add-Ins.**

Paste Function

Function category:	Function name:
Most Recently Used	QUARTILE
All	QUOTIENT
Financial	RADIANS
Date & Time	RAND
Math & Trig	RANDBETWEEN
Statistical	RANK
Lookup & Reference	RATE
Database	RECEIVED
Text	REGISTER.ID
Logical	REPLACE
Information	REPT

RANDBETWEEN(bottom,top)

Returns a random number between the numbers you specify.

OK   Cancel

Select RANDBETWEEN and then fill in the bottom and top numbers. Alternatively, you may simply type =RANDBETWEEN(bottom,top) in the formula bar, with numbers in place of bottom and top.

The random number generators of Excel have the characteristic that whenever a command is entered anywhere in the active workbook, the random numbers change because they are recalculated. To prevent this from happening, change the recalculate mode from automatic to manual. Select ➤ **Tools** ➤ **Options,** and then click on the tab labeled Calculation. Select Manual calculation, then press Enter.

Options

Tabs: Transition | Custom Lists | Chart | Color | View | Calculation | Edit | General

Calculation
- ○ Automatic
- ○ Automatic except tables
- ● Manual
- ☑ Recalculate before save

Calc Now (F9)
Calc Sheet

☐ Iteration
Maximum iterations: 100   Maximum change: 0.001

Workbook options
- ☑ Update remote references
- ☐ Precision as displayed
- ☐ 1904 date system
- ☑ Save external link values
- ☑ Accept labels in formulas

OK   Cancel

With automatic recalculation disabled, you can still recalculate by pressing the **Shift-F9** key combination. Let us see this in an example, where we select a list of random numbers in a designated range and sort the list in ascending order.

Example

There are 175 students enrolled in a large section of introductory statistics. Draw a random sample of fifteen of the students.

We assign each of the students a distinct number between 1 and 175. To find the numbers of the fifteen students to be included in the sample, we do the following steps.

(a) Change the Calculation mode to Manual.

(b) Type the label Sample in Cell A1.

(c) Select Cell A2.

(d) Type =RANDBETWEEN(1,175) in the formula bar and press Enter.

(e) Position the mouse pointer in the lower right corner of Cell A2 until it becomes a + sign, and click-drag downward until you reach Cell A16. Release. Then press **Shift-F9.**

X Microsoft Excel - Book1									
File Edit View Insert Format Tools Data Window Help									
A2		=	=RANDBETWEEN(1,175)						
	A	B	C	D	E	F	G	H	I
1	Sample								
2	119								
3	67								
4	106								
5	83								
6	175								
7	102								
8	99								
9	141								
10	23								
11	129								
12	156								
13	63								
14	89								
15	119								
16	154								
17									
18									

Sheet1 / Sheet2 / Sheet3 /

Ready   NUM

(f) Use one of the Sort buttons to sort the data, so you can easily check for repetitions. If there are repetitions, press **Shift-F9** again and re-sort. Below, with the data sorted, we can verify that there are no repetitions.

Sometimes we will want to sample from data already in our worksheet. In such case, we can use the **Sampling** dialog box. To reach the **Sampling** dialog box, use the man menu toolbar and select Sampling under ➤ **Tools** ➤ **Data Analysis.**

Example

Enter the even numbers from 0 through 200 in Column A. Then take a sample of size ten, without replacement, from the population of even numbers 0 through 200, and place the results in Column B.

First we need to enter the even numbers 0 through 200 in Column A. Let's type the label Even # in Cell A1. The easiest way to generate the even numbers from 0 through 200 is to use the **Fill** menu selection. To do this, we

(a) Place the value 0 into Cell A2, and finish with Cell A2 highlighted.

(b) From the main menu, select ➤ **Edit** ➤ **Fill** ➤ **Series.** In the dialog box select Series in Columns, Type Linear. Enter 2 as the Step value and 200 as the Stop value. Press OK.

Now Column A should contain the even numbers from 0 to 200.

Now we will use the Sampling dialog box to select a sample of size ten from Column A, and we will place the sample in Column B. Notice that we labeled Column B as Sample. To draw the sample,

(a) From the main menu select ➤**Tools**➤**Data Analysis**➤**Sampling.**

(b) In the dialog box, designate the data range from which we are sampling A1:A102. Also specify that the range contains a label. Select Random and enter 10 as the Number of Samples. Finally select the output range and type the destination B2:B11. Note that B1 already contains the label. Press Enter.

The worksheet now shows the random sample in Column B.

Note: After you finish the random number examples and the lab activities, you may want to set the calculation mode back to Automatic, especially if you are using a school computer.

LAB ACTIVITIES FOR RANDOM SAMPLES

1. Out of a population of 8173 eligible county residents, select a random sample of fifty for prospective jury duty. (Should you sample with or without replacement?) Use the RANDBETWEEN command with bottom value 1 and top value 8173. Then sort the data to check for repetitions. Note: Be sure that Calculation mode is set to manual and use the **Shift-F9** key combination to generate the sample in Rows 2 through 51 of Column A.

2. Retrieve the Excel worksheet Svls02.xls from the data CD-ROM. This file contains weights of a random sample of linebackers on professional football teams. The data is in column form. Use the SAMPLING dialog box to take a random sample of ten of these weights. Print the ten weights included in the sample.

Simulating experiments in which outcomes are equally likely is another important use of random numbers.

3. We can simulate dealing bridge hands by numbering the cards in a bridge deck from 1 to 52. Then we draw a random sample of thirteen numbers without replacement from the population of 52 numbers. A bridge deck has four suits: hearts, diamonds, clubs, and spades. Each suit contains thirteen cards; those numbered 2 through 10, a jack, a queen, a king, and an ace. Decide how to assign the numbers 1 through 52 to the cards in the deck.

   (a) Use RANDBETWEEN to generate the numbers of the thirteen cards in one hand. Translate the numbers to specific cards and tell what cards are in the hand. For a second game, the cards would be collected and reshuffled. Use the computer to determine the hand you might get in a second game.

   (b) Generate the numbers 1–52 in Column A, and then use the SAMPLING dialog box to sample thirteen cards. Put the results in Column B, Label Column B as "My hand" and print the results. Repeat this process to determine the hand you might get in a second game.

   (c) Compare the four hands you have generated. Are they different? Would you expect this result?

4. We can also simulate the experiment of tossing a fair coin. The possible outcomes resulting from tossing a coin are heads or tails. Assign the outcome heads the number 2 and the outcome tails the number 1. Use RANDBETWEEN(1,2) to simulate the act of tossing a coin ten times. Simulate the experiment another time. Do you get the same sequence of outcomes? Do you expect to? Why or why not.

CHAPTER 2 ORGANIZING DATA

BAR GRAPHS, CIRCLE GRAPHS, AND TIME PLOTS (SECTION 2.1 OF *UNDERSTANDABLE STATISTICS*)

Excel has a Chart Wizard that produces a wide variety of charts. To access the Chart Wizard, use the Standard toolbar and select the button with a picture of a chart on it:

When you press the Chart Wizard button, a dialog box like the following is displayed.

Chart Wizard - Step 1 of 4 - Chart Type

Standard Types | Custom Types

Chart type:
- Column
- Bar
- Line
- Pie
- XY (Scatter)
- Area
- Doughnut
- Radar
- Surface
- Bubble
- Stock

Chart sub-type:

Clustered Column. Compares values across categories.

Press and hold to view sample

Cancel | < Back | Next > | Finish

As you can see, you choose from a variety of chart types and select the chart subtype you want.

Bar Graphs

You have the option of making a vertical bar graph (called a column graph in Excel) or a horizontal bar graph (called a bar graph in Excel).

Before making a chart, you must enter the necessary data in a worksheet, in rows or columns with appropriate row and column headers.

Example

If you are out hiking, and the air temperature is 50°F with no wind, a light jacket will keep you comfortable. However, if a wind comes up, you will feel cold, even though the temperature has not changed. This is called wind chill. In the following spreadsheet, wind speeds and equivalent temperatures as a result of wind chill are given for a calm-air temperature of 50°F.

Notice that the Columns A and B are widened, and the labels are typed in bold. Also, each wind speed is followed by a label. This will cause Excel to treat the wind speeds as row headers, rather than as numerical values. Now, after making sure that a cell in or touching the data blocks is selected, call up the Chart Wizard, choose the first option, and view a sample chart. The row headers, i.e., the wind speeds, give us the labels on the horizontal axis, while the values in Column B, the equivalent temperatures, appear as bar heights.

In the second step of the Chart Wizard you see this dialog box:

Since the data values are in a column (Column B), we select Series in Columns. Check that the data range is correct. If you click on the Series tab, you can set other parameters. See what happens when you select Series in Rows instead of Series in Columns under the Data Range tab. Click Next. Then click Back and reselect Series in Columns. Again, click Next.

Notice again that there are several tabs. Click on the various tabs to see a variety of options for labeling the axes, setting grids, labeling the heights of the bars, etc. Try different buttons and options. If you do not like the results, deselect the options and return to the Titles tab. Notice that we typed in a title for the graph and also a label for the horizontal axis. Click Next.

The last dialog box asks where we want the graph to appear. If we select "As new sheet," we will get the chart on a worksheet all by itself. Selecting "As object in" places the chart on the designated sheet. It will appear with your data. The chart will have handles on it. When you click inside the chart and hold down the buttons, you can drag the chart to a new location on the worksheet. You can also use the handles to resize the graph. Click outside the chart to remove the chart handles.

Circle Graphs

Another option available from the Chart Wizard is Pie Chart format.

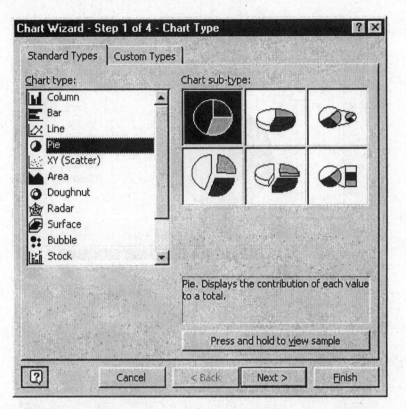

Again, enter the data in the worksheet first.

Example

Where do we hide the mess when company comes? According to *USA Today,* a survey showed that 68% of the respondents hide the mess in the closet, 23% put things under the bed, 6% put things in the bathtub, and 3% put things in the freezer. Make a circle graph for these data. We will put labels in Column A and the percents in Column B.

Select the basic pie chart on the Chart Wizard. In the Data Range dialog box, select
Series in Columns, then click Next.

For the title, type "Where We Hide the Mess." Click Next. Select "As object in" and then Finish.

Time Plots

You can make time charts by selecting Line Charts in the Chart Wizard. In step three, select the Axes tab and then choose Time-scale.

LAB ACTIVITIES FOR BAR GRAPHS, CIRCLE GRAPHS, AND TIME PLOTS

1. According to a survey of chief information officers at large companies, the technology skills most in demand are: Networking 33%; Internet/intranet development, 21%; Applications development 18%; Help desk/user support, 8%; Operations, 6%; Project management, 6%, Systems analysis, 5%; Other 3%.

   (a) Make a bar graph displaying this data.

   (b) Make a circle graph displaying this data.

2. In survey where respondents could name more than one choice, on-line Internet users were asked where they obtained news about current events. The results are: Search engine/directory sites, 49%; Cable news site, 41%; On-line service; 40%; Broadcast news site, 40%; Local newspapers, 30%; National newspaper site; 24%; Other, 13%; National newsweekly site 12%; Haven't accessed news on-line, 11%.

   (a) Make a horizontal bar graph displaying this information.

   (b) Is this information appropriate for a circle graph display? Why or why not?

3. What percentage of its income does the average household spend on food, and how may workdays are devoted to earning the money spent on food in an average household? The American Farm Bureau Federation gave the following information, by year: In 1930, 25% of a household's budget went to food, and it took 91 workdays to earn the money. In 1960, 17% of the budget was for food, and the money took 64 workdays to earn. In 1990, food was 12% of the budget, earned in 43 workdays. For the year 2000, it was projected that the food budget would 11% of total income and that it would take 40 workdays to earn the money.

   (a) Enter these data in an Excel worksheet so you can create graphs.

   (b) Make bar charts for both the percent of budget for food, by year, and for the workdays required.

   (c) Use the Chart Wizard to make a "double" bar graph that shows side-by-side bars, by year, for the percent of budget and for the number of days of work. (You may need to change the format of the first column of numbers to something other than percent.)

   (d) Are these data suitable for a time plot? If so, use the Line graph option in the Chart Wizard to create a time plot that shows both the percent of budget and number of work days consumed on household food.

FREQUENCY DISTRIBUTIONS AND HISTOGRAMS (SECTION 2.2 OF *UNDERSTANDABLE STATISTICS*)

Excel's Histogram dialog box is found under ➤**Tools**➤**Data Analysis**➤**Histogram**.

Example

Let's make a histogram with four classes, using the data we stored in the workbook Book1ads (created in Chapter 1). Use ➤**File**➤**Open** to locate the workbook, and click on it.

The number of ads per hour of TV is in Column A; we will represent these values in the histogram. We also need to specify class boundaries, and for this we will use Column C. Using methods shown in the text *Understandable Statistics*, we see that the upper class boundaries for four classes are 16.5, 20.5, 24.5, and 28.5.

Label Column C as Ad Count, and below the label enter these values, smallest to largest. The horizontal axis of the histogram will carry the label Ad Count. Note: Excel follows the convention that a data value is counted in a class if the value is less than or equal to the upper boundary (upper bin value) of the class.

Now select ➤**Tools**➤**Data Analysis**➤**Histogram**. Check the Labels option, and select New workbook for the output option. Finally, check Chart output for the histogram, and check Cumulative percentage to produce an ogive. Note that you can select both options, or just one of the options. Your dialog box should look similar to the one at the top of the next page.

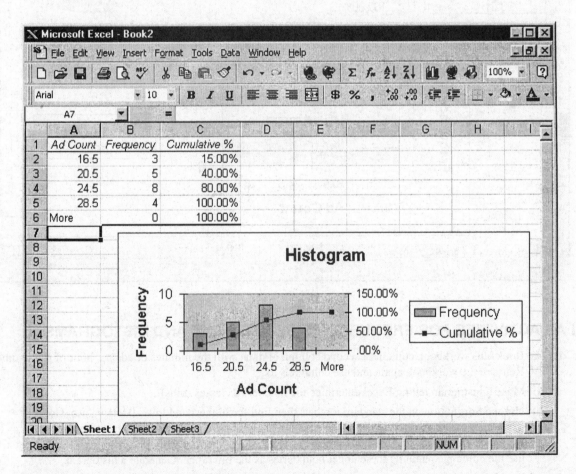

In the resulting worksheet, we moved and resized the chart window. Notice that the class boundaries are not shown directly under the tick marks. The first bar represents all the values less than or equal to 16.5, the second bar is for values greater than 16.5 and less than or equal to 20.5, and so on. There are no values above 28.5.

If you do not specify the cells containing the bin range, Excel automatically creates enough bins (classes) to show the data distribution.

Adjusting the Histogram

Excel automatically supplies the "more" category (which will be empty if you specify upper class boundaries). To remove the "more" category, select the row of the frequency table that contains "more". Then, on the main menu bar select ➤**Edit**➤**Delete**➤**Shift cells up.**

To make the bars touch, *right-click* on one of the bars of the histogram. Then click **Format Data Series** and click the Options tab. Set Gap width to 0. Then press Enter.

The results should be similar to this:

LAB ACTIVITIES FOR FREQUENCY DISTRIBUTIONS AND HISTOGRAMS

1. The Book1ads workbook contains a second column of data, with the minutes of ads per hour of prime time TV. Retrieve the workbook again and use column B to

(a) Make a histogram letting Excel determine the number of classes (bins).

(b) Use the Sort button on the standard toolbar, then find the highest and lowest data values. Use the techniques of the text to find the upper class boundaries for five classes. Make a column in your worksheet that contains these boundaries, and label the column Min/Hr. In the Histogram dialog box, use the column containing these upper boundaries as the bin range. Generate a histogram.

2. As a project for her finance class, Linda gathered data about the number of cash requests made at an automatic teller machine located in the student center between the hours of 6 P.M. and 11 P.M. She made a count every day for four weeks. The results follow.

25	17	33	47	22	32	18	21	12	26	43	25
19	27	26	29	39	12	19	27	10	15	21	20
32	24	17	181								

(a) Enter the data.

(b) Repeat part (b) of Problem 1.

3. Choose one of the following workbooks from the Excel data disk.

DISNEY STOCK VOLUME: **Svls01.xls**
WEIGHTS OF PRO FOOTBALL PLAYERS: **Svls02.xls**
HEIGHTS OF PRO BASKETBALL PLAYERS: **Svls03.xls**
MILES PER GALLON GASOLINE CONSUMPTION: **Svls04.xls**
FASTING GLUCOSE BLOOD TESTS: **Svls05.xls**
NUMBER OF CHILDREN IN RURAL CANADIAN FAMILIES: **Svls06.xls**

(a) Make a histogram letting Excel scale it

(b) Make a histogram using five classes. Use the method of part (b) of Problem 1.

4. Histograms are not effective displays for some data. Consider the data

1	2	3	6	7	9	8	4	12	10
1	9	1	12	12	11	13	4	6	206

Enter the data and make a histogram letting Excel do the scaling. Now drop the high value, 206, from the data set. Do you get more refined information from the histogram by eliminating the high and unusual data value?

CHAPTER 3 AVERAGES AND VARIATION

CENTRAL TENDENCY AND VARIATION OF UNGROUPED DATA (SECTIONS 3.1 AND 3.2 OF *UNDERSTANDABLE STATISTICS*)

Sections 3.1 and 3.2 of *Understandable Statistics* describe some of the measures used to summarize the character of a data set. Excel supports these descriptive measures.

On the standard toolbar, click the Paste Function button.

In the dialog box that appears, select Statistical in the left column, and in the right column select the measure of interest. To find the arithmetic mean of a set of numbers, for instance, we use the function AVERAGE.

Paste Function	? ✕

Function category:
- Most Recently Used
- All
- Financial
- Date & Time
- Math & Trig
- Statistical
- Lookup & Reference
- Database
- Text
- Logical
- Information

Function name:
- AVEDEV
- AVERAGE
- AVERAGEA
- BETADIST
- BETAINV
- BINOMDIST
- CHIDIST
- CHIINV
- CHITEST
- CONFIDENCE
- CORREL

AVERAGE(number1,number2,...)

Returns the average (arithmetic mean) of its arguments, which can be numbers or names, arrays, or references that contain numbers.

[?]   OK   Cancel

Notice that the bottom of the dialog box contains a brief explanation of how the selected function works.

Of the descriptive measures discussed in Sections 3.1 and 3.2 of the text, Excel supports the following:

AVERAGE(data range)

returns the arithmic mean $\bar{x}$ or μ of the data values in the designated cells.

COUNT(data range)

returns the number of data values in the designated cells.

MEDIAN(data range)

returns the median of the data values in the designated cells.

MODE(data range)

returns the mode of the data values in the designated cells.

STDEV(data range)

returns the sample standard deviation of the data values in the designated cells.

STDEVP(data range)

returns the population standard deviation of the data values in the designated cells.

TRIMMEAN(data range, percent as a decimal)

returns a trimmed mean based on the *total* percentage of data removed from both the bottom and top of the ordered data values. If you want a 5% trimmed mean, implying that 5% of the bottom data and 5% of the top data will be removed, then enter 0.10 for the percent in Excel.

Formatting the Worksheet to Display the Summary Statistics

It is a good idea to create a column in which you type the name of each descriptive measure you use, next to the column where the corresponding computations are performed.

Example

Let's again use the data about ads during prime time TV. We will retrieve Book1ads and find summary statistics for Column B, the time taken up with ads. And we will use Column C as our column of names for the descriptive statistics measures. Notice that we widened the column to accommodate the names. We label Column C to remind us that the summary statistics apply to the number of minutes per hour that ads consume.

Ad Count	Min/Hour	Measures for Min/Hr	D
25	11.5	Mean	11.675
23	10.7	Median	
28	10.2	Mode	
15	9.3	StDev s	
13	11.3	StDev p	
24	11	5% trim mean	
27	15	Count of data n	
22	12		
17	10		
19	10.5		
20	14.3		
22	11.7		
18	14.9		
19	10.7		
23	12.3		
13	10.1		
23	11.2		
21	10.8		
22	10.3		
25	15.7		

Formula bar: D2 = =AVERAGE(B2:B21)

Notice that Cell D2 is highlighted, and the formula bar shows the command = AVERAGE(B2:B21). The value in Cell D2 is the mean of the data in the Cells B2 through B21.

To compute the other measures, we enter the appropriate formulas in Column D and identify the measures used in Column C, as shown. Notice that you can type the commands directly in the formula bar (don't forget to put = before the command), or you can use the Paste Function button and retrieve the function from the category Statistical.

```
X Microsoft Excel - Book1ads                                    _ □ X
 File  Edit  View  Insert  Format  Tools  Data  Window  Help    _ 日 X
 [toolbar icons]                                        100%
 Arial          10   B I U                   $ %  ,
    D7                 =   =TRIMMEAN(B2:B21,0.1)
       A        B            C              D       E    F    G    H
 1  Ad Count  Min/Hour  Measures for Min/Hr
 2    25       11.5    Mean               11.675
 3    23       10.7    Median             11.1
 4    28       10.2    Mode               10.7
 5    15        9.3    StDev s            1.848719
 6    13       11.3    StDev p            1.801909
 7    24        11     5% trim mean       11.58333
 8    27        15     Count of data n    20
 9    22        12
 10   17        10
 11   19       10.5
 12   20       14.3
 13   22       11.7
 14   18       14.9
 15   19       10.7
 16   23       12.3
 17   13       10.1
 18   23       11.2
 19   21       10.8
 20   22       10.3
 21   25       15.7
 22
   Sheet1 / Sheet2 / Sheet3
 Ready                                              NUM
```

Don't forget that you can control the number of digits displayed after the decimal by using the buttons on the standard toolbar to increase or decrease the number of decimal places.

Another way to obtain a table of some descriptive statistics is to use the menu choices ➤**Tools**➤**Data Analysis**➤**Descriptive Statistics**. A dialog box appears. If you check Summary statistics, an output table containing the mean, median, mode, and other measures appears.

Descriptive Statistics

Input

Input Range: `$A$1:$B$21`

Grouped By: ● Columns ○ Rows

☑ Labels in First Row

Output options

● Output Range: `$D$1`

○ New Worksheet Ply:

○ New Workbook

☑ Summary statistics

☐ Confidence Level for Mean: `95` %

☐ Kth Largest: `1`

☐ Kth Smallest: `1`

[OK] [Cancel] [Help]

For the Ad count and Min/Hr data the results are.

Microsoft Excel - Book1ads

File Edit View Insert Format Tools Data Window Help

Arial 10 B I U

I26 =

	A	B	C	D	E	F	G
1	Ad Count	Min/Hour		*Ad Count*		*Min/Hour*	
2	25	11.5					
3	23	10.7		Mean	20.95	Mean	11.675
4	28	10.2		Standard Error	0.944443584	Standard Error	0.413386209
5	15	9.3		Median	22	Median	11.1
6	13	11.3		Mode	23	Mode	10.7
7	24	11		Standard Deviation	4.223680111	Standard Deviation	1.848719329
8	27	15		Sample Variance	17.83947368	Sample Variance	3.417763158
9	22	12		Kurtosis	-0.34256161	Kurtosis	0.166618753
10	17	10		Skewness	-0.43164497	Skewness	1.12401684
11	19	10.5		Range	15	Range	6.4
12	20	14.3		Minimum	13	Minimum	9.3
13	22	11.7		Maximum	28	Maximum	15.7
14	18	14.9		Sum	419	Sum	233.5
15	19	10.7		Count	20	Count	20
16	23	12.3					
17	13	10.1					
18	23	11.2					
19	21	10.8					
20	22	10.3					
21	25	15.7					

Sheet1 / Sheet2 / Sheet3 /

Ready NUM

LAB ACTIVITIES FOR CENTRAL TENDENCY AND VARIATION OF UNGROUPED DATA

1. A random sample of twenty people were asked to dial thirty telephone numbers each. The incidence of numbers misdialed by these people follow:

   3 2 0 0 1 5 7 8 2 6
   0 1 2 7 2 5 1 4 5 3

   Enter the data and use the appropriate commands to find the mean, median, mode, sample standard deviation, population standard deviation, 10% trimmed mean, and data count.

2. Consider the test scores of thirty students in a political science class.

   85 73 43 86 73 59 73 84 100 62
   75 87 70 84 97 62 76 89 90 83
   70 65 77 90 94 80 68 91 67 79

   (a) Use the appropriate commands to find the mean, median, mode, sample standard deviation, 10% trimmed mean, and data count.

   (b) Suppose that Greg, a student in a political science course, missed a several classes because of illness. Suppose he took the final exam anyway and made a score of 30 instead of 85 as listed in the data set. Change the 85 (first entry in the data set) to 30 and use the appropriate commands to find the new mean, median, mode, sample standard deviation, 10% trimmed mean, and data count. Compare the new mean, median and standard deviation with the ones in part (a). Which average was most affected: median or mean? What about the standard deviation?

3. Consider the ten data values

4	7	3	15	9	12	10	2	9	10

   (a) Use the appropriate commands to find the sample standard deviation and the population standard deviation. Compare the two values.

   (b) Now consider these fifty data values in the same general range:

7	9	10	6	11	15	17	9	8	2
2	8	11	15	14	12	13	7	6	9
3	9	8	17	8	12	14	4	3	9
2	15	7	8	7	13	15	2	5	6
2	14	9	7	3	15	12	10	9	10

   Again use the appropriate commands to find the sample standard deviation and the population standard deviation. Compare the two values.

   (c) Compare the results of parts (a) and (b). As the sample size increases, does it appear that the difference between the population and sample standard deviations decreases? Why would you expect this result from the formulas?

4. In this problem we will explode the effects of changing data values by multiplying each data value by a constant, or by adding the same constant to each data value.

   (a) Clear your workbook or begin a new one. Then enter the following data into Column A, with the column label "Original" in Cell A1.

   1 8 3 5 7 2 10 9 4 6 3

(b) Now label Column B as A * 10. Select Cell B2. In the formula bar, type

=A2*10

and press Enter. Select Cell B2 again and move the cursor to the lower right corner of Cell B2. A small black + should appear. Click-drag the + down the column so that Column B contains all the data of Column A, but with each value in Column A multiplied by 10.

(c) Now suppose we add 30 to each data value in Column A and put the new data in Column C.

First label Column E as A + 30. Then select Cell C2. In the formula bar type

=A2+30

and press Enter. Then select Cell C2 again and position the cursor in the lower right corner. The cursor should change shape to a small black +. Click-drag down Column C to generate all the entries of Column A increased by 30.

(d) Predict how you think the mean and the standard deviation of Columns B and C will be similar to or different from those values for column A. Use the ➤**Tools**➤**Data Analysis**➤**Descriptive Statistics** dialogue box to find the mean and standard deviation for all three columns. Note: use all three columns as input columns in the dialogue box. Compare the actual results to your predictions. What do you predict will happen to these descriptive statistics values if you multiply each data value of Column A by 50? If you add 50 to each data value in Column A?

BOX-AND-WHISKER PLOTS (SECTION 3.4 OF *UNDERSTANDABLE STATISTICS*)

Excel does not have any commands or dialogue boxes that produce box-and-whisker plots directly. Macros can be written to accomplish the task. However, Excel does have commands to produce the five-number summary, and you can then draw a box-and-whisker plot by hand.

The commands for the five-number summary can be found in the dialogue box obtained by pressing the Paste Function (or function wizard) key on the standard tool bar.

MIN(data range) which returns the minimum data value from the designated cells.

QUARTILE(data range, 1) which returns the first quartile for the data in the designated cells.

MEDIAN(data range) which returns the median of the data in the designated cells.

QUARTILE(data range, 3) which returns the third quartile for the data in the designated cells.

MAX(data range) which returns the maximum data value from the designated cells.

Example

Generate the five-number summary for the number of minutes of ads per hour on commercial TV, using the data in Book1ads.

	A	B	C	D	E	F	G	H
	Ad Count	Min/Hour	**Measures for Min/Hr**					
1	Ad Count	Min/Hour	**Measures for Min/Hr**					
2	25	11.5	Minimum	9.3				
3	23	10.7	Quartile 1	10.45				
4	28	10.2	Median	11.1				
5	15	9.3	Quartile 2	12.075				
6	13	11.3	Maximum	15.7				
7	24	11						
8	27	15						
9	22	12						
10	17	10						
11	19	10.5						
12	20	14.3						
13	22	11.7						
14	18	14.9						
15	19	10.7						
16	23	12.3						
17	13	10.1						
18	23	11.2						
19	21	10.8						
20	22	10.3						
21	25	15.7						
22								

Microsoft Excel - Book1ads

File Edit View Insert Format Tools Data Window Help

Arial   10

D5   =QUARTILE(B2:B21,3)

Sheet1 / Sheet2 / Sheet3

Ready   NUM

In computing quartiles, Excel uses a slightly different process from the one adopted in *Understandable Statistics*. However, the results will generally be nearly the same.

CHAPTER 4 ELEMENTARY PROBABILITY THEORY

SIMULATIONS

Excel has several random number generators. Recall from Chapter 1 that RANDBETWEEN(Bottom,Top) puts out a random integer between (and including) the bottom and top numbers. Again, the Analysis ToolPak needs to be included as an Add-In to make RANDBETWEEN available. To find the RANDBETWEEN function, click the Paste Function or Function Wizard button on the standard tool bar. Then select All in the left column and scroll down in the right column until you find RANDBETWEEN. You can also type the command directly in the formula bar, but again, remember to type = first.

We can use the random number generator to simulate experiments such as tossing coins or rolling dice.

Example

Simulate the experiment of tossing a fair coin 200 times. Look at the percent of heads and the percent of tails. How do these compare with the expected 50% for each?

Assign the outcome heads to digit 1 and tails to digit 2. We will draw a random sample of size 200 from the distributions of integers from a minimum of 1 to a maximum of 2. When using a random number generator, you are best off setting recalculation to manual. To do this, go to ➤**Tools**➤**Options** and, under Calculation, choose the option Manual Calculation.

Now put the label Coin Toss in Cell A1, and enter = RANDBETWEEN(1,2) in Cell A2. Then press Enter. Reselect Cell A2, move the cursor to the lower right corner until the + symbol appears, and drag down through Cell A201. Since calculation is set to manual, press **Shift-F9** to apply the random integer generation command to each of the selected cells. Column A should now have 200 entries.

	Microsoft Excel - Book1								
	A	B	C	D	E	F	G	H	I
1	Coin Toss								
2	2								
3	2								
4	1								
5	1								
6	1								
7	1								
8	1								
9	1								
10	1								
11	2								
12	1								

A2 = =RANDBETWEEN(1,2)

Sheet1 / Sheet2 / Sheet3

Count the number of heads and the number of tails

Next we want to count the number of 1's and 2's in column A. We will set up a table in Columns C and D to display the counts. Label Cell C1 as Outcome. Type Heads in Cell C2 and Tails in Cell C3. Then Label Cell D1 as Frequency.

We will use the COUNTIF command to count the 1's and 2's. The syntax for the COUNTIF command is

$$\text{COUNTIF(data range,condition)}$$

Recall that we assigned the number 1 to the outcome heads and the number 2 to the outcome tails. Select Cell D2, and enter

$$= \text{COUNTIF(A1:A201,1)}$$

This will return the number of 1's, or heads, in the designated cell range. Next select Cell D3, and enter = COUNTIF(A1:A200,2). This will return the number of 2's, or tails, in the same designated cell range.

Compute the relative frequency of each outcome

Let's use column E to display the probability of each outcome. Label Cell E1 as Rel Freq. Select Cell E2. In the formula bar, type

$$=\text{D2/200}$$

and press Enter. Then select Cell E3 and enter =D3/200. The display should be similar to the one shown.

Notice that the percentages of heads and of tails are each close to 50%. This is what the Law of Large Numbers predict. Of course, each time you repeat the simulation, you will in general get slightly different results.

	A	B	C	D	E	F	G	H	I
1	Coin Toss		Outcome	Frequenc	Rel Freq				
2	1		Heads	105	0.525				
3	2		Tails	95	0.475				
4	1								
5	1								
6	1								
7	1								
8	1								
9	1								
10	1								
11	2								
12	1								

LAB ACTIVITIES FOR SIMULATIONS

1. Use the RANDBETWEEN command to simulate 50 tosses of a fair coin. Make a table showing the frequency of the outcomes and the relative frequency. Compare the results with the theoretical expected percents (50% heads, 50% tails). Repeat the process for 500 trials. Are these outcomes closer to the results predicted by theory?

2. Use RANDBETWEEN to simulate 50 rolls of a fair die. Use the number 1 for the bottom value and 6 for the top. Make a table showing the frequency of each outcome and the relative frequency. Compare the results with the theoretical expected percents (16.7% for each outcome). Repeat the process for 500 tosses. Are these outcomes closer to the results predicted by theory?

CHAPTER 5
THE BINOMIAL DISTRIBUTION AND RELATED TOPICS

THE BINOMIAL PROBABILITY DISTRIBUTION (SECTIONS 5.1, 5.2 OF *UNDERSTANDABLE STATISTICS*)

The binomial probability distribution is discussed in Chapter 5 of *Understandable Statistics*. It is a discrete probability distribution controlled by the number of trials, *n,* and the probability of success on a single trial, *p.*

The Excel function that generates binomial probabilities is

$$\text{BINOMDIST(r,n,p,cumulative)}$$

where *r* represents the number of successes. Using TRUE for cumulative returns the cumulative probability of obtaining no more than *r* successes in *n* trials, and using FALSE returns the probability of obtaining exactly *r* successes in *n* trials.

You can type the command directly in the formula bar (don't forget the preceding equal sign), or you can call up the dialog box by pressing the Paste Function (Function Wizard) button on the standard menu bar, then selecting Statistical in the left column and BINOMDIST in the right column.

To compute the probability of exactly three successes out of four trials where the probability of success on a single trial is 0.50, we would enter the following information into the dialog box. When we press Enter, the formula result 0.25 will appear in the active cell.

```
┌─BINOMDIST ─────────────────────────────────────────┐
│      Number_s │3                        │ = 3         │
│        Trials │4                        │ = 4         │
│  Probability_s│0.5                      │ = 0.5       │
│    Cumulative │False                    │ = FALSE     │
│                                                      │
│                                      = 0.25          │
│ Returns the individual term binomial distribution probability. │
│                                                      │
│     Cumulative is a logical value: for the cumulative distribution function, use TRUE; for │
│              the probability mass function, use FALSE. │
│  [?]   Formula result =              [  OK  ]  [ Cancel ] │
└──────────────────────────────────────────────────────┘
```

Example

A surgeon regularly performs a certain difficult operation. The probability of success for any one such operation is $p = 0.73$. Ten similar operations are scheduled. Find the probability of success for 0 through 10 successes out of these operations.

First let's put information regarding the number of trials and probability of success on a single trial into the worksheet. We type n = 10 in Cell A1 and p = 0.73 in Cell B1.

Next, we will put the possible values for the number of successes, *r,* in Cells A3 through A13 with the label **r** in Cell A2. We can also duplicate A2 through A13 in D2 through D13 for easier reading of the finished table. Now place the label **P(r)** in Cell B2 and the label **P(X ≤ r)** in Cell C2. We will have Excel generate the probabilities of the individual number of successes **r** in Cells B3 through B13 and the corresponding cumulative probabilities in Cells C3 through C13.

Generate _P(r)_ values and adjust format

Select Cell B3 as the active cell. In the formula bar enter

=BINOMDIST(A3,10,0.73,false)

and press Enter. Then select Cell B3 again, and move the cursor to the lower right corner of the cell. When the small black + appears, drag through Cell B13. This process generates the probabilities for each value of **r** in Cells A3 through A13.

The probabilities are expressed in scientific notation, where the value after the E indicates that we are to multiply the decimal value by the given power of 10. To reformat the probabilities, select them all and press the comma button on the formatting tool bar. Then press the button to move the decimal point until you see four digits after it.

Generate Cumulative Probabilities _P(X ≤ r)_ and adjust format

Select Cell C3 as the active cell. In the formula bar enter

=BINOMDIST(A3,10,0.73,true)

and press Enter. Then select Cell C3 again, move the cursor to the lower right corner of the cell, and drag down through Cell C13. This generates the cumulative probabilities for each value of **r** in Cells A3 through A13. Again, reformat the probabilities to show four decimal places.

	A	B	C	D
1	n = 10	p = 0.73		
2	r	P(r)	P(X ≤ r)	r
3	0	0.0000	0.0000	0
4	1	0.0001	0.0001	1
5	2	0.0007	0.0007	2
6	3	0.0049	0.0056	3
7	4	0.0231	0.0287	4
8	5	0.0750	0.1037	5
9	6	0.1689	0.2726	6
10	7	0.2609	0.5335	7
11	8	0.2646	0.7981	8
12	9	0.1590	0.9570	9
13	10	0.0430	1.0000	10
14				

(B3 = =BINOMDIST(A3,10,0.73,FALSE))

Next, let's use the Chart Wizard to create a bar graph showing the probability distribution.

Press the Chart Wizard button on the standard tool bar. Select Column and the first sub-type. Press Enter.

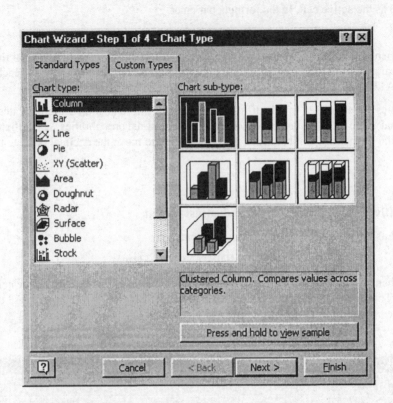

For the data range, activate Cells B2 through B13 and press Enter.

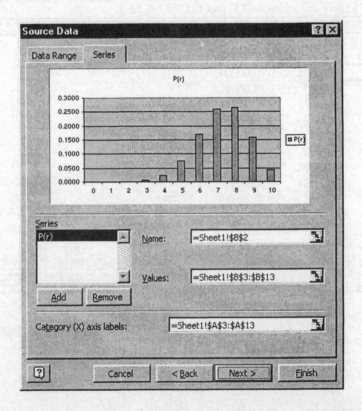

Select the Series Tab, and go to Category (X) axis labels. Select Cells A3 through A13 on the worksheet.

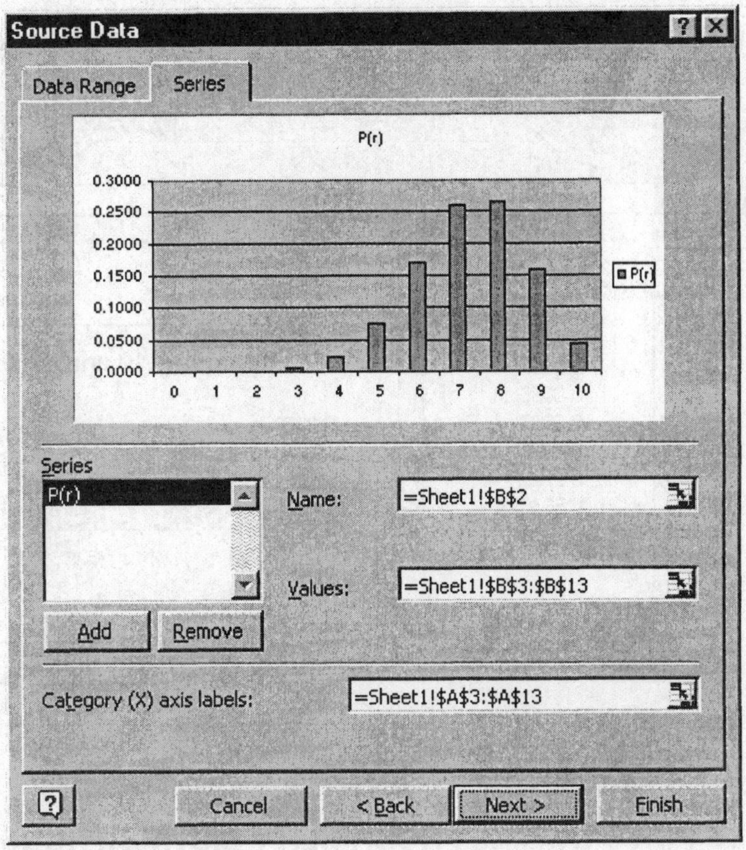

Press Enter or click Next. In the dialog box, fill in the Titles options as shown below and press Finish. The graph will appear on your worksheet.

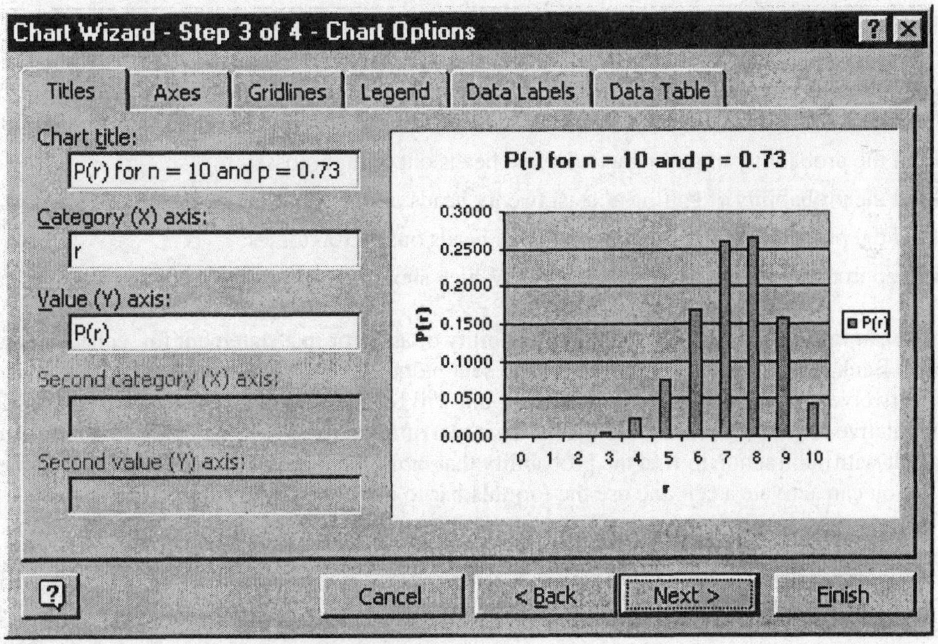

Move the graph so it doesn't block the data columns, and size it as you wish.

LAB ACTIVITIES FOR BINOMIAL PROBABILITY DISTRIBUTIONS

1. You toss a coin n times. Call heads success. If the coin is fair, the probability of success p is 0.5. Use BINOMDIST command with False for cumulative to find each of the following probabilities.

   (a) Find the probability of getting exactly five heads out of eight tosses.

   (b) Find the probability of getting exactly twenty heads out of 100 tosses.

   (c) Find the probability of getting exactly forty heads out of 100 tosses.

2. You toss a coin n times. Call heads success. If the coin is fair, the probability of success p is 0.5. Use BINOMDIST command with True for cumulative to find each of the following probabilities

   (a) Find the probability of getting at least five heads out of eight tosses.

   (b) Find the probability of getting at least twenty heads out of 100 tosses.

   (c) Find the probability of getting at least forty heads out of 100 tosses.

   Hint: Keep in mind how BINOMDIST works. Which should be larger, the value in (b) or the value in (c)?

3. A bank examiner's record shows that the probability of an error in a statement for a checking account at Trust Us Bank is 0.03. The bank statements are sent monthly. What is the probability that exactly two of the next twelve monthly statements for our account will be in error? Now use the BINOMDIST with True for cumulative to find the probability that *at least* two of the next twelve statements contain errors. Use this result with subtraction to find the probability that *more than* two of the next twelve statements contain errors. You can activate a cell and use the formula bar to do the required subtraction.

4. Some tables for the binomial distribution give values only up to 0.5 for the probability of success p. There is a symmetry between values of p greater than 0.5 and values of p less than 0.5.

   (a) Consider the binomial distribution with $n = 10$ and $p = .75$. Since there are anywhere from 0 to 10 successes possible, put the numbers 0 through 10 in Cells A2 through A12. Use Cell A1 for the label **r.** Use BINOMDIST with cumulative False option to generate the probabilities for $r = 0$ through 10. Store the results in Cells B2 through B12. Use Cell B1 for the label **p = 0.75.**

   (b) Now consider the binomial distribution with $n = 10$ and $p = .25$. Use BINOMDIST with cumulative False option to generate the probabilities for $r = 0$ through 10. Store the results in Cells C2 through C12. Use Cell C1 for the label **p = 0.25.**

   (c) Now compare the entries in Columns B and C. How does $P(r = 4$ successes with $p = .75)$ compare to $P(r = 6$ successes with $p = .25)$?

5. (a) Consider a binomial distribution with fifteen trials and probability of success on a single trial $p = 0.25$. Create a worksheet showing values of r and corresponding binomial probabilities. Generate a bar graph of the distribution.

   (b) Consider a binomial distribution with fifteen trials and probability of success on a single trial $p = 0.75$. Create a worksheet showing values of r and the corresponding binomial probabilities. Generate a bar graph.

   (c) Compare the graphs of parts (a) and (b). How are they skewed? Is one symmetric with the other?

CHAPTER 6 NORMAL DISTRIBUTIONS

GRAPHS OF NORMAL PROBABILITY DISTRIBUTIONS (SECTION 6.1 OF *UNDERSTANDABLE STATISTICS*)

A normal distribution is a continuous probability distribution governed by the parameters μ (the mean) and σ (the standard deviation), as discussed in Section 6.1 of *Understandable Statistics*. The Excel command that generates values for a normal distribution is

NORMDIST(x,mean,standard deviation,cumulative)

When we set the value of cumulative to FALSE, the command gives the values of the normal probability density function for the corresponding x value. You can type the command directly into the formula bar, or find it by using the Paste Function button on the standard toolbar, selecting Statistical in the left column and scrolling to NORMDIST in the right column to call up the dialog box below. We filled in the entries that we will use in the next example.

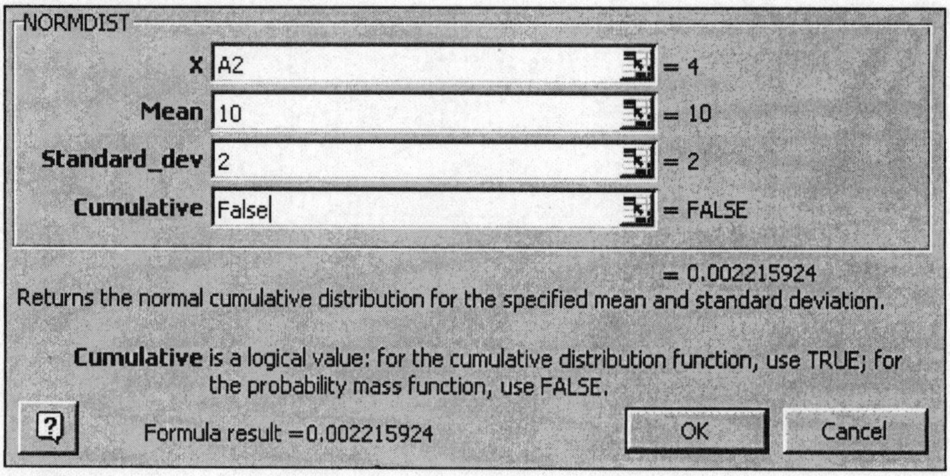

```
┌─NORMDIST────────────────────────────────────────────────┐
│            X  │A2                          │ = 4          │
│         Mean  │10                          │ = 10         │
│  Standard_dev │2                           │ = 2          │
│    Cumulative │False                       │ = FALSE      │
│                                                           │
│                                         = 0.002215924     │
│  Returns the normal cumulative distribution for the       │
│  specified mean and standard deviation.                   │
│                                                           │
│   Cumulative is a logical value: for the cumulative       │
│   distribution function, use TRUE; for                    │
│       the probability mass function, use FALSE.           │
│                                                           │
│  [?]   Formula result =0.002215924     [ OK ]  [ Cancel ] │
└───────────────────────────────────────────────────────────┘
```

Example

Graph the normal distribution with mean $\mu = 10$ and standard deviation $\sigma = 2$.

Since most of the normal curve occurs over the values $\mu - 3\sigma$ to $\mu + 3\sigma$, we will start the graph at $10 - 3(2) = 4$ and end it at $10 + 3(2) = 16$. We will let Excel set the scale on the vertical axis automatically.

Generate the column of x values

To graph a normal distribution, we must have a column of x values and a column of corresponding y values. We begin by generating a column of x values ranging 4 to 16, with an increment of 0.25. To do this, select Cell A2 and enter the number 4. Select Cell A2 again and use the ➤**Edit**➤**Fill**➤**Series** to open the dialogue box shown next. Set the options in the dialogue box as shown and then press Enter.

You will see that Column A now contains the numbers 4, 4.25, 4.50, ... all the way up to 16.

Generate the column of y values

Select Cell B2, and enter

$$=NORMDIST(A2,10,2,False)$$

Press Enter. Select Cell B2 again and move the cursor to the lower right corner of the cell. When the cursor changes to a small black +, hold down the left mouse button and drag down the column until each Column-A entry has a corresponding Column-B entry. Release the left mouse button. All the y values should now appear.

Create the graph of a normal distribution

We will use the Chart Wizard to create the graph of a normal distribution. On the standard toolbar, press the Chart Wizard button. Select Line Graph and choose the first chart sub-type. Click Next.

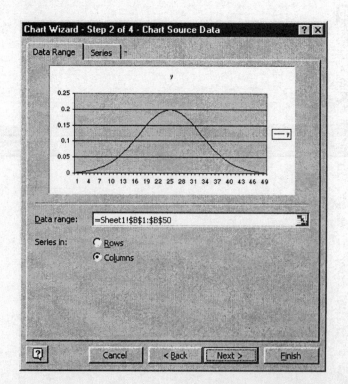

Enter all of Column B as the Data range.

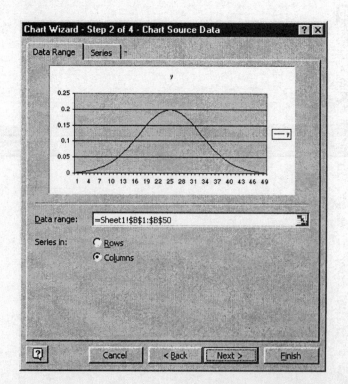

Click on the series tab. Click to place the cursor at Category (*x*) axis labels, and then, in the worksheet, select the cells in Column A, starting with Cell A2. Click Next.

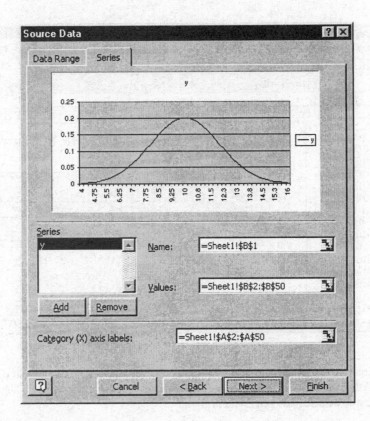

Next we will add a title and variable label to the *x*-axis. In the Chart Title, type

Normal Distribution, Mean = 10, St Dev = 2

For Category (X) axis, type x. Click Finish.

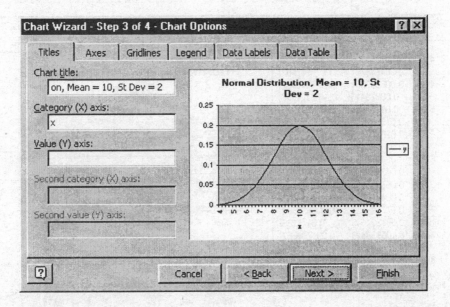

Now your worksheet will contain the graph of the described normal distribution. Move the graph and size it to your liking. Notice that as you make the graph wider or taller, the labels shown on the *x*-axis might change.

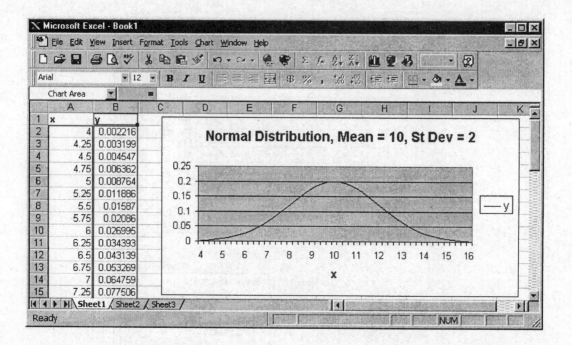

We can also graph two (or more) normal distributions on the same graph. In the next display we generated *y* values for a normal distribution with mean 10 and standard deviation 1 in Column C. Then, in the Data Range box of the Chart Wizard, we selected both Column B and C.

	A	B	C
1	x	y2 (st dev = 2)	y1 (st dev = 1)
2	4	0.002215924	6.07588E-09
3	4.25	0.00319906	2.63924E-08
4	4.5	0.004546781	1.07698E-07
5	4.75	0.006362091	4.12847E-07
6	5	0.00876415	1.48672E-06
7	5.25	0.01188595	5.02951E-06
8	5.5	0.015869826	1.59837E-05
9	5.75	0.020860493	4.77186E-05
10	6	0.026995483	0.00013383
11	6.25	0.034393138	0.000352596
12	6.5	0.043138659	0.000872683
13	6.75	0.053269134	0.002029048
14	7	0.064758798	0.004431848
15	7.25	0.077506133	0.009093563
16	7.5	0.091324543	0.0175283
17	7.75	0.105938323	0.031739652
18	8	0.120985362	0.053990967
19	8.25	0.136027499	0.086277319
20	8.5	0.150568716	0.129517596

STANDARD SCORES AND NORMAL PROBABILITIES

Excel has several built-in functions relating to normal distributions.

STANDARDIZE(x,mean,standard deviation) which returns the z score for the given x value from a distribution with the specified mean and standard deviation.

NORMDIST(x_1,mean,standard deviation,cumulative) which when cumulative is TRUE, returns probability that a random value x selected from this distribution is $\leq x_1$, i.e. it returns $P(x \leq x_1)$. This is the same as the area to the left of the specified x_1 value under the described normal distribution. When cumulative is FALSE, it returns the height of the normal probability density function evaluated at x_1. We used this function to graph a normal distribution.

NORMINV(probability,mean,standard deviation) which returns the inverse of the normal cumulative distribution. In other words, when a probability is entered, the command returns the value from the normal distribution with specified mean and standard deviation so that the area to the left of that value is equal to the designated probability.

NORMSDIST(z_1) which returns the probability that a randomly selected z score is less or equal the specified value of z_1, i.e., it returns $P(z \leq z_1)$. This is the same as the area to the left of the specified z_1 value under the standard normal distribution. This command is equivalent to NORMDIST(x_1,0,1,true).

NORMSINV(probability) which returns the value such that the area to its left under the standard normal distribution is equal to the specified probability. This command is equivalent to NORMINV(probability,0,1).

Each of these commands can be typed directly into the formula bar for an active cell or accessed by using the Paste Function button.

Examples

(a) Consider a normal distribution with mean 100 and standard deviation 15. Find the z score corresponding to $x = 90$ and find the area to the left of 90 under the distribution.

First we place some headers and labels on the worksheet. Then,

in Cell B3, enter =STANDARDIZE(90,100,15), and

in Cell C3, enter =NORMDIST(90,100,15,true).

(b) Find the z score so that 10% of the area under the standard normal distribution is to the left of z.

Again, we put some labels on the worksheet. Then, since we are working with a standard normal distribution, we use NORMSINV(0.1).

X Microsoft Excel - 06-13								_ □ ×
File Edit View Insert Format Tools Data Window Help								_ ⊟ ×

B3   = =NORMSINV(0.1)

	A	B	C	D	E	F	G	H	I
2	P(z <= z_1)	z_1							
3	0.1	-1.28155							
4									
5									

Sheet1 / Sheet2 / Sheet3 /

Ready   NUM

To find areas under normal curves between two values, we do simple arithmetic with the cumulative areas provided by Excel. For instance, to find the area under a standard normal distribution between –2 and 3, we would use NORMDIST(3) to find the cumulative area to the left of 3 and then subtract the cumulative area to the left of –2, found using NORMDIST(-2).

X Microsoft Excel - Book1								_ □ ×
File Edit View Insert Format Tools Data Window Help								_ ⊟ ×

C2   = =A2 - B2

	A	B	C	D	E	F	G	H
1	P(z <= 3)	P(z <= -2)	P(-2 <= z <= 3)					
2	0.998650033	0.022750062	0.975899971					
3								
4								

Sheet1 / Sheet2 / Sheet3 /

Ready   NUM

To find areas under normal curves to the right of a specified value, we subtract the cumulative area to the left of the value from 1. For instance, consider the normal distribution with mean 50 and standard deviation 5. At the top of the next page is a worksheet in which the area to the right of 60 is found.

```
X Microsoft Excel - Book1                                                    _ □ X
 File  Edit  View  Insert  Format  Tools  Data  Window  Help               _ 日 X
 [toolbar icons]                                          100%  ▼  🔁
 Arial            ▼ 10  ▼  B  I  U  ≡ ≡ ≡ 圉  $ % ,  .0 .00  ⌐ ⌐  □ ▼ ⌐ ▼ A ▼
        C3           ▼         =  =1 - B3
        A           B         C         D      E      F      G      H      I
  2  x₁           P(x <= x₁) P(x > x₁)
  3        60     0.97725   0.02275
  4
  5
 [◄ ◄ ► ►] \Sheet1 / Sheet2 / Sheet3 /              |◄|                    ►|
 Ready                                                          NUM
```

LAB ACTIVITIES FOR NORMAL DISTRIBUTIONS

1. (a) Use the Chart Wizard to sketch a graph of the standard normal distribution with a mean of 0 and a standard deviation of 1. Generate x values in column A ranging from –3 to 3 in increments of 0.5. Use NORMDIST to generate the y values in column B.

    (b) Use the Chart Wizard to sketch a graph of a normal distribution with a mean of 10 and a standard deviation of 1. Generate x values in Column A ranging from 7 to 13 in increments of 0.5. Use NORMDIST to generate the y values in column B. Compare the graphs of parts (a) and (b). Do the height and spread of the graphs appear to be the same? What is different? Why would you expect this difference.

    (c) Sketch a graph of a normal distribution with a mean of 0 and a standard deviation of 2. Generate x values in Column A ranging from –6 to 6 in increments of 0.5. Use NORMDIST to generate the y values in Column B. Compare that graph to that of part (a). Do the height and spread of the graphs appear to be the same? What is different? Why would you expect this difference? Note, to really compare the graphs, it is best to graph them using the same scales. Redo the graph of part (a) using x from –6 to 6. Then redo the graph in this part using the same x values as in part (a) and y values ranging from 0 to the high value in part (a).

2. Use NORMDIST or NORMSDIST plus arithmetic to find the specified area.

    (a) Find the area to the left of 2 on a standard normal distribution.

    (b) Find the area to the left of –1 on a standard normal distribution.

    (c) Find the area between –1 and 2 on a standard normal distribution.

    (d) Find the area to the right of 2 on a standard normal distribution.

    (e) Find the area to the left of 40 on a normal distribution with $\mu = 50$ and $\sigma = 8$.

    (f) Find the area to the left of 55 on a normal distribution with $\mu = 50$ and $\sigma = 8$.

    (g) Find the area between 40 and 55 on a normal distribution with $\mu = 50$ and $\sigma = 8$.

    (h) Find the area to the right of 55 on a normal distribution with $\mu = 50$ and $\sigma = 8$.

3. Use NORMINV or NORMSINV to find the specified x or z value.

    (a) Find the z value so that 5% of the area under the standard normal curve falls to the left of z.

    (b) Find the z value so that 15% of the area under the standard normal curve falls to the left of z.

    (c) Consider a normal distribution with mean 10 and standard deviation 2. Find the x value so that 5% of the area under the normal curve falls to the left of x.

    (d) Consider a normal distribution with mean 10 and standard deviation 2. Find the x value so that 15% of the area under the normal curve falls to the left of x.

CHAPTER 7
INTRODUCTION TO SAMPLING DISTRIBUTIONS

Excel has no commands that directly support the demonstration of sampling distributions. To do the class project in Section 7.3 of *Understandable Statistics,* use Random Number Generation (under ➤**Tools**➤**Data Analysis**) and the AVERAGE and STDEV functions.

CHAPTER 8 ESTIMATION

CONFIDENCE INTERVALS FOR THE MEAN – WHEN σ IS KNOWN (SECTION 8.1 OF *UNDERSTANDABLE STATISTICS*)

Excel's function for computing confidence intervals for a population mean assumes a normal distribution, regardless of sample size. Therefore, as stated in the text, if the population distribution is not a normal distribution, then a large sample should be used. The command syntax is

CONFIDENCE(alpha, standard deviation, sample size)

where alpha equals 1 minus the confidence level. In other words, an alpha of 0.05 indicates a 95% confidence interval. To generate a 99% confidence interval, use alpha = 0.01. The standard deviation is the population standard deviation σ.

Recall from the discussion in *Understandable Statistics* that a confidence interval for the population mean has the form

$$\bar{x} - E \leq \mu \leq \bar{x} + E$$

Excel uses the formula

$$E = z_c \frac{\sigma}{\sqrt{n}}$$

where z_c is the critical value for the chosen confidence level. For $c = 95\%$, $z_c = 1.96$.

CONFIDENCE returns the value of E. To find the lower boundary of the confidence interval, you must subtract E from the sample mean $\bar{x}$ of your data; to find the upper boundary, you add E to the sample mean.

Example

Lucy decided to try to estimate the average number of miles she drives each day. For a three-month period, she selected a random sample of 35 days and kept a record of the distance driven on each of those sample days. The sample mean was 28.3 miles, and also assume that the standard deviation was $\sigma = 5$ miles. Find a 95% confidence interval for the population mean of miles Lucy drove per day in the three-month period.

For a 95% confidence level, alpha is 0.05. In our Excel worksheet, after putting in some information and labels, we enter =CONFIDENCE(0.05,5,35) in Cell B3 to represent E. In Cell D4, we type =28.3–B3 to compute the lower value of the confidence interval, and we place the upper value of the confidence interval in Cell F4 using =28.3+B3.

Confidence Intervals for the Mean – When σ is unknown (Section 8.2 of *Understandable Statistics*)

When the population standard deviation σ is unknown, the sample standard deviation s is used to replace σ, and the confidence intervals are computed using Student's t distribution.

The confidence interval is computed as

$$\overline{x} - E \quad \text{to} \quad \overline{x} + E$$

where $E = t_c \dfrac{s}{\sqrt{n}}$.

Excel has two commands for Student's t distribution under the statistical options of Paste Function.

TDIST(x, degrees of freedom, tails)

returns the area in the tail of Student's t distribution beyond the specified value of x, for the specified number of degrees of freedom and number of tails (1 or 2).

TINV(probability, degrees of freedom)

returns the t_c value such that the area in the two tails beyond the t_c value equals the specified probability for the specified degrees of freedom.

We can use the TINV command to find the t_c value to use in the confidence interval. For instance, if we have a sample of size 10 with mean $\overline{x} = 6$ and $s = 1.2$, then the t_c value we use in the computation of a 95% confidence interval is 2.262. Notice that for a 95% confidence interval, 5% of the area is in the two tails. The number of degrees of freedom is $10 - 1 = 9$.

We can also find the E value in the Describe Statistics dialog box found by selecting ➤**Tools**➤**Data Analysis**➤**Descriptive Statistics.** If you have data entered into a worksheet, you can use these menu selections to automatically compute the sample mean $\overline{x}$ and the sample standard deviation s for the data, as well as the value of E for the confidence interval on the basis of Student's t distribution no matter what the sample size is.

Example

Open the Excel workbook Svls02.xls. This workbook contains the weights in pounds of fifty randomly selected professional football linebackers.

Use the menu selection ➤**Tools**➤**Data Analysis**➤**Descriptive Statistics** to access the following dialog box. Since the weights are in Column A, we select the cells containing the range and use these for the input range. We place the upper left corner of the output in Cell C1 and check that we want summary statistics and a 95% confidence interval.

Next we widen cells in Column C and D to accommodate the output. Notice that the sample mean is in Cell D3 and the value of *E* for a 95% confidence interval (again, based on Student's *t* distribution, not the standard normal distribution) is in Cell D16. We put the confidence interval in Cells F2 to H2, by entering =D3–D16 in Cell F2 and =D3+D16 in Cell H2. The display should be similar to the one shown at the top of the next page.

X Microsoft Excel - Svls02.xls								
File Edit View Insert Format Tools Data Window Help								
Arial ▾ 10 ▾ **B** *I* U								
H2 = =D3 + D16								

	A	B	C	D	E	F	G	H
1	WEIGHTS		*WEIGHTS*			**95% Confidence Interval**		
2	225					236.1748	to	241.2252
3	230		Mean	238.7				
4	235		Standard Error	1.256574547				
5	238		Median	240				
6	232		Mode	240				
7	227		Standard Deviation	8.885323832				
8	244		Sample Variance	78.94897959				
9	222		Kurtosis	-0.741626333				
10	250		Skewness	-0.115643512				
11	226		Range	33				
12	242		Minimum	222				
13	253		Maximum	255				
14	251		Sum	11935				
15	225		Count	50				
16	229		Confidence Level(95.0%)	2.525179561				
17	247							
18	239							
19	223							
20	233							
21	222							

Sheet1 / Sheet2 / Sheet3 /

Ready   NUM

LAB ACTIVITIES FOR CONFIDENCE INTERVALS FOR THE MEAN

1. Retrieve the worksheet Svls03.xls. This contains the heights (in feet) of 65 randomly selected pro basketball players. Use the Descriptive Statistics dialog box to get summary statistics for the data and to create a 90% confidence interval.

2. Retrieve the worksheet Svls01.xls from the Excel data disk. This worksheet contains the number of shares of Disney Stock (in hundreds of shares) sold for a random sample of sixty trading days in 1993 and 1994. Use the Descriptive Statistics dialog box to get summary statistics for the data and to create the following confidence intervals.

   (a) Find a 99% confidence interval for the population mean volume.

   (b) Find a 95% confidence interval for the population mean volume.

   (c) Find a 90% confidence interval for the population mean volume.

   (d) Find an 85% confidence interval for the population mean volume.

   (e) What do you notice about the lengths of the intervals as the confidence level decreases?

3. There are many types of errors that will cause a computer program to terminate or give incorrect results. One type of error is punctuation. For instance, if a comma is inserted in the wrong place, the program might not run. A study of programs written by students in a beginning programming course showed that 75 out of 300 errors selected at random were punctuation errors. Find a 99% confidence interval for the proportion of errors made by beginning programming students that are punctuation errors. Next find a 90% confidence interval. Use the CONFIDENCE command to find the interval. Is this interval longer or shorter?

CHAPTER 9 HYPOTHESIS TESTING

TESTING A SINGLE POPULATION MEAN – WHEN σ IS KNOWN (SECTIONS 9.2 OF *UNDERSTANDABLE STATISTICS*)

Chapter 9 of *Understandable Statistics* introduces tests of hypotheses. Testing involving a single mean is found in Sections 9.2. Hypothesis tests in this section test the value of the population mean, μ, against some specified value, denoted by k.

When population standard deviation σ is known, one-sample z-tests are appropriate for testing the null hypothesis $H_0: \mu = k$ against one of the three alternative hypotheses $H_1: \mu > k$, $H_1: \mu < k$, or $H_1: \mu \neq k$ when (1) the data in the sample are known to be from a normal distribution (in which case any sample size will do) or when (2) the data distribution is unknown or the data are believed to be from a non-normal distribution, but the sample size, n, is large ($n \geq 30$).

In Excel, the **ZTEST** function finds the P value for an upper- or right-tailed test, used to decide between the hypotheses $H_0: \mu = k$ and $H_1: \mu > k$. The null hypothesis says that value of the population mean μ is k. A right-tailed test is used when the sample mean $\overline{x}$ is greater than k, suggesting that μ may in fact be greater than k, as the alternative hypothesis states. **Note: the Excel documentation for ZTEST should be ignored. It mistakenly says that ZTEST gives the result of a two-tailed test.**

The syntax is

ZTEST(array, x, sigma)

The array is the list of sample values; what Excel calls x is k, and sigma is the known value of σ, the population standard deviation. If the syntax used is ZTEST(array, x), i.e., if there is no sigma value given, then Excel calculates the sample standard deviation, s, from the sample data in array and uses that in place of σ.

The P value returned by ZTEST is the probability, given that the null hypothesis is true, of getting results at least as extreme as those observed in the sample. More precisely, ZTEST gives the probability of obtaining a sample mean greater than or equal to the observed sample mean, $\overline{x}$. When this probability is small, it means that the data in the observed sample would be surprising if H_0 were true. This is a reason to reject H_0.

ZTEST can also be used to apply a left-tailed test ($H_0: \mu = k$ versus $H_1: \mu < k$) or a two-tailed test ($H_0: \mu = k$ versus $H_1: \mu \neq k$). To apply a left-tailed test, for the case where the sample mean x is less than k, simply apply a right-tailed test and then subtract the result from 1. (When $\overline{x} < k$, the area found by ZTEST in the upper "tail" will be greater than 0.5, and 1 minus that area will be the area in the lower tail.) To apply a two-tailed test, either double the P value from a right-tailed test (when $\overline{x} > k$) or double the P value from a left-tailed test (when $\overline{x} < k$).

To call up the ZTEST dialog box, click the Paste Function button on the standard tool bar, select Statistical in the left column, and scroll to ZTEST in the right column. The dialog box should be similar to the one shown on the top of the next page.

Enter the cell range containing the sample values in the Array blank, and in the blank for X, enter the mean given by the null hypothesis. In the Sigma blank, enter the value of the population standard deviation σ if it is known. Again: sigma is optional; if this box is left blank, Excel will compute the sample standard deviation for the data in the specified Array and use s instead of σ in the computation for z. Recall that if we are dealing with large samples, s and σ are fairly close, so this approximation produces reliable results. Finally, when you click on OK, the P value of the right-tailed test of x is computed.

You can also type the command directly into the formula bar using

ZTEST(data range, X, sigma).

Once the P value is computed, the user can then compare it with α, the level of significance of the test. If

P value $\leq \alpha$, we reject the null hypothesis.

P value $> \alpha$, we do not reject the null hypothesis

Example

ZTEST requires the use of large samples (size 30 or greater) when the population distribution is unknown. Let's use some data from the data CD. The Excel workbook Svls03.xls contains heights in feet of 65 randomly selected professional basketball players.

Assume that twenty years ago the average height of professional basketball players was 6.3 feet (that translates to 6 feet 3.6 inches). Let's use the sample data in Svls03.xls to consider whether the current population mean height of professional basketball players is greater than it was twenty years ago. The null hypothesis will be that their average height is the same. Given our alternative hypothesis ("greater than"), we will apply a right-tailed test.

Open the workbook Svls03.xls found on the data CD. The data will appear in Column A. Cell A1 contains the label Heights. The data are in cells A2:A66.

After typing in some labeling information, we want to display the P value provided by ZTEST in Cell F3. Activate Cell F3 and, in the Paste Function dialog box, select Statistical in the left column and ZTEST in the right column. The dialog box should be similar to the one shown at the top of the next page. **Ignore the dialog box statement that this is a two-tailed test. It is right-tailed.**

We will use the cell range A2:A66 for the Array and 6.3 as the value of X. The population standard deviation is not given in this case. To demonstrate the use of ZTEST, let's use the value of s, the sample standard deviation for the value of σ, and let Excel compute s from the sample data. Notice that with the space for Sigma left blank, the dialog box tells us that the P value is 2.12826E-05. We interpret this as the probability that 65 data values could come out with a mean greater than or equal to that of the sample, given that they were taken from a normal distribution with a mean of 6.3.

So the P value is about 0.00002. Since this is less than even the very restrictive $\alpha = 0.01$, we reject the null hypothesis and conclude that the population mean height of pro basketball players now is greater than it was twenty years ago.

For completeness, we also used the **➤Tools➤Data Analysis➤Descriptive statistics** menu choice and dialog box to generate the descriptive statistics for our data. We selected output range beginning in cell C5, and then widened the columns to fit the display.

LAB ACTIVITIES FOR TESTING A SINGLE POPULATION MEAN

1. Open or retrieve the worksheet Svls04.xls from the data CD. The data in Column A of this worksheet represent the miles per gallon gasoline consumption (highway) for a random sample of 55 makes and models of passenger cars (source: Environmental Protection Agency).

30	27	22	25	24	25	24	15
35	35	33	52	49	10	27	18
20	23	24	25	30	24	24	24
18	20	25	27	24	32	29	27
24	27	26	25	24	28	33	30
13	13	21	28	37	35	32	33
29	31	28	28	25	29	31	

Test the hypothesis that the population mean mile per gallon gasoline consumption for such cars is greater than 25 mpg.

(a) Do we know σ for the mpg consumption? If not, use the value of s for the value of σ (sometimes this is done in practice when sample size is large.) Can we use the normal distribution for the hypothesis test?

(b) State the null and alternate hypothesis, and type them on your worksheet.

(c) Use ZTEST with Sigma omitted.

(d) Look at the P value in the output. Compare it to α. Do we reject the null hypothesis or not? Does it depend on the level of significance?

(e) Use the Descriptive Statistics dialog box to generate the summary statistics for the data, and place the results on the worksheet.

TESTS INVOLVING PAIRED DIFFERENCES – DEPENDENT SAMPLES (SECTION 9.4 OF *UNDERSTANDABLE STATISTICS*)

The test for difference of means of dependent samples is presented in Section 9.4 of *Understandable Statistics*. Dependent samples arise from before-and-after studies, some studies of data taken from the same subjects, and some studies on identical twins.

In Excel there are two functions that produce the P value for a one- or two-tailed test of paired differences. The first command is **TTEST,** found using Paste Function with Statistical in the left column and TTEST in the right column. You can also activate a cell and type the command in the formula bar. This command returns only the P value for the test. The syntax is

TTEST(data range of sample 1, data range of sample 2, tails, type)

If tails = 1, then TTEST returns the P value for a one-tailed test, and if tails = 2, then TTEST returns the two-tailed value. For the parameter called type, there are three choices:

type	Test performed using Student's t distribution
1	Paired difference test
2	Difference of means test for two samples with equal variances
3	Difference of means test for two samples with unequal variances

The other Excel command, ➤**Tools**➤**Data Analysis**➤**t-Test: Paired Two Sample for Means,** gives much more information than TTEST. We will use this command in the next example.

Example

Promoters of a state lottery decided to advertise the lottery heavily on television for one week during the middle of one of the lottery games. To see if the advertising improved ticket sales, they surveyed a random sample of 8 ticket outlets and recorded weekly sales for one week before the television campaign and for one week after the campaign. The results follow (in ticket sales) where row A gives sales prior to the campaign and row B gives sales afterward.

| A | 3201 | 4529 | 1425 | 1272 | 1784 | 1733 | 2563 | 3129 |
| B | 3762 | 4851 | 1202 | 1131 | 2172 | 1802 | 2492 | 3151 |

Test the claim that the television campaign increased lottery ticket sales at the 0.05 level of significance.

We enter the data in Columns A and B, with appropriate headers. Next, open the dialog box below, using ➤**Tools**➤**Data Analysis**➤**t-Test: Paired Two Sample for Means.**

Notice that we use Column A cells for Variable 1 Range, Column B cells for Variable 2 range, and we check the Labels box. The null hypothesis is H_o: $\mu = 0$, so we enter 0 as the value for the Hypothesized Mean difference. We select Cell D8 as the upper left cell for the Output Range, and we widen the output columns to fit the display.

	A	B	C	D	E	F	G
1	**Before**	**After**		**Paired Difference Test**			
2	3201	3762		**Null Hypothesis: Mean of differences = 0**			
3	4529	4851					
4	1425	1202					
5	1272	1131					
6	1784	2172					
7	1733	1802					
8	2563	2492		t-Test: Paired Two Sample for Means			
9	3129	3151					
10					*Before*	*After*	
11				Mean	2454.5	2570.375	
12				Variance	1250835	1665406	
13				Observations	8	8	
14				Pearson Correlation	0.983465		
15				Hypothesized Mean Difference	0		
16				df	7		
17				t Stat	-1.17846		
18				P(T<=t) one-tail	0.138559		
19				t Critical one-tail	1.894578		
20				P(T<=t) two-tail	0.277118		
21				t Critical two-tail	2.364623		
22							

Notice that we get $P = 0.1386$ for a one-tailed test. Since this value is larger than the level of significance, we do not reject the null hypothesis. The same output gives the P value for a two-tailed test as well. In addition, we see the sample t value of -1.17846, together with the critical values for a one- or two-tailed test using $\alpha = 0.05$.

LAB ACTIVITIES FOR TESTS INVOLVING PAIRED DIFFERENCES

1. Open or retrieve the worksheet Tvds01.xls from the data CD-ROM. The data are pairs of values where the entries in Column A represents average salary ($1000/yr) for male faculty members at an institution and those in Column B represent the average salary for female faculty members ($1000/yr) at the same institution. A random sample of 22 U.S. colleges and universities was used (source: Academe, Bulletin of the American Association of University Professors).

    (34.5, 33.9)   (30.5, 31.2)   (35.1, 35.0)   (35.7, 34.2)   (31.5, 32.4)
    (34.4, 34.1)   (32.1, 32.7)   (30.7, 29.9)   (33.7, 31.2)   (35.3, 35.5)
    (30.7, 30.2)   (34.2, 34.8)   (39.6, 38.7)   (30.5, 30.0)   (33.8, 33.8)
    (31.7, 32.4)   (32.8, 31.7)   (38.5, 38.9)   (40.5, 41.5)   (25.3, 25.5)
    (28.6, 28.0)   (35.8, 35.1)

    (a) The data are in Columns A and B.

    (b) Use the ➤Tools➤Data Analysis➤t-Test Paired Two Sample for Means dialog box to test the hypothesis that there is a difference in salary. What is the P value of the sample test statistic? Do we reject or fail to reject the null hypothesis at the 5% level of significance? What about at the 1% level of significance?

    (c) Use the ➤Tools➤Data Analysis➤t-Test Paired Two Sample for Means dialog box to test the hypothesis that female faculty members have a lower average salary than male faculty members. What is the test conclusion at the 5% level of significance? At the 1% level of significance?

2. An audiologist is conducting a study on noise and stress. Twelve subjects selected at random were given a stress test in a room that was quiet. Then the same subjects were given another stress test, this time in a room with high-pitched background noise. The results of the stress tests were scores 1 through 20 with 20 indicating the greatest stress. The results follow, where A represents the score of the test administered in the quiet room and B represents the scores of the test administered in the room with the high-pitched background noise.

Subject	1	2	4	5	6	7	8	9	10	11	12
A	13	12	16	19	7	13	9	15	17	6	14
B	18	15	14	18	10	12	11	14	17	8	16

Test the hypothesis that the stress level was greater during exposure to high-pitched background noise. Look at the P value. Should you reject the null hypothesis at the 1% level of significance? At the 5% level?

TESTING OF DIFFERENCES OF MEANS (SECTION 9.5 OF *UNDERSTANDABLE STATISTICS*)

Tests of difference of means for independent samples are presented in Section 9.5 of *Understandable Statistics*. We consider the $\bar{x}_1 - \bar{x}_2$ distribution. The null hypothesis is that there is no difference between means, so $H_0 : \mu_1 = \mu_2$ or $H_0 : \mu_1 - \mu_2 = 0$.

When σ_1 and σ_2 are known

When we are testing the difference of means with known σ_1 and σ_2, the z-test based on normal distribution is used if either the population has a normal distribution or the sample size is large. The values of population variance, which is the square of population standard deviation, are actually needed in the Excel testing procedure dialog box. Some examples used in this section do not provide the values of population standard deviations. In order to demonstrate the use of the testing procedure, we use the sample variance values for the population variance values. In an Excel worksheet, use Paste Function, and select Statistical in the left column, VAR in the right column. We did this to get the variances shown in the next worksheet.

The menu selection ➤**Tools**➤**Data Analysis**➤**z-Test Two Sample for Means** provides the sample z statistic, P values for a one- or two-tailed test, and critical z values for a one- or two-tailed test at the specified level of significance.

We used the Excel worksheet Tvis01.xls to generate the worksheet that follows. This data gives the heights of a random sample of football players (Column A) and the heights of a random sample of basketball players (Column B).

	A	B	C	D	E	F
1	HtFt	HtBk		**Sample Variance for Football Players**	0.134028	
2	6.33	6.08		**Sample Variance for Basketball Players**	0.098735	
3	6.5	6.58				
4	6.5	6.25				
5	6.25	6.58		z-Test: Two Sample for Means		
6	6.5	6.25				
7	6.33	5.92			*HtFt*	*HtBk*
8	6.25	7		Mean	6.178889	6.45325
9	6.17	6.41		Known Variance	0.134028	0.098735
10	6.42	6.75		Observations	45	40
11	6.33	6.25		Hypothesized Mean Difference	0	
12	6.42	6		z	-3.71752	
13	6.58	6.92		P(Z<=z) one-tail	0.000101	
14	6.08	6.83		z Critical one-tail	1.644853	
15	6.58	6.58		P(Z<=z) two-tail	0.000201	
16	6.5	6.41		z Critical two-tail	1.959961	
17	6.42	6.67				
18	6.25	6.67				
19	6.67	5.75				
20	5.91	6.25				
21	6	6.25				

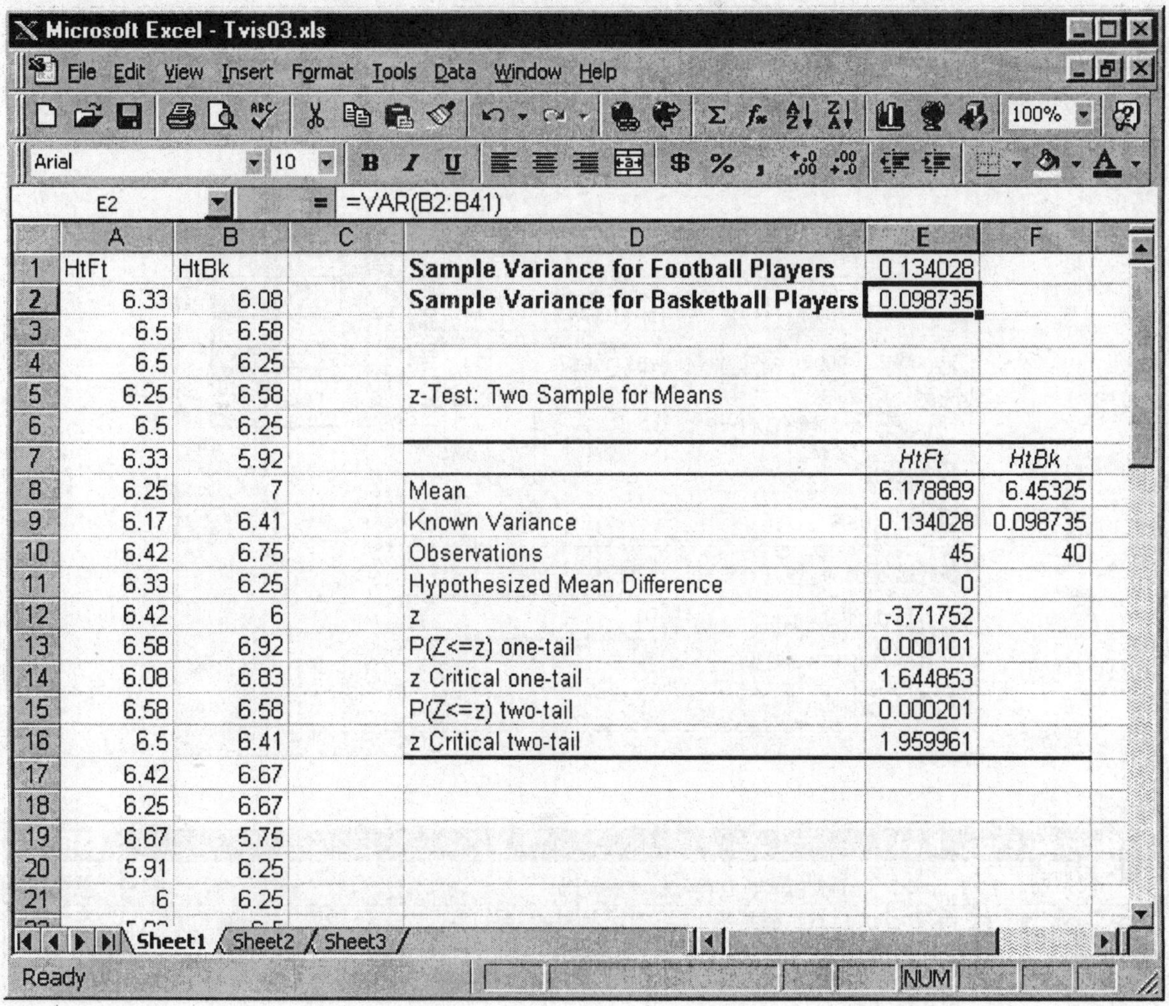

When σ_1 and σ_2 are unknown

To do a test of difference of sample means when σ_1 and σ_2 are unknown, t-tests are used if either the population has approximately a normal distribution or the sample size is large. If we assume that the samples come from populations with the same standard deviation, use **➤Tools➤Data Analysis➤t-Test: Two-Sample Assuming Equal Variances.** If we believe that population standard deviations are not equal, use **➤Tools➤Data Analysis➤t-Test: Two-Sample Assuming Unequal Variances.**

Example

Sellers of microwave French fry cookers claim that their process saves cooking time. The McDougle Fast Food Chain is considering the purchase of these new cookers, but wants to test the claim.

Six batches of French fries were cooked in the traditional way. The cooking times (in minutes) were

15 17 14 15 16 13

Six batches of French fries of the same weight were cooked using the new microwave cooker. These cooking times (in minutes) were

11 14 12 10 11 15

Let us assume that both populations are approximately normal with equal standard deviation. Test the claim that the microwave process takes less time. Use $\alpha = 0.05$.

We will enter the traditional data in Column A and the new data in Column B. Then we use ➤**Tools**➤**Data Analysis**➤**t-Test: Two-Sample Assuming Equal Variances.** The hypothesized mean difference is zero.

t-Test: Two-Sample Assuming Equal Variances

Input
Variable 1 Range: $A$1:$A$7
Variable 2 Range: $B$1:$B$7
Hypothesized Mean Difference:
☑ Labels
Alpha: 0.05

Output options
◉ Output Range: D1
○ New Worksheet Ply:
○ New Workbook

Microsoft Excel - Book1

	A	B	C	D	E	F	G
1	Traditional	New		t-Test: Two-Sample Assuming Equal Variances			
2	15	11					
3	17	14			Traditional	New	
4	14	12		Mean	15	12.16667	
5	15	10		Variance	2	3.766667	
6	16	11		Observations	6	6	
7	13	15		Pooled Variance	2.883333		
8				Hypothesized Mean Difference	0		
9				df	10		
10				t Stat	2.890087		
11				P(T<=t) one-tail	0.008052		
12				t Critical one-tail	1.812462		
13				P(T<=t) two-tail	0.016105		
14				t Critical two-tail	2.228139		

We see that the P value for a one-tail test is 0.00805. Since this value is less than 0.05, we reject the null hypothesis and conclude that the new method takes less time on average.

LAB ACTIVITIES FOR TESTING DIFFERENCES OF MEANS

1. Calm Cough Medicine is testing a new ingredient to see if its addition will lengthen the effective cough relief time of a single dose. A random sample of fifteen doses of the standard medicine was tested, and the effective relief times (in minutes) were

42	35	40	32	30	26	51	39	33	28
37	22	36	33	41					

   Then a random sample of twenty doses with the new ingredient was tested. The effective relief times (in minutes) were

43	51	35	49	32	29	42	38	45	74
31	31	46	36	33	45	30	32	41	25

   Assume that the standard deviations of the relief times are equal for the two populations. Also assume that both populations are approximately normal. Test the claim that the effective relief time is longer when the new ingredient is added. Use $\alpha = 0.01$.

2. Retrieve the worksheet Tvis06.xls from the data CD-ROM. The data represent numbers of cases of red fox rabies for a random sample of sixteen areas in each of two different regions of southern Germany.

   Number of Cases in Region 1

10	2	2	5	3	4	3	3	4	0	2	6	4	8	7	4

   Number of Cases in Region 2

1	1	2	1	3	9	2	2	4	5	4	2	2	0	0	2

   Test the hypothesis that the average number of cases in Region 1 is greater than the average number of cases in Region 2. Use a 1% level of significance. Assume the both populations are approximately normal and have equal standard deviations.

3. Retrieve the Excel worksheet Tvis02.xls from the data CD-ROM. The data represent the petal length (in centimeters) for a random sample of 35 *Iris Virginica* plants and for a random sample of 38 *Iris Setosa* plants (source: Anderson, E., Bulletin of American Iris Society).

   Petal Length of *Iris Virginica*

5.1	5.8	6.3	6.1	5.1	5.5	5.3	5.5	6.9	5.0	4.9	6.0	4.8	6.1	5.6	5.1
5.6	4.8	5.4	5.1	5.1	5.9	5.2	5.7	5.4	4.5	6.1	5.3	5.5	6.7	5.7	4.9
4.8	5.8	5.1													

   Petal Length of *Iris Setosa*

1.5	1.7	1.4	1.5	1.5	1.6	1.4	1.1	1.2	1.4	1.7	1.0	1.7	1.9	1.6	1.4
1.5	1.4	1.2	1.3	1.5	1.3	1.6	1.9	1.4	1.6	1.5	1.4	1.6	1.2	1.9	1.5
1.6	1.4	1.3	1.7	1.5	1.7										

   Test the hypothesis that the average petal length for *Iris Setosa* is shorter than the average petal length for *Iris Virginica*. Assume that population standard deviations are unequal.

CHAPTER 10 REGRESSION AND CORRELATION

LINEAR REGRESSION – TWO VARIABLES (SECTIONS 10.1–10.3 OF *UNDERSTANDABLE STATISTICS*)

Chapter 10 of *Understandable Statistics* introduces linear regression. The formula for the correlation coefficient r is given in Section 10.1. Formulas to find the equation of the least squares line,

$$y = a + bx$$

are given in Section 10.2. This section also contains the formula for the coefficient of determination r^2. The equation for the standard error of estimate as well as the procedure to find a confidence interval for the predicted value of y are given in Section 10.3.

Excel supports several functions related to linear regression. To use these, first enter the paired data values in two columns. Put the explanatory variable in a column labeled with x, or an appropriate descriptive name, and put the response variable in a column labeled with y, or an appropriate descriptive name.

The functions and corresponding syntax are

LINEST(y range, x range) which returns the slope b and y-intercept a of the least-squares line, in that order. Although this command can be found under the Paste Function button menu, it is best to type it in because it is an array formula. (The full LINEST function involves two more, optional parameters, but we will ignore these.) To use LINEST,

1. Activate two cells, the first to hold the slope b, the second for the intercept a.
2. In the formula bar, type =LINEST(y range,x range) with the appropriate cell ranges in place of x range and y range.
3. Instead of pressing Enter, press **Ctrl+Shift+Enter.** This key combination activates the array formula features so that you get the outputs for both b and a. Otherwise, you will get only the slope b of the least-squares line.

The other functions are employed in the usual way, by activating a cell and then either typing the command directly in the formula bar, followed by Enter, or by using the Paste Function dialog box.

SLOPE(y range, x range) which returns the slope b of the least-squares line.

INTERCEPT(y range, x range) which returns the intercept a of the least-squares line.

CORREL(y range, x range) which returns the correlation coefficient r.

FORECAST(x value, y range, x range) which returns the predicted y value for the specified x value, using extrapolation from the given pairs of x and y values. Note that you need to use a new FORECAST command for each different x value.

STEYX(y range, x range) which returns the standard error of estimate, S_e.

We will now use the Chart Wizard to generate a scatter diagram and then add the results of the least-squares regression.

Example

In retailing, merchandise loss due to shoplifting, damage, and other causes is called shrinkage. The managers at H.R. Merchandise think that there is a relationship between shrinkage and the number of clerks on duty. To explore this relationship, a random sample of seven weeks was selected. During each week, the staffing level of sales clerks was held constant and the dollar value (in hundreds of dollars) of the shrinkage was recorded.

X	10	12	11	15	9	13	8	Staffing level
Y	19	15	20	9	25	12	31	(in hundreds)

Open a worksheet. Place the X values in Column A with a corresponding label and the Y values in Column B.

Create a Scatter Diagram

Click on the Chart Wizard and select XY (Scatter). Choose the first subtype, which shows only points. Click Next.

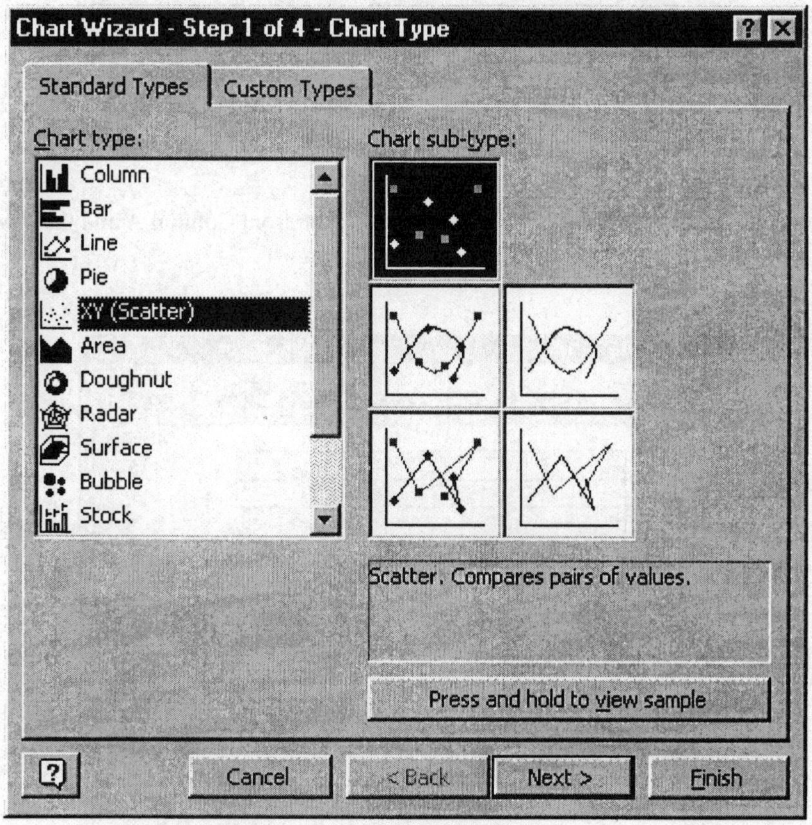

Select the X and Y value cells (no label) for the data range.

Click the Series tab to verify that the X values are being taken from Column A and the Y values from Column B.

Give the chart a title and give the axes labels. Then click on the Legend tab and turn off Legend.

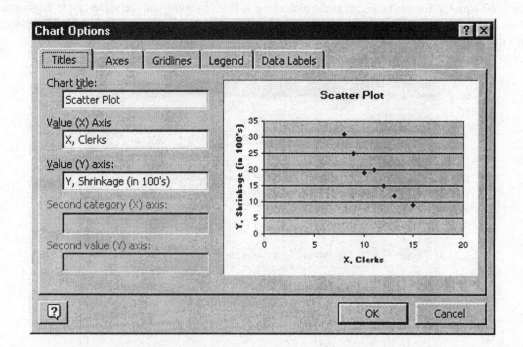

Click Finish to see the scatter plot on your worksheet.

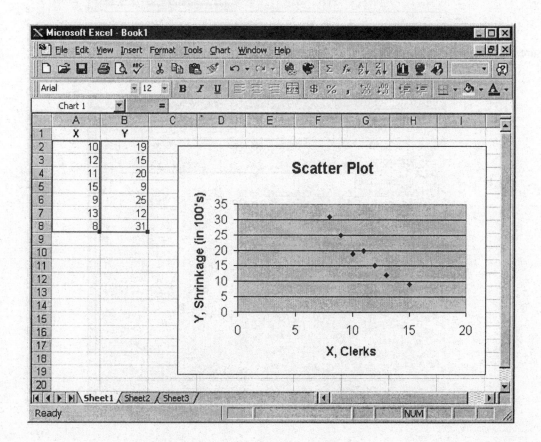

Add least-squares results to the plot

The least-squares line can be added to the plot, along with its equation and the value of r^2. Right-click on one of the data points shown in the scatter diagram. A drop-down menu will appear. Select Add Trendline to call up a dialog box.

Be sure Linear is selected as the Type. Click on the Options tab.

Check Display equation on chart and Display R-squared on chart, and then click OK. Now our scatter diagram shows the graph of the least squares-line and the equation. You can move the equation out of the way of the graph by clicking on it and dragging the resulting box to a convenient location.

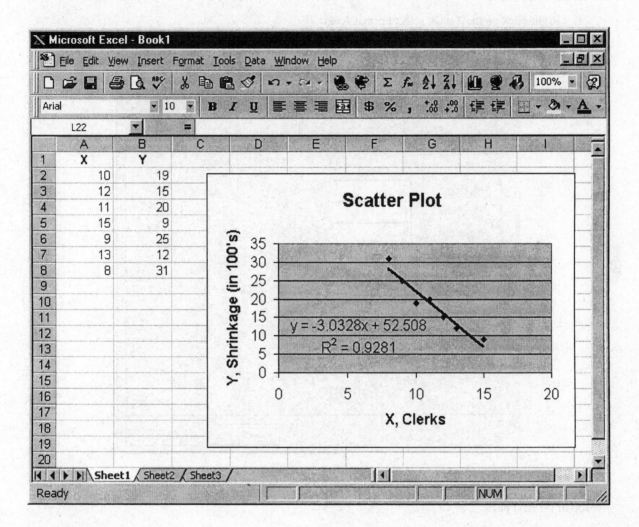

Scale the Axes

Sometimes all the points of a scatter diagram are in a corner, and we want to rescale the axes to reflect the data range. To do so,

1. Right-click on the *X*-axis, select Format Axes.
2. Click on the Scale tab. Change the minimum and maximum values for X. You may retain 0 as the value for Value (Y) axis Crosses at.

Go through a similar procedure to rescale the *Y*-axis. The result of our changes appears as shown at the top of the next page.

Forecast a value

Let's predict the shrinkage when fourteen clerks are available, using FORECAST. Click Paste Function on the standard tool bar, then select Statistical in the left column and Forecast in the right column. Fill in the dialog box as shown.

The predicted *Y* value is about 10.05. Since the unit for *Y* is hundreds of dollars, this represents a predicted shrinkage of $1005 when fourteen clerks are on duty.

Find the standard error of estimate and the value of *r*

We use Paste Function, Statistical and select STEYX and CORREL to find the standard error and the correlation coefficient.

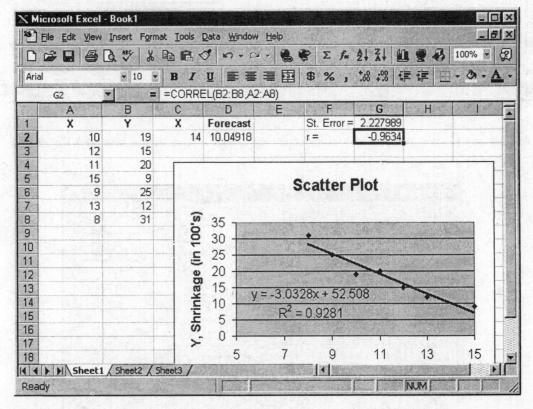

The worksheet shows the results.

If we want to obtain the values of *b* and *a* in the least-squares line without using the Chart Wizard, we can use LINEST(y range, x range). Activate *two* cells on the worksheet. Type =LINEST(B2:B8,A2:A8) in the formula bar, and press **Ctrl+Shift+Enter** to generate the values of both *b* and *a*.

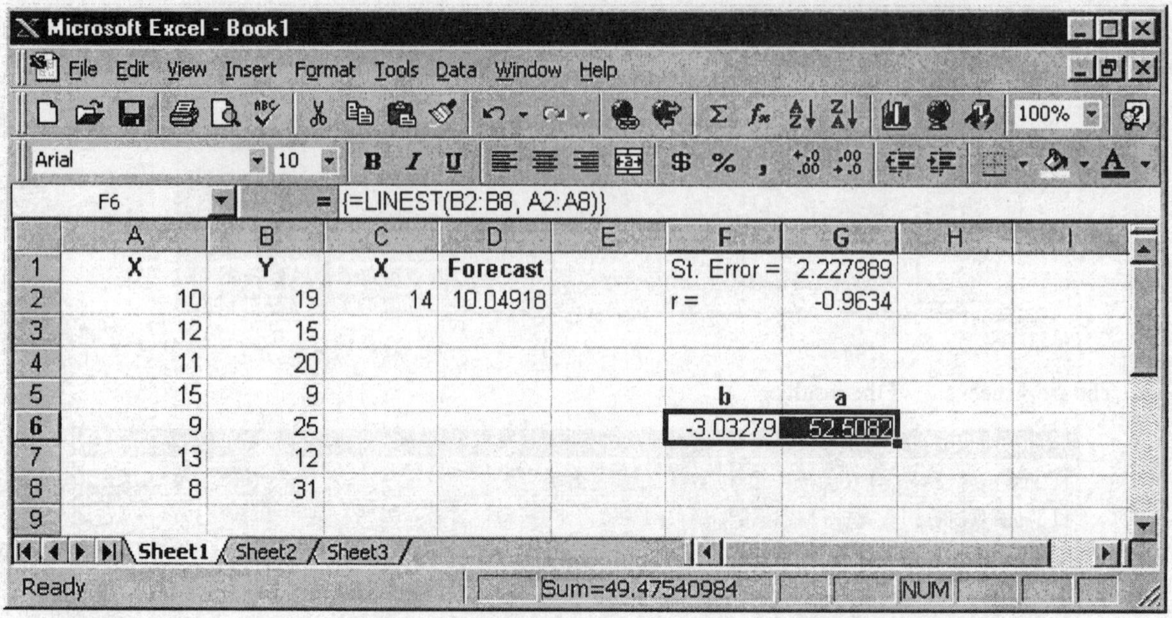

```
X Microsoft Excel - Book1                                              _ □ ×

 S  File  Edit  View  Insert  Format  Tools  Data  Window  Help        _ 8 ×

 D  ⯊  🖫  🖨  🔍 ABC  ✂  🗎 🖺 ✐  ⟳ ▾ ⟲ ▾  🌐 🌐  Σ  ƒ× ᴬ�ↆ ᶻↆ  🏛 🌐 ⌖  100% ▾  ▨

 Arial          ▾ 10 ▾  B  I  U  ≡ ≡ ≡ 🖽  $  %  ,  ⁺⁰.₀₀ .₀₀⁺⁰  🧾 🧾  ▦ ▾ ▨ ▾ A ▾

        F6          ▾     = {=LINEST(B2:B8, A2:A8)}
```

	A	B	C	D	E	F	G	H	I
1	X	Y	X	Forecast		St. Error =	2.227989		
2	10	19	14	10.04918		r =	-0.9634		
3	12	15							
4	11	20							
5	15	9				b	a		
6	9	25				-3.03279	52.5082		
7	13	12							
8	8	31							
9									

```
 I◀ ◀ ▶ ▶I \ Sheet1 / Sheet2 / Sheet3 /           I◀                    ▶I

 Ready                              │ Sum=49.47540984 │    │    NUM    │
```

Another way to get a great deal of information at once is with ➤**Tools**➤**Data Analysis**➤**Regression.**

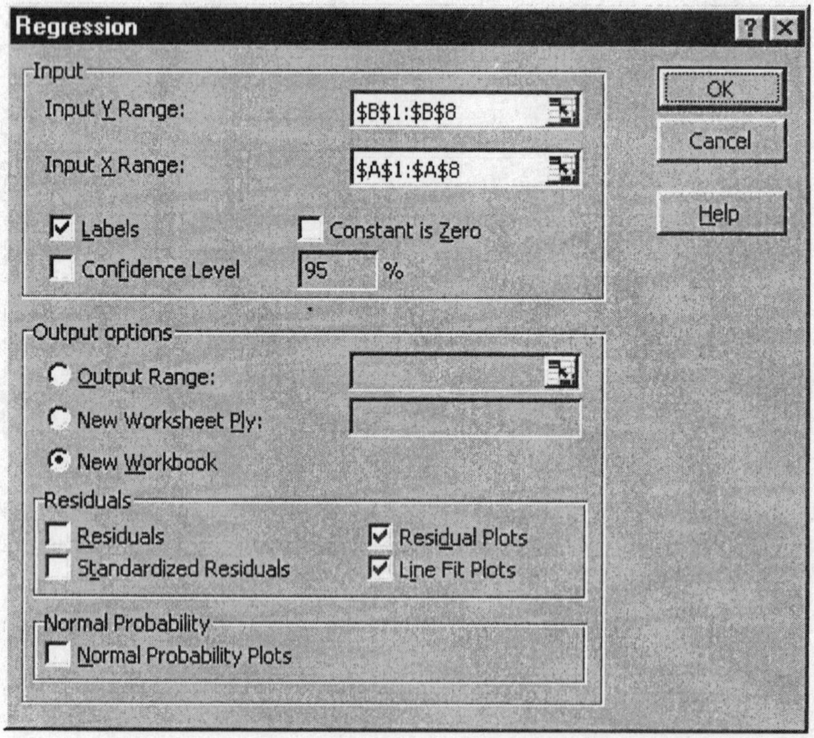

Regression   ? ×

┌─Input───────────────────────────────┐   ┌──────────┐
│ Input Y Range: $B$1:$B$8 🔢 │   │   OK   │
│   │   └──────────┘
│ Input X Range: $A$1:$A$8 🔢 │   ┌──────────┐
│   │   │ Cancel │
│ ☑ Labels   ☐ Constant is Zero   │   └──────────┘
│ ☐ Confidence Level   95   %   │   ┌──────────┐
│   │   │   Help   │
│   │   └──────────┘
├─Output options──────────────────────┤
│ ○ Output Range: [   ] 🔢 │
│ ○ New Worksheet Ply: [   ]   │
│ ● New Workbook   │
│ ┌─Residuals──────────────────────┐ │
│ │ ☐ Residuals   ☑ Residual Plots │ │
│ │ ☐ Standardized Residuals   │ │
│ │   ☑ Line Fit Plots │ │
│ └─────────────────────────────────┘ │
│ ┌─Normal Probability─────────────┐ │
│ │ ☐ Normal Probability Plots   │ │
│ └─────────────────────────────────┘ │
└──────────────────────────────────────┘

We checked Residuals Plot and Line Fit Plot. Notice that the output goes to a new worksheet. We used
>Format>Column>AutoFit Selection to adjust the column widths of the output.

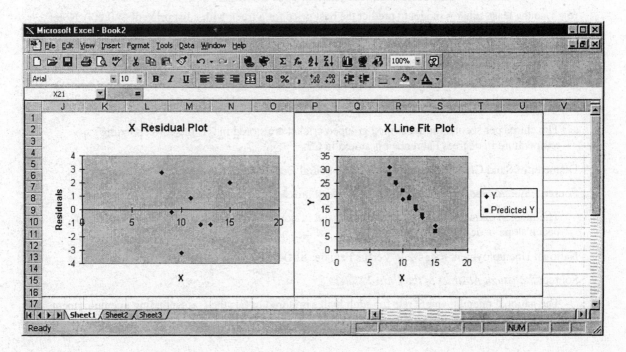

Scrolling down would show the rest of the residual output. Scrolling right shows the charts.

LAB ACTIVITIES FOR TWO-VARIABLE LINEAR REGRESSION

1. Open or retrieve the worksheet Slr01.xls from the data CD-ROM. This worksheet contains the following data, with the list price in Column C1 and the best price in Column C2. The best price is the best price negotiated by a team from the magazine.

LIST PRICE VERSUS BEST PRICE FOR A NEW GMC PICKUP TRUCK

In the following data pairs (X, Y),

X = List Price (in $1000) for a GMC Pickup Truck

Y = Best Price (in $1000) for a GMC Pickup Truck

Source: Consumers Digest, February 1994

        (12.400), 11.200)   (14.300, 12.500)   (14.500, 12.700)

        (14.900, 13.100)   (16.100, 14.100)   (16.900, 14.800)

        (16.500, 14.400)   (15.400, 13.400)   (17.000, 14.900)

        (17.900, 15.600)   (18.800, 16.400)   (20.300, 17.700)

        (22.400, 19.600)   (19.400, 16.900)   (15.500, 14.000)

        (16.700, (14.600)   (17.300, 15.100)   (18.400, 16.100)

        (19.200, 16.800)   (17.400, 15.200)   (19.500, 17.000)

        (19.700, 17.200)   (21.200, 18.600)

  (a) Use the Chart Wizard to create a scatter plot for the data

  (b) Right-click on a data point and use the Add Trendline option to show the least-squares line on the scatter diagram, along with its equation and the value of r^2.

  (c) What is the value of the standard error of estimate? (Use STEYX.)

  (d) What is the value of the correlation coefficient r? (Use CORREL.)

  (e) Use the least-squares model to predict the best price for a truck with a list price of $20,000. Note: Enter this value as 20, since X is assumed to be in thousands of dollars. (Use FORECAST.)

2. Other Excel worksheets appropriate to use for simple linear regression are

Cricket Chirps Versus Temperature: Slr02.xls

Source: *The Song of Insects* by Dr. G.W. Pierce, Harvard Press

  The chirps per second for the striped grouped cricket are stored in C1; the corresponding temperature in degrees Fahrenheit is stored in C2.

Diameter of Sand Granules Versus Slope on a Natural Occurring Ocean Beach: Slr03.xls

Source *Physical Geography* by A.M. King, Oxford press

  The median diameter (MM) of granules of sand is stored in C1; the corresponding gradient of beach slope in degrees is stored in C2.

National Unemployment Rate Male Versus Female: Slr04.xls

Source: *Statistical Abstract of the United States*

  The national unemployment rate for adult males is stored in C1; the corresponding unemployment rate for adult females for the same period of time is stored in C2.

Select these worksheets and repeat Parts (a)–(d) of Problem 1, using Column A as the explanatory variable and Column B as the response variable.

3. A psychologist studying the correlation between interruptions and job stress rated a group of jobs for interruption level. She selected a random sample of twelve people holding jobs from among those rated, and analyzed the people's stress level. The results follow, with X being interruption level of the job on a scale of 1 (fewest interruptions) to 20 and Y the stress level on a scale of 1 (lowest stress) to 50.

Person	1	2	3	4	5	6	7	8	9	10	11	12
X	9	15	12	18	20	9	5	3	17	12	17	6
Y	20	37	45	42	35	40	20	10	15	39	32	25

   (a) Enter the X values into Column A and the Y values into Column B.

   (b) Follow parts (a) through (d) of Problem 1 using the X values as the explanatory data values and the Y values as response data values.

   (c) Redo Part (b). This time change the X values to the response data values and the Y values to the explanatory data values (i.e. exchange headers for the X and Y columns) How does the scatter diagram compare? How does the least-squares equation compare? How does the correlation coefficient compare? How does the standard error of estimate compare? Does it seem to make a difference which variable is the response variable and which is the explanatory?

4. The researcher of Problem 3 was able to add to her data. Another random sample of eleven people had their jobs rated for interruption level and were then evaluated for stress level.

Person	13	14	15	16	17	18	19	20	21	22	23
X	4	15	19	13	10	9	3	11	12	15	4
Y	20	35	42	37	40	23	15	32	28	38	12

Add this data to the data in problem 3, and repeat Parts (a) and (b). Be sure the label column A as the X values and column B as the Y values. Compare the new standard error of estimate with the old one. Does more data tend to reduce the value of standard error of estimate? What about the value of r?

MULTIPLE REGRESSION

An introduction to multiple regression is presented in Section 10.4 of *Understandable Statistics*. ➤**Tools**➤**Data Analysis**➤**Regression** calls up a dialog box that supports multiple regression (and simple regression as well). The data must be entered in adjacent columns, so that one data range includes the data for all the explanatory variables. Data for the response variable can be entered in any other column or row.

Example

Bowman Brothers, a large sporting goods store in Denver, has a giant ski sale every October. The chief executive officer at Bowman Brothers is studying the following variables regarding the ski sale

   x_1 = Total dollar receipts from October ski sale

   x_2 = Total dollar amount spend advertising ski sale on local TV

   x_3 = Total dollar amount spend advertising ski sale on local radio

   x_4 = Total dollar amount spend advertising ski sale in Denver newspapers

Data for the past eight years is shown below (in thousands of dollars).

Year	1	2	3	4	5	6	7	8
x_1	751	768	801	832	775	718	739	780
x_2	19	23	27	32	25	18	20	24
x_3	14	17	20	24	19	9	10	19
x_4	11	15	16	18	12	5	7	14

Enter the data

Since we will be using x_1 as the response variable, let's enter the x_1 data in Column A and the other data in Columns B through D.

Look at descriptive statistics for each column of data

Use ➤**Tools**➤**Data Analysis**➤**Descriptive Statistics** to open the Descriptive Statistics dialog box. Specify the Input Range as shown and check the box labeled Summary Statistics.

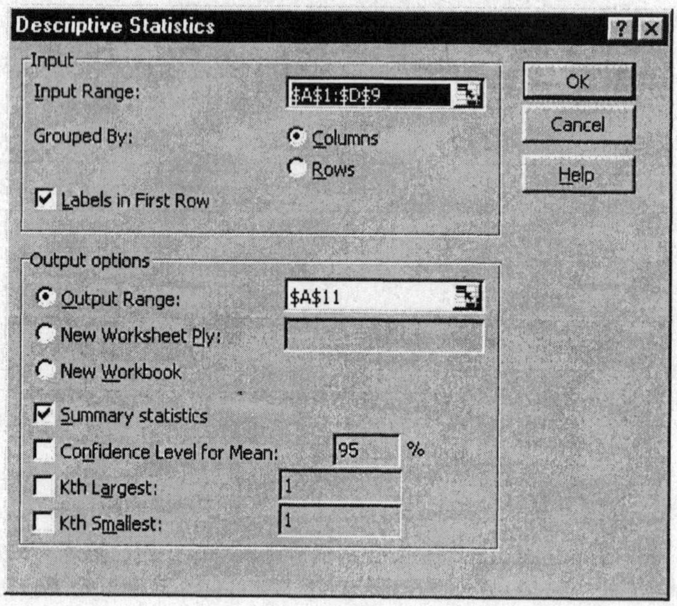

The following worksheet shows the data and the descriptive statistics. The column width was adjusted to show all the measurements on one page.

	X1		X2		X3		X4	
	A	B	C	D	E	F	G	H
1	X1	X2	X3	X4				
2	751	19	14	11				
3	768	23	17	15				
4	801	27	20	16				
5	832	32	24	18				
6	775	25	19	12				
7	718	18	9	5				
8	739	20	10	7				
9	780	24	19	14				
10								
11	X1		X2		X3		X4	
12								
13	Mean	770.5	Mean	23.5	Mean	16.5	Mean	12.25
14	Standard Error	12.64769883	Standard Error	1.636634177	Standard Error	1.822478689	Standard Error	1.578312842
15	Median	771.5	Median	23.5	Median	18	Median	13
16	Mode	#N/A	Mode	#N/A	Mode	19	Mode	#N/A
17	Standard Deviation	35.77309444	Standard Deviation	4.629100499	Standard Deviation	5.154748158	Standard Deviation	4.464142855
18	Sample Variance	1279.714286	Sample Variance	21.42857143	Sample Variance	26.57142857	Sample Variance	19.92857143
19	Kurtosis	0.10476719	Kurtosis	0.21504	Kurtosis	-0.830385016	Kurtosis	-0.685027171
20	Skewness	0.326784965	Skewness	0.691279008	Skewness	-0.300380746	Skewness	-0.553996223
21	Range	114	Range	14	Range	15	Range	13
22	Minimum	718	Minimum	18	Minimum	9	Minimum	5
23	Maximum	832	Maximum	32	Maximum	24	Maximum	18
24	Sum	6164	Sum	188	Sum	132	Sum	98
25	Count	8	Count	8	Count	8	Count	8

Examine the correlation between each pair of columns of data

To make a table of correlation coefficients between columns of data, use ➤**Tools**➤**Data Analysis**➤ **Correlation.**

Correlation

Input

Input Range: $A$1:$D$9

Grouped By: ⦿ Columns
◯ Rows

☑ Labels in First Row

Output options

⦿ Output Range: $F$1
◯ New Worksheet Ply:
◯ New Workbook

OK
Cancel
Help

The results are shown in the workbook on the top of the next page.

	A	B	C	D	E	F	G	H	I	J
	X1	X2	X3	X4			*X1*	*X2*	*X3*	*X4*
1	**X1**	**X2**	**X3**	**X4**						
2	751	19	14	11		X1	1			
3	768	23	17	15		X2	0.973964	1		
4	801	27	20	16		X3	0.96761	0.933948	1	
5	832	32	24	18		X4	0.9366	0.864126	0.949834	1
6	775	25	19	12						
7	718	18	9	5						
8	739	20	10	7						
9	780	24	19	14						
10										

Find the linear regression formula with x_1 as the response variable and x_2, x_3, x_4 as the explanatory variables.

We use ➤**Tools**➤**Data Analysis**➤**Regression** to find the coefficients of the linear regression formula as well as some other useful statistics relative to the model.

Regression

Input

Input Y Range: $A$1:$A$9

Input X Range: $B$1:$D$9

☑ Labels   ☐ Constant is Zero

☐ Confidence Level   95 %

Output options

○ Output Range:

◉ New Worksheet Ply:

○ New Workbook

Residuals

☐ Residuals   ☐ Residual Plots

☐ Standardized Residuals ☐ Line Fit Plots

Normal Probability

☐ Normal Probability Plots

OK   Cancel   Help

We use ➤**Format**➤**Column**➤**Auto Fit Selection** to adjust the column width.

```
Microsoft Excel - Book1
File  Edit  View  Insert  Format  Tools  Data  Window  Help

Arial        10   B  I  U

L23
```

	A	B	C	D	E	F	G	H	I
1	SUMMARY OUTPUT								
2									
3	*Regression Statistics*								
4	Multiple R	0.992286921							
5	R Square	0.984633333							
6	Adjusted R Square	0.973108333							
7	Standard Error	5.866314876							
8	Observations	8							
9									
10	ANOVA								
11		*df*	*SS*	*MS*	*F*	*Significance F*			
12	Regression	3	8820.345399	2940.115133	85.43456199	0.000440478			
13	Residual	4	137.6546009	34.41365022					
14	Total	7	8958						
15									
16		*Coefficients*	*Standard Error*	*t Stat*	*P-value*	*Lower 95%*	*Upper 95%*	*Lower 95.0%*	*Upper 95.0%*
17	Intercept	617.7229808	14.92113647	41.39919113	2.03468E-06	576.2951787	659.1507829	576.2951787	659.1507829
18	X2	4.6982481	1.369369847	3.430956297	0.026512392	0.896260015	8.500236184	0.896260015	8.500236184
19	X3	0.652174846	1.978850052	0.329572645	0.758258915	-4.842005074	6.146354767	-4.842005074	6.146354767
20	X4	2.580188072	1.622720787	1.59003822	0.187031511	-1.925216447	7.08559259	-1.925216447	7.08559259
21									

```
Sheet4 / Sheet1 / Sheet2 / Sheet3
Ready                                              NUM
```

The coefficients of the regression equation are given in the last box. The equation is

$$x_1 = 617.72 + 4.70x_2 + 0.65x_3 + 2.58x_4.$$

Notice that the value of coefficient of multiple determination R-square is 0.9846, and the standard error of estimate is 5.866. The P value for each coefficient is given. Remember, we are testing the null hypothesis H_0: $\beta = 0$ against the alternate hypothesis H_1: $\beta \neq 0$. Finally, a 95% confidence interval is given for each coefficient.

Note that we do not use the results of the Analysis of Variance.

Forecast a value

Supply some labels on the worksheet, and then for the forecast value, type in the linear regression equation using the specified values for x_2, x_3, x_4. For instance, to forecast the total receipts for October (x_1) given that $21,000 dollars is spent on local TV ads, $11,000 is spent on local radio ads, and $8,000 is spent on newspaper ads, we use the corresponding values in the regression equation.

Microsoft Excel - Book1

Cell reference: H16

	A	B	C	D	E	F	G	H	I	J
1	X1	X2	X3	X4			*X1*	*X2*	*X3*	*X4*
2	751	19	14	11		X1	1			
3	768	23	17	15		X2	0.973964	1		
4	801	27	20	16		X3	0.96761	0.933948	1	
5	832	32	24	18		X4	0.9366	0.864126	0.949834	1
6	775	25	19	12						
7	718	18	9	5						
8	739	20	10	7						
9	780	24	19	14						
10										
11	X2	X3	X4	Forecast X1						
12	21	11	8	744.21						
13										

Sheet4 \ **Sheet1** / Sheet2 / Sheet3

Ready   NUM

Change the response variable

To change the response variable to x_2, and use x_1, x_3, x_4 as explanatory variables, we first need to move the x_2 column to the far left and put the x_1 column next to the other explanatory variables. Use ➤**Edit**➤**Cut** and ➤**Edit**➤**Paste** to move the data columns.

Microsoft Excel - Book1

Cell reference: L14

	A	B	C	D	E	F	G	H	I	J
1		X2	X3	X4	X1					
2		19	14	11	751					
3		23	17	15	768					
4		27	20	16	801					
5		32	24	18	832					
6		25	19	12	775					
7		18	9	5	718					
8		20	10	7	739					
9		24	19	14	780					
10										
11										
12										
13										

Sheet4 \ **Sheet1** / Sheet2 / Sheet3

Ready   NUM

LAB ACTIVITIES FOR MULTIPLE REGRESSION

For 1–6, use the data provided on Excel worksheets stored on the data CD-ROM to explore the relationships among the variables. Activities 1–4 come from Section 10.5 Problems 3–6 respectively. Activities 5 and 6 are two additional case studies on the data CD-ROM.

1. Systolic Blood Pressure Data
   Excel CD-ROM worksheet Mlr02.xls

2. Test Scores for General Psychology
   Excel CD-ROM worksheet Mlr03.xls

3. Hollywood Movies
   Excel CD-ROM worksheet Mlr04.xls

4. All Greens Franchise
   Excel CD-ROM worksheet Mlr05.xls

5. Crime
   This data is a case study of education, crime, and police funding for small cities in ten Eastern and Southeastern states. The states are New Hampshire, Connecticut, Rhode Island, Maine, New York, Virginia, North Carolina, South Carolina, Georgia, and Florida. The data are for a sample of 50 small cities in these states.

   x_1 = Total overall reported crime rate per 1 million residents

   x_2 = Reported violent crime rate per 100,000 residents

   x_3 = Annual police funding in dollars per resident

   x_4 = Percent of people 25 years and older that have had four years of high school

   x_5 = Percent of 16- to 19-year-olds not in high school and not high school graduates

   x_6 = Percent of 18- to 24-year-olds enrolled in college

   x_7 = Percent of people 25 years and older with at least four years of college

   Excel data CD-ROM worksheet Mlr06.xls

6. Health
   This data is a case study of public health, income, and population density for small cities in eight Midwestern states: Ohio, Indiana, Illinois, Iowa, Missouri, Nebraska, Kansas, and Oklahoma. The data are for a sample of 53 small cities in these states.

   x_1 = Death rate per 1000 residents

   x_2 = Doctor availability per 100,000 residents

   x_3 = Hospital availability per 1000,000 residents

   x_4 = Annual per capita income in thousands of dollars

   x_5 = Population density people per square mile

   Excel data CD-ROM worksheet Mlr07.xls

CHAPTER 11 CHI-SQUARE AND *F* DISTRIBUTIONS

CHI-SQUARE TEST OF INDEPENDENCE (SECTION 11.1 OF *UNDERSTANDABLE STATISTICS*)

Use of the chi-square distribution to test independence is discussed in Section 11.1 of *Understandable Statistics*. In such tests we use hypotheses

H_0: The variables are independent

H_1: The variables are not independent

In Excel, the applicable command (accessed using the Paste Function button) is

CHITEST(set of observed values,set of expected values)

which returns the *P* value of the sample X^2 value, where the sample X^2 value is computed as

$$X^2 = \sum \frac{(O-E)^2}{E}$$

Here, E stands for the expected count in a cell, and O stands for the observed count in that same cell. The sum is taken over all cells.

In our Excel worksheet, we first enter in the contingency table of observed values. If the table does not contain column sums or row sums, use the Sum button on the tool bar to generate the sums. We need to create the table of expected values, where the expected value E for a cell is

$$E = (\text{column total}) \left(\frac{\text{row total}}{\text{grand total}} \right)$$

By careful use of absolute and relative cell references, we can type the formula once and then copy it to different positions in the contingency table of expected values. Recall that an absolute cell reference has $ symbols preceding the column and row designators.

Example

A computer programming aptitude test has been developed for high school seniors. The test designers claim that scores on the test are independent of the type of school the student attends: rural, suburban, urban. A study involving a random sample of students from these types of institutions yielded the following contingency table. Use the CHITEST command to compute the *P* value of the sample chi-square value. Then determine if type of school and test score are independent at the $\alpha = 0.05$ level of significance.

	School Type		
Score	Rural	Suburban	Urban
200-299	33	65	83
300-399	45	79	95
400-500	21	47	63

First we enter the table into a worksheet and use the sum button $\boxed{\Sigma}$ on the standard toolbar to generate the column, row, and grand total sums.

Microsoft Excel - Book1

File Edit View Insert Format Tools Data Window Help

E5 = =SUM(B5:D5)

	A	B	C	D	E	F	G	H	I
1	Observed Values								
2	Score	Rural	Suburban	Urban	Row Total				
3	200-299	33	65	83	181				
4	300-399	45	79	95	219				
5	400-500	21	47	63	131				
6	Col Total	99	191	241	531				
7									
8									
9									

Sheet1 / Sheet2 / Sheet3 /

Ready   NUM

Next we create the contingency table of expected values, where the expected value for Cell B3 will go in Cell H3. Notice that in the formula

$$E = (\text{row total})\left(\frac{\text{row total}}{\text{grand total}}\right)$$

the grand total stays the same for each expected value. The grand total is in Cell E6. Since we want this to be an absolute address used in each computation, we use the cell label $E$6. (Alternatively, we could just type in the grand total of 531 in the formula bar.) The column totals are all in Row 6, so when we refer to a column total, we will fix the row address by using $6 and let the column names vary. The row totals are all in Column E, so we will fix the column address as $E and let the row address vary when we use row totals. So the formula as entered in Cell H3 should be

$$= B\$6*(\$E3/\$E\$6)$$

Now move the cursor to the lower right corner of Cell H3. When the small + appears, drag it to the lower right corner of cell J5. The calculations for all the cells will automatically be done.

We now have both the observed values and the expected values.

Microsoft Excel - Book 1

File Edit View Insert Format Tools Data Window Help

H3 = =B$6*($E3/$E$6)

	A	B	C	D	E	F	G	H	I	J
1	Observed Values						Expected Values			
2	Score	Rural	Suburban	Urban	Row Total		Score	Rural	Suburban	Urban
3	200-299	33	65	83	181		200-299	33.74576	65.10546	82.14878
4	300-399	45	79	95	219		300-399	40.83051	78.77401	99.39548
5	400-500	21	47	63	131		400-500	24.42373	47.12053	59.45574
6	Col Total	99	191	241	531					
7										

Sheet1 / Sheet2 / Sheet3 /

Ready   NUM

Now use the CHITEST command. After clicking the Paste Function button on the tool bar, select the CHITEST function and, in the dialog box that appears, enter the required ranges of values as shown.

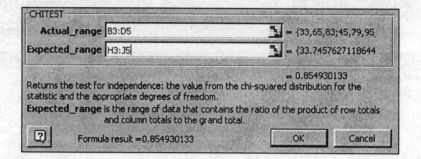

The resulting *P* value is 0.8549. Place appropriate labels on the worksheet

Observed Values						Expected Values			
Score	Rural	Suburban	Urban	Row Total		Score	Rural	Suburban	Urban
200-299	33	65	83	181		200-299	33.74576	65.10546	82.14878
300-399	45	79	95	219		300-399	40.83051	78.77401	99.39548
400-500	21	47	63	131		400-500	24.42373	47.12053	59.45574
Col Total	99	191	241	531					
P-value for observed chi square is				0.85493					

Cell E8: =CHITEST(B3:D5,H3:J5)

Since the *P* value is greater than $\alpha = 0.05$, we do not reject the null hypothesis. There is insufficient evidence to conclude that school type and test scores are not independent. Intuitively, we can confirm this by observing that the values we would expect, assuming independence (the values in the right table) are not very different from the values observed (the values in the left table).

LAB ACTIVITIES FOR CHI-SQUARE TEST OF INDEPENDENCE

In each activity, enter the contingency tables into a worksheet, then use the Sum button to generate the required row and column sums. Create a table of expected values. Finally, use CHITEST to find the P value of the sample statistic and draw the appropriate conclusion.

1. We Care Auto Insurance had its staff of actuaries conduct a study to see if vehicle type and loss claim are independent. A random sample of auto claims over the first six months gives the information in this contingency table.

Type of Vehicle	Total Loss Claims per Year per Vehicle			
	$0–999	$1000–2999	$3000–5999	$6000+
Sports Car	20	10	16	8
Truck	16	25	33	9
Family Sedan	40	68	17	7
Compact	52	73	48	12

Test the claim that car type and loss claim are independent. Use $\alpha = 0.05$.

2. An educational specialist is interested in comparing three methods of instruction:

SL – standard lecture with discussion

TV – video taped lectures with no discussion

IM – individualized method with reading assignments and tutoring, but no lectures.

The specialist conducted a study of these three methods to see if they were independent. A course was taught using each of the three methods, and a standard final exam given at the end. Students were put into the different method sections at random. The course type and test results are shown in the contingency table below.

Course Type	Final Exam Score				
	<60	60–69	70–79	80–89	90–100
SL	10	4	70	31	25
TV	8	3	62	27	23
IM	7	2	58	25	22

Test the claim that the instruction method and final exam test scores are independent, using $\alpha = 0.01$.

Note: If you have raw data entered in columns, you can use the Pivot Wizard to construct a contingency table. See the Excel help menu or other manuals for details.

TESTING TWO VARIANCES (SECTION 11.4 OF *UNDERSTANDABLE STATISTICS*)

In Excel there are two commands that give the *P* value for testing two variances.

FTEST(data range of first variable,data range of second variable) which returns the *P* value for a test with the alternate hypothesis $H_1: \sigma_1^2 \neq \sigma_2^2$. To find this command, use the Paste Function button.

F-Test: Two Sample for Variances which returns the *P* value for a one-tailed test as well as some summary statistics. The dialog box for this command is found using ➤**Tools**➤**Data Analysis**➤**F-Test**.

Example

Two processes for manufacturing paint are under study. The useful life (in years) of the paint for a random sample produced by method A is

      5.2   5.9   6.2   4.8   5.3   5.2

With method B, the useful life of a random sample of test patches is

      5.1   4.9   4.6   5.8   6.7   5.2

Use a 5% level of significance to test if the variances are equal. First enter the data and use the **FTEST** command from Paste Function. The results are

	Method A	Method B		P-value is	0.420341
2	5.2	5.1			
3	5.9	4.9			
4	6.2	4.6			
5	4.8	5.8			
6	5.3	6.7			
7	5.2	5.2			

(Excel screen: cell E1 = FTEST(A2:A7,B2:B7))

Since the *P* value is greater than 0.05, we do not reject H_0, and we conclude that there is not enough evidence to judge the variances unequal.

Next use the F-Test dialog box found under ➤**Tools**➤**Data Analysis**➤**F-Test**.

	A	B	C	D	E	F	G	H
1	**Method A**	**Method B**		F-Test Two-Sample for Variances				
2	5.2	5.1						
3	5.9	4.9			*Method A*	*Method B*		
4	6.2	4.6		Mean	5.433333333	5.383333333		
5	4.8	5.8		Variance	0.266666667	0.573666667		
6	5.3	6.7		Observations	6	6		
7	5.2	5.2		df	5	5		
8				F	0.46484602			
9				P(F<=f) one-tail	0.210170305			
10				F Critical one-tail	0.198006944			
11								

Notice that this output gives the P value for a one-tailed test. Double the value to get the results for a two-tailed test. Again, the P value is too large to reject the null hypothesis.

LAB ACTIVITIES FOR TESTING TWO VARIANCES

Select several exercises from 11.4 of the text *Understandable Statistics*.

ANALYSIS OF VARIANCE – ONE-WAY (SECTION 11.5 OF *UNDERSTANDABLE STATISTICS*)

Section 11.5 of *Understandable Statistics* introduces single-factor analysis of variance (also called one-way ANOVA). We consider several populations that are assumed to follow a normal distribution. The standard deviations of the populations are assumed to be approximately equal. ANOVA provides a method to compare several different populations to see if the means are the same. Let population 1 have mean μ_1, population 2 have mean μ_2, and so on. The hypotheses of ANOVA are

$$H_0 : \mu_1 = \mu_2 = ... = \mu_n$$
H_1 : not all the means are equal

In Excel we use ➤**Tools**➤**Data Analysis**➤**Anova: Single Factor.** The output provides summary statistics of the groups, and a summary table giving the degrees of freedom, sum of squares, mean squares, the value of F, and the P value of F.

Example

A psychologist has developed a series of tests to measure a person's level of depression. The composite scores range from 50 to 100, with 100 representing the most severe depression level. A random sample of twelve patients with approximately the same depression level (as measured by the tests) was divided into three different treatment groups. Then, one month after treatment was competed, the depression of level of each patient was again evaluated. The after-treatment depression levels are as given:

Treatment 1   70   65   82   83   71

Treatment 2   75   62   81

Treatment 3   77   60   80   75

Enter the data in Columns A, B, and C. Then access the ANOVA Single Factor dialog box by using the menu selections ➤**Tools**➤**Data Analysis**➤**Anova: Single Factor.**

The output is

LAB ACTIVITIES FOR ANALYSIS OF VARIANCE – ONE-WAY

1. A random sample of twenty overweight adults was randomly divided into four equal groups. Each group was given a different diet plan, and the weight loss for each individual (in pounds) was measured after three months:

Plan 1	18	10	20	25	17
Plan 2	28	12	22	17	16
Plan 3	16	20	24	8	17
Plan 4	14	17	18	5	16

Test the claim that the population mean weight loss is the same for the four diet plans, at the 5% level of significance.

2. A psychologist is studying the time it takes rats to respond to stimuli after being given doses of different tranquilizing drugs. A random sample of eighteen rats were divided into three groups. Each group was given a different drug. The time of response to stimuli was measured (in seconds).

Drug A	3.1	2.5	2.2	1.5	0.7	2.4
Drug B	4.2	2.5	1.7	3.5	1.2	3.1
Drug C	3.3	2.6	1.7	3.9	2.8	3.5

Test the claim that the population mean response times for the three drugs is the same at the 5% level of significance.

3. A research group is testing various chemical combinations designed to neutralize and buffer the effects of acid rain on lakes. Eighteen lakes of similar size in the same region have all been affected in the same way by acid rain. The lakes are divided into four groups, and each group of lakes is sprayed with a different combination of chemicals. An acidity index is then taken after treatment. The index ranges from 60 to 100, with 100 indicating the greatest acid rain pollution. The results follow.

Combination I	63	55	72	81	75
Combination II	78	56	75	73	82
Combination III	59	72	77	60	
Combination IV	72	81	66	71	

Test the claim that the population mean acidity index after each of the four treatments is the same, at the 0.01 level of significance.

ANALYSIS OF VARIANCE – TWO-WAY (SECTION 11.6 OF *UNDERSTANDABLE STATISTICS*)

Excel has two commands for two-way ANOVA under the ➤**Tools**➤**Data Analysis** menu choices.

Anova: Two-Factor With Replication is used when two or more sample measurements are available for each factor combination. Each factor combination must contain the same number of data values.

Anova: Two-Factor Without Replication is used when there is exactly one sample measurement for each factor combination.

Example (with replication)

This data set is from Example 7 of Section 11.6 of the text. The example gives customer satisfaction data with regard to time of day and type of customer contact. The data are listed on the spreadsheet.

Open the **Anova: Two-Factor With Replication** dialog box and fill in the designated boxes.

The output covers two pages:

Microsoft Excel - Book1

	A	B	C	D	E	F	G	H
1		**Automated**	**Bank Rep**		Anova: Two-Factor With Replication			
2	**Morning**	6	8					
3		5	7		SUMMARY	Automated	Bank Rep	Total
4		8	9		*Morning*			
5		4	9		Count	4	4	8
6	**Afternoon**	3	9		Sum	23	33	56
7		5	10		Average	5.75	8.25	7
8		6	6		Variance	2.916666667	0.916666667	3.428571429
9		5	8					
10	**Evening**	5	9		*Afternoon*			
11		5	10		Count	4	4	8
12		7	10		Sum	19	33	52
13		5	9		Average	4.75	8.25	6.5
14					Variance	1.583333333	2.916666667	5.428571429
15								
16					*Evening*			
17					Count	4	4	8
18					Sum	22	38	60
19					Average	5.5	9.5	7.5
20					Variance	1	0.333333333	5.142857143
21								
22					*Total*			
23					Count	12	12	

Sheet1 / Sheet2 / Sheet3

Ready

Microsoft Excel - Book1

	E	F	G	H	I	J	K
22	*Total*						
23	Count	12	12				
24	Sum	64	104				
25	Average	5.333333333	8.666666667				
26	Variance	1.696969697	1.515151515				
27							
28							
29	ANOVA						
30	*Source of Variation*	*SS*	*df*	*MS*	*F*	*P-value*	*F crit*
31	Sample	4	2	2	1.24137931	0.312576642	3.55456109
32	Columns	66.66666667	1	66.66666667	41.37931034	4.70341E-06	4.413863053
33	Interaction	2.333333333	2	1.166666667	0.724137931	0.498336378	3.55456109
34	Within	29	18	1.611111111			
35							
36	Total	102	23				
37							

Sheet1 / Sheet2 / Sheet3

Ready

The "Columns" data is for type of contact, and the "Sample" data is for time of contact. We see the sample F values for each factor and for the interaction and the corresponding P value. The F critical values are critical values for $\alpha = 0.05$ (as specified in the dialog box).

Example (without replication)

For this example, let's look at the example in Guided Exercises 12 of Section 11.6 of *Understandable Statistics*. In this example, we have the average grams of fat in 3-oz servings of potato chips, as measured by three labs from three brands of chips.

The data are entered in the worksheet. Since there is only one measurement for each factor combination, we use ➤**Tools**➤**Data Analysis**➤**Anova: Two-Factor without Replication.**

The output is

ANOVA output table:

Source of Variation	SS	df	MS	F	P-value	F crit
Rows	108.7022222	2	54.35111111	968.6336634	4.2457E-06	6.944276265
Columns	0.142222222	2	0.071111111	1.267326733	0.374692378	6.944276265
Error	0.224444444	4	0.056111111			
Total	109.0688889	8				

SUMMARY:

SUMMARY	Count	Sum	Average	Variance
Texas	3	98.4	32.8	0.13
Great	3	113.4	37.8	0.01
Chip Oh	3	88	29.33333333	0.043333333
Lab 1	3	99.4	33.13333333	19.76333333
Lab 2	3	100.2	33.4	17.29
Lab 3	3	100.2	33.4	17.41

Data:

	Lab 1	Lab 2	Lab 3
Texas	32.4	33.1	32.9
Great	37.9	37.7	37.8
Chip Oh	29.1	29.4	29.5

The "Columns" data involve the laboratory factor, and the "Rows" data involve the chip brand factor. We see the sample F for each factor and the corresponding P value. The F critical values are for the designated level of significance, $\alpha = 0.05$.

LAB ACTIVITIES FOR ANALYSIS OF VARIANCE – TWO-WAY

See *Understandable Statistics* Section 11.6 Problems. Enter the data for the problems into an Excel worksheet and use the appropriate two-factor command to generate the computer output. Use the computer-generated output to draw conclusions.

TABLE OF EXCEL FUNCTIONS

DESCRIPTIVE STATISTICS

Can be entered by using the Paste Function button on the standard toolbar, or by directly typing the function in the formula bar.

=AVERAGE(cells containing data) returns the arithmetic mean of data

=COUNT(cells containing data) returns the number of data items

=COUNTIF(cells containing data,criteria) returns the number of data items satisfying the criteria

=MAX(cells containing data) returns the maximum value in the data

=MEDIAN(cells containing data) returns the median value of the data

=MIN(cells containing data) returns the minimum value in the data

=MODE(cells containing data) returns the mode of the data

=QUARTILE(cells containing data,quartile) returns the indicated quartile of the data

=STDEV(cells containing data) returns the sample standard deviation of the data

=STDEVP(cells containing data) returns the population standard deviation of the data

=TRIMMEAN(cells containing data,percent) returns the mean of the data trimmed by the total indicated percent, half from the bottom and half from the top.

=VAR(cells containing data) returns the sample variance of the data

=VARP(cells containing data) returns the population variance of the data

Dialog boxes found using the menu selections **➤Tools➤Data Analysis➤Descriptive Statistics** dialog box with Summary Statistics checked returns the mean standard error of mean, median, mode, standard deviation, variance, kurtosis, skewness, range, minimum, maximum, sum, count of the data. Checking the confidence level box returns the error E for a confidence interval $\overline{x} - E$ to $\overline{x} + E$ at the designated confidence level.

GRAPHICS

Use the Chart Wizard on the standard tool bar to create graphs for column charts (bar graphs), pie charts (circle graphs), line graphs (for time series), or scatter diagrams. Right-click a data point and click Add trend line to show a regression line and formula. Use **➤Tools➤Data Analysis➤Histogram** to open the histogram dialog box.

RANDOM SAMPLES

Use the Paste Function button on the standard tool bar or type directly

=RAND( ) returns a random number between 0 and 1.

=RANDBETWEEN(bottom,top) returns a random number between the designated values.

Use the menu selection **➤Tools➤Data Analysis** to access the dialog box.

Sampling returns a random sample from a designated cell range.

Random Number Generator returns a random sample from a designated distribution (uniform, normal, binomial, poisson, patterned, discrete).

PROBABILITY

Use the Paste Function button on the standard tool bar or type directly.

=BINOMDIST(# of success,# of trails,probability of success,cumulative)

returns the probability of exactly r success if the value of cumulative is false the probability of at most *r* successes if the value of cumulative is true.

=CHIDIST(x degrees of freedom)

returns the area to the right of *x* under a chi-square distribution with the designated degrees of freedom. The area is the probability that a value is greater than or equal to *x*.

=CHIINV(probability,degrees of freedom)

returns the chi-square value x such that the probability a value falls to the right of *x* is the designated probability.

=FDIST(x,degrees of freedom numerator,degrees of freedom denominator)

returns area in the right tail of the distribution beyond *x*.

=FINV(probability,degrees of freedom numerator,degrees of freedom denominator)

returns the value *x* such that the probability of a value falling to the right of *x* is the designated probability.

=HYPGEOMDIST(# of success in sample,size of sample,number of successes in population,population size)

returns the probability that the designated number of successes in the sample occur.

=NORMDIST(x,mean,standard deviation,cumulative)

returns the probability that a value is less than or equal x for true as the value of cumulative.

returns the density function value of *x* for false as the value of cumulative.

=NORMINV(probability,mean,standard deviation)

returns the *x* value such that the area under the normal distribution to the left of *x* equals the designated probability.

=NORMSDIST(z)

returns the area to the left of *z* under the standard normal.

=NORMSINV(probability)

returns the *z* value such that the area to the left of *z* is equal to the specified probability.

=POISSON(x,mean,cumulative)

returns the probability that the number of random events occurring will be between zero and *x* inclusive when the value of cumulative is true.

the density function value that the number of events occurring will be exactly *x* when the value of cumulative is false.

=TDIST(x,degrees of freedom,tails)

returns the area in the tail to the right of x when the value of tails is 1. When the value of tails is 2, returns the total area in the tail to the right of *x* and the tail to the left of –*x*.

=TINV(probability,degree of freedom)

returns the *x* value that the total area to the right of *x* and that to the left of –*x* equals the designated probability.

CONFIDENCE INTERVAL

Use Paste Function or type directly

=CONFIDENCE(alpha,standard deviation,size)

returns the value of E in the confidence interval for a mean $\overline{x} - E$ to $\overline{x} + E$. The confidence level equals $100(1 - \alpha)\%$. The value of E is computed using a normal distribution.

Confidence intervals for other parameters are included in some of the outputs for certain dialog boxes contained in the ➤**Tools**➤**Data Analysis** choices.

HYPOTHESIS TESTING

Use the Paste Function on the standard tool bar or type directly

=CHITEST(observed value data cells,expected value data cells)

returns the P value of the sample chi-square statistics for a test of independence.

=FTEST(cells containing the data of first sample,cells containing data of second sample)

returns P value for a two-tailed test of variances.

=TTEST(cells containing first sample,cells containing second sample,tails,type)

returns the P value of a one-tailed test if the value of tails is 1 and a two-tailed test if the value of tails is 2.

The type of test performed is as follows:

type = 1: a paired test is performed.
type = 2: a two-sample equal variance test is performed.
type = 3: a two-sample unequal variance test is performed.

=ZTEST(cells containing data,x,sigma)

returns the P value for a two tailed test where $H_0: \mu = x$ is the null hypothesis and $H_0: \mu - x$ is the alternate hypothesis. For sigma, use the population standard deviation, if it is known. Otherwise leave out sigma, and Excel will use the sample standard deviation computed from the data. The normal distribution is the sampling distribution used.

These dialog boxes are under the menu choice ➤**Tools**➤**Data Analysis**.

F-Test: Two Sample for Variances

returns the P value for a one-tailed test as well as some summary statistics for the data.

t-Test: Paired Two Sample for Means

returns P values for a one-tailed test and P values for a two-tailed test, the t value of the sample test statistic, critical one-tail or two-tail values, and summary statistics for the data.

t-Test: Two-Sample Assuming Equal Variances

returns P values for a one-tailed test and P values for a two-tailed test, the t value of the sample test statistic, critical one-tailed or two-tailed values, and summary statistics for the data.

t-Test: Two-Sample Assuming Unequal Variances

returns P values for a one-tailed test and P values for a two-tailed test, the t value of the sample test statistic, critical one-tailed or two-tailed values, and summary statistics for the data.

z-Test: Two-Sample for Means

returns P values for a one-tailed test and P values for a two-tailed test, the z value of the sample test statistic, critical one-tailed or two-tailed values, and summary statistics for the data.

ANALYSIS OF VARIANCE

These dialog boxes are under the menu choice ➤**Tools**➤**Data Analysis.**

Anova: Single Factor

Anova: Two-Factor With Replication

(equal number greater than 1 of measurements per factor combination)

Anova: Two-Factor Without Replication

(one measurement per factor combination)

Each of these dialog boxes returns summary statistics for the factors and a summary table giving degrees of freedom, sum of squares, mean squares, F values, P values.

LINEAR REGRESSION

Use Paste Function on the standard tool bar or type the command directly into the formula bar.

=CORREL(y range,x range)

returns the correlation coefficient r between two data sets.

=FORECAST(x value,y range,x range)

returns the y value for the specified x as computed using the least-squares regression equation.

=INTERCEPT(y range,x range)

returns the y-intercept of the least -squares regression line.

=LINEST(y range,x range)

returns the slope and y-intercept of the least-squares line. (We are ignoring two optional parameters.) Since this is an array formula, two cells need to be activated to receive the output. The command needs to be typed into the formula bar and entered with the **Ctrl+Shift+Enter** combination.

=PEARSON(y range,x range)

returns the Pearson product moment correlation coefficient r for the two data sets.

=RSQ(y range,x range)

returns r^2 (the square of the Pearson product moment correlation coefficient).

=SLOPE(y range,x range)

returns the slope of the least-squares line.

=STEYX(y range,x range)

returns the standard error of estimate for the predicted y value.

These dialog boxes are under the menu choice ➤**Tools**➤**Data Analysis.**

Correlation

returns the correlation between all variables in a multiple linear regression model.

Regression

returns the coefficients for a simple or multiple linear regression model. Other options provide for a table of residuals, a residual plot a line fit plot, a normal probability plot, and an ANOVA table. An output table including coefficients, standard error of y estimates, r^2 values, number of observations, and standard error of coefficients is provided.

MISCELLANEOUS

Under the Paste Function button on the standard menu bar

=STANDARDIZE(x,mean,standard deviation)

returns the z value for the specified x value.

On the tool bars

Σ **button** gives the sum of the selected cells.

Sort buttons AZ↓ or ZA↑ sort data in increasing or decreasing order.

Chart Wizard button opens the chart-making tools.

$ button formats numbers in selected cells in dollar and cent format.

% button formats numbers in selected cells to percents.

, button formats numbers in selected cells to comma format.

←.0 button formats numbers in selected cells with one more digit after the decimal each time the button is pushed.

→.00 button formats numbers in selected cells with one less digit after the decimal each.

PART III

MINITAB GUIDE

FOR

UNDERSTANDABLE STATISTICS

EIGHTH EDITION

CHAPTER 1 GETTING STARTED

GETTING STARTED WITH MINITAB

In this chapter you will find

(a) general information about MINITAB

(b) general directions for using the Windows style pull-down menus

(c) general instructions for choosing values for dialog boxes

(d) how to enter data

(e) other general commands.

General Information

MINITAB is a command driven software package with more than 350 commands available. However, in Windows versions of MINITAB, menu options and dialog boxes can be used to generate the appropriate commands. After using the menu options and dialog boxes, the actual commands are shown in the Session window along with the output of the desired process. Data are stored and processed in a table with rows and columns. Such a table is similar to a spreadsheet and is called a worksheet. Unlike electronic spreadsheets, a MINITAB worksheet can contain only numbers and text, not formulas. Constraints are also stored in the worksheet, but are not visible.

MINITAB will accept words typed in upper or lower case letters as well as a combination of the two. Comments elaborating on the commands may be included. In this *Guide* we will follow the convention of typing the essential parts of a command in upper case letters and optional comments in lower case letters:

COMMAND with comments

Note that only the *first four letters* of a command are essential. However, we usually give the entire command name in examples.

Numbers must be typed *without* commas. Be sure you type the number zero instead of the letter "O" and that you use the number one (1) instead of a lower case letter "l." Exponential notation is also acceptable. For instance

127.5   1.257E2   1.257E+2

are all acceptable in MINITAB and have the same value.

The MINITAB *worksheet* contains columns, rows, and constants. The rows are designated by numbers. The columns are designated by the letter **C** followed by a number. **C1, C2, C3** designate columns 1, 2, and 3. Constants require the letter **K,** and may be followed by a number if there are several constants. **K1, K2** designate constant 1 and constant 2, respectively.

Starting and Ending MINITAB

The steps you use to start MINITAB will differ according to the computer equipment you are using. You will need to get specific instructions for your installation from your professor or computer lab manager. Use this space to record the details of logging onto your system and accessing MINITAB. For Windows versions, you generally click on the MINITAB icon to begin the program.

_____

_____

_____

_____

_____

_____

The first screen will look similar to the image displayed in the center of the following screen:

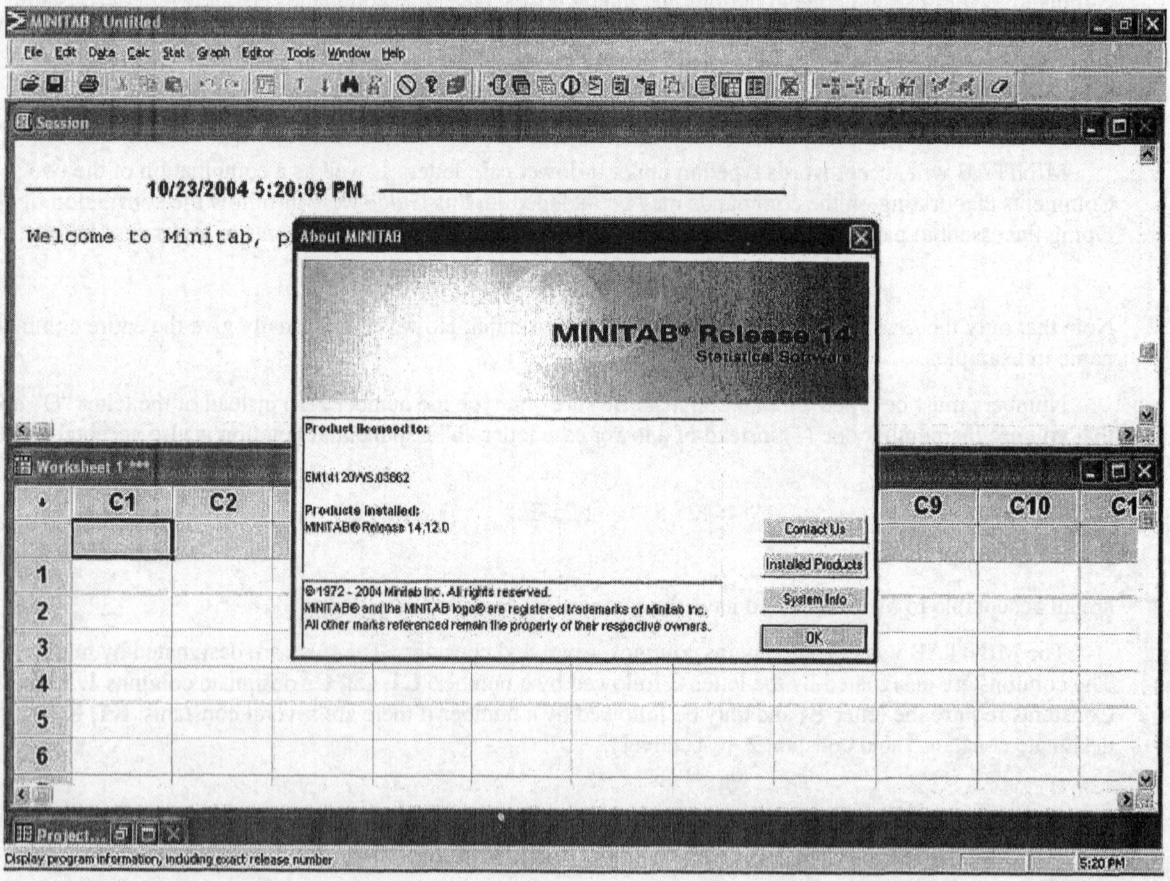

After a pause, a split screen appears. The session window is at the top, and the data window is at the bottom.

Notice the main menu items:

      File   Edit   Data   Calc   Stat   Graph   Editor   Tools   Window   Help

The toolbar contains icons for frequently used operations.

We enter data in the individual cells in the Data window. We can enter commands directly in the Session window, or use the menu options together with dialog boxes to generate the commands.

To end MINITAB: Click on the File option. Select Exit and click on it or press ENTER.

    Menu selection summary: ➤**File**➤**Exit**

Entering Data

One of the first tasks you do when you begin a MINITAB session is to insert data into the worksheet. The easiest way to enter data is to enter it directly into the Data worksheet. Notice that the active cell is outlined by a heavier box.

To enter a number, type it in the active box and then press ENTER or TAB. The data value is entered and the next cell is activated. Data for a specific variable are usually entered by column. Notice that there is a cell for a column label above row number 1.

To change a data value in a cell, click on the cell, correct the data, and press ENTER or TAB.

Example

Open a new worksheet by selecting ➤**File**➤**New.**

Let's create a new worksheet that has data in it regarding ads on TV. A random sample of 20 hours of prime time viewing on TV gave information about the number of TV ads in each hour as well as the total time consumed in the hour by ads. We will enter the data into two columns: one column representing the number of ads and the other the time per hour devoted to ads.

Notice that we typed a name for each column. To switch between the Data window and the Session window, click on <u>W</u>indow and select the window you want. The Data window is called the Worksheet window.

Working with Data

There are several commands for inserting or deleting rows or cells. One way to access these commands is to use the D<u>a</u>ta menu option or the <u>E</u>dit menu option.

Click on the D<u>a</u>ta menu item. You will see these cascading options in the pull-down menu.

A useful item is **Change Data Type.** If you accidentally typed a letter instead of a number, you have changed the data type to text. To change it back to numeric, use ➤**Data**➤**Change Data Type** and fill in the dialog box. The same process can be used to change back to text.

If you want to see the data displayed in the session window, select ➤**Data**➤**Display Data** and select the columns you want to see displayed.

Click on the Edit menu item. You will see these cascading options in the pull-down menu.

To print the worksheet, click the cell in the upper-left corner of the Data window. Press [Shift] + [Ctrl] + [End] to highlight the entire Data window containing all your data. Then click on the printer icon on the toolbar. You can also select ➤**File**➤**Print Worksheet** from the menus.

Manipulating Data

You can also do calculations with entire columns. Click on the Calc menu item and select Calculator (➤**Calc**➤**Calculator**). The dialog box appears:

You can store the results in a new column, say C3. To multiply each entry from C1 by 3 and add 4, type 3, click on the multiply key * on the calculator, type C1, click on the + key on the calculator, type 4. Click on OK. The results of this arithmetic will appear in column C3 of the data sheet.

Saving a Worksheet

Click on the File menu and select Save Current Worksheet As. A dialog box similar to the following appears. (Select ➤**File**➤**Save Current Worksheet As.**)

For most computer labs, you will save your file on a 3½ inch floppy. Insert it in the appropriate drive. Scroll down the Save in button until you find 3½ Floppy (A:). Then select a file name. In most cases you will save the file as a MINITAB file. If you change versions of MINITAB or systems, you might select MINITAB portable.

Example

Let's save the worksheet created in the previous example (information about ads on TV).

If you added Column C3 as described under Manipulating the Data, highlight all the entries of the column and press the Del key. Your worksheet should have only two columns. Use ➤**File**➤**Save Current Worksheets as.** Insert a diskette in drive A. Scroll down the Save in box and select 3½ Floppy (A:). Name the file Ads. Click on Save. The worksheet will be saved as Ads.mtw.

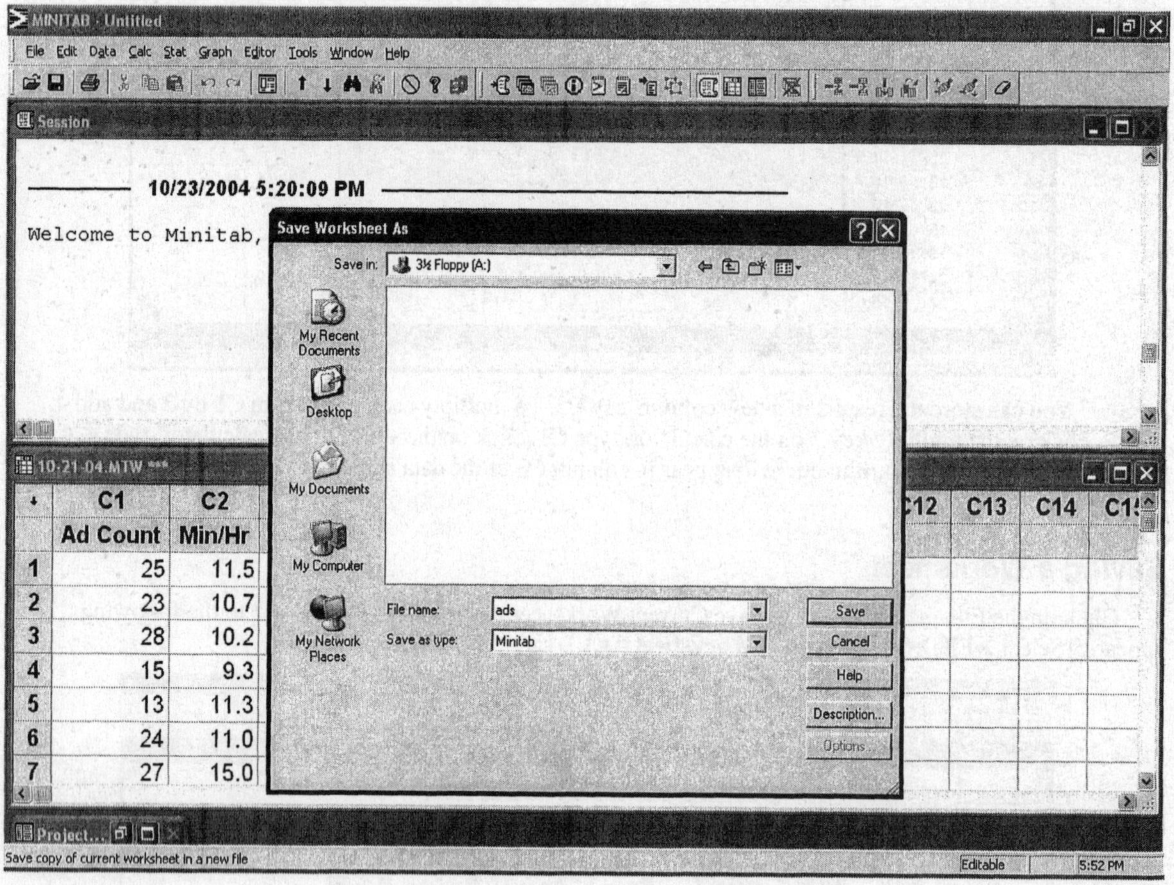

LAB ACTIVITIES FOR GETTING STARTED WITH MINITAB

1. Go to your computer lab (or use your own computer) and learn how to access MINITAB.

2. (a) Use the data worksheet to enter the data:

Enter the data

| 1 | 3.5 | 4 | 10 | 20 | in Column C1. |

| 3 | 7 | 9 | 8 | 12 | in Column C2. |

(b) Use ➤**Calc**➤**Calculator** to create C3. The data in C3 should be 2*C1 + C2. Check to see that the first entry in C3 is 5. Do the other entries check?

    (c) Name C1 First, C2 Second, C3 Result.

    (d) Save the worksheet as Prob 2 on a floppy diskette.

    (e) Retrieve the worksheet by selecting ➤**File**➤**Open Worksheet.**

    (f) Print the worksheet. Use either the Print Worksheet button or select ➤**File**➤**Print Worksheet.**

RANDOM SAMPLES (SECTION 1.2 OF *UNDERSTANDABLE STATISTICS*)

In MINITAB you can take random samples from a variety of distributions. We begin with one for the simplest: random samples from a range of consecutive integers under the assumption that each of the integers is equally likely to occur. The basic command RANDOM draws the random sample, and subcommands refer to the distribution being sampled. To sample from a range of equally likely integers, we use subcommand INTEGER. The menu selection options are:

➤**Calc**➤**Random Data**➤**Integer**

Dialog Box Responses

    Generate _____ rows of data: Enter the sample size.

    Store in: Enter the column number C# in which you wish to store the sample numbers.

    Minimum: Enter the minimum integer value of your population.

    Maximum: Enter the maximum integer value of your population.

The random sample numbers are given in the order of occurrence. If you want them in ascending order (so you can quickly check to see if any values are repeated), use the SORT command.

➤**Data**➤**Sort**

Dialog Box Responses

    Sort columns: Enter the column number C# containing the data you wish to sort.

    Store sorted column in: Choose where you want to store the sorted data. You may choose to store it in the original column that contains the original unsorted data, or in another column in the current worksheet, or in a new worksheet.

    Sort by column: Enter the same column number C# that contains the original data. Leave the rest of the sort by columns options empty.

Example

There are 175 students enrolled in a large section of introductory statistics. Draw a random sample of 15 of the students.

We number the students from 1 to 175, so we will be sampling from the integers 1 to 175. We don't want any student repeated, so if our initial sample has repeated values, we will continue to sample until we have 15 distinct students. We sort the data so that we can quickly see if any values are repeated.

First, generate the sample.

Next, sort the data.

Switch to the Data window and type the name Sample as the header to C1. To display the data, use the command ➤**Data**➤**Display Data.** The results are shown. (Your results will vary.)

We see that no data are repeated. If you have repetitions, keep sampling until you get 15 distinct values.

Random numbers are also used to simulate activities or outcomes of a random experiment such as tossing a die. Since the six outcomes 1 through 6 are equally likely, we can use the RANDOM command with the INTEGER subcommand to simulate tossing a die any number of times. When outcomes are allowed to occur repeatedly, it is convenient to tally, count, and give percents of the outcomes. We do this with the TALLY command and appropriate subcommands.

➤**Stat**➤**Table**➤**Tally Individual Variables**

Dialog Box Responses

Variables: Column number C# or column name containing data

Option to check: counts, percents, cumulative counts, cumulative percents.

Example

Use the RANDOM command with INTEGER A = 1 to B = 6 subcommand to simulate 100 tosses of a fair die. Use the TALLY command to give a count and percent of outcomes.

Generate the random sample using the menu selection **Calc➤Random Data➤Integer**, with generate at 100, min at 1 and max at 6. Type the name Outcome as the header for C1. Then use **➤Stat➤Tables➤Tally Individual Variables** with counts and percents checked.

The results are shown. (Your results will vary.)

If you have a finite population, and wish to sample from it, you may use the command SAMPLE. This command requires that your population already be stored in a column.

➤Calc➤Random Data➤Sample from Columns

Dialog Box Responses

Sample _____ rows from columns: Provide sample size and list column number C# containing population

Store sample in: Provide column number C# where you want to store the sample items.

Example

Take a sample of size 10 without replacement from the population of numbers 1 through 200.

First we need to enter the numbers 1 through 200 in column C1. The easiest way to do this is to use the patterned data option.

➤Calc➤Make Patterned Data➤Simple Set of Numbers

Dialog Box Responses

Store patterned data in: List column number

From first number: 1 for this example

To last value: 200 for this example

In steps of: 1 for this example

Tell how many times to list each value or sequence.

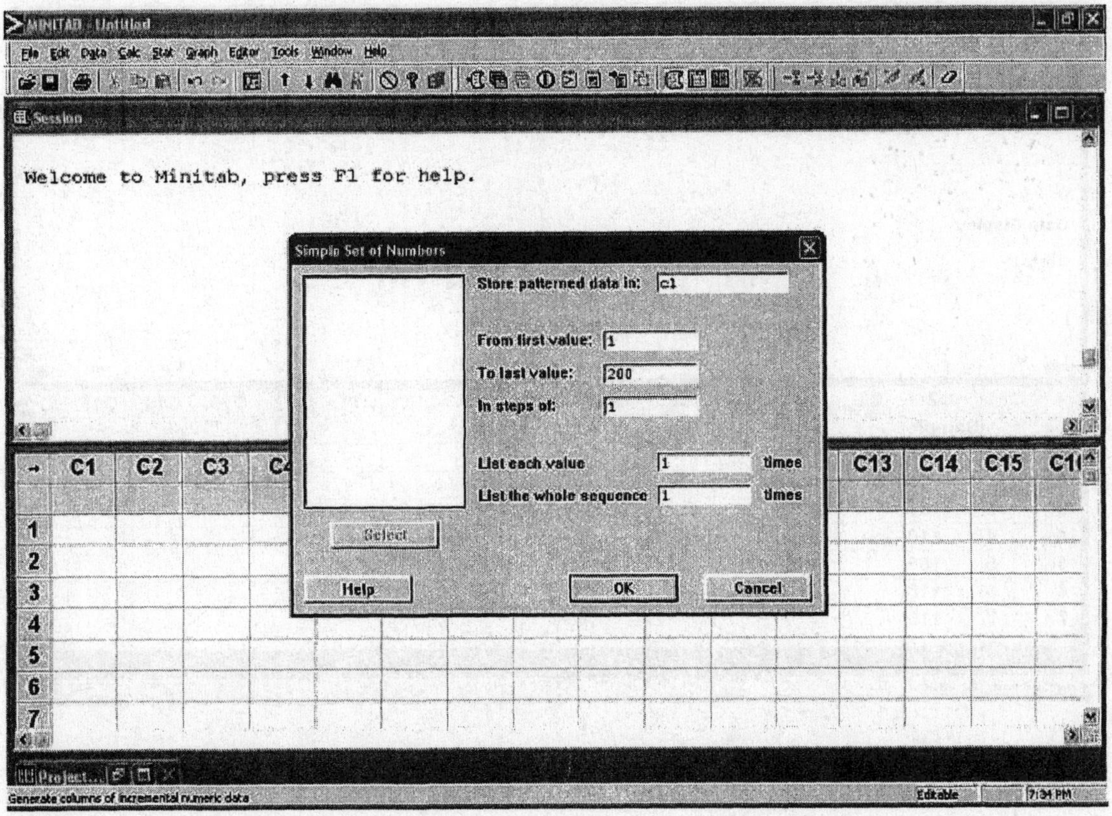

Next we use the **>Calc>Random Data>Sample from Columns** choice to take a sample of 10 items from C1 and store them in C2.

Finally, go to the Data window and label C2 as Sample. Use **>Data>Display Data.** The results are shown. (Your results will vary.)

LAB ACTIVITIES FOR RANDOM SAMPLES

1. Out of a population of 8173 eligible count residents, select a random sample of 50 for prospective jury duty. Should you sample with or without replacement? Use the RANDOM command with subcommand INTEGER A = 1 to B = 8173. Use the SORT command to sort the data so that you can check for repeated values. If necessary, use the RANDOM command again to continue sampling until you have 50 different people.

2. Retrieve the MINITAB worksheet Svls02.mtp on the CD-ROM. This file contains weights of a random sample of linebackers on professional football teams. The data is in Column 1. Use the SAMPLE command to take a random sample of 10 of these weights. Print the 10 weights included in the sample.

Simulating experiments in which outcomes are equally likely is another important use of random numbers.

3. We can simulate dealing bridge hands by numbering the cards in a bridge deck from 1 to 52. Then we draw a random sample of 13 numbers without replacement from the population of 52 numbers. A bridge deck has 4 suits: hearts, diamonds, clubs, and spades. Each suit contains 13 cards: those numbered 2 through 10, a jack, a queen, a king, and an ace. Decide how to assign the numbers 1 through 52 to the cards in the deck.

   (a) Use the RANDOM command with INTEGER subcommand to get the numbers of the 13 cards in one hand. Translate the numbers into cards and tell what cards are in the hand. For a second game, the cards would be collected and reshuffled. Use the computer to determine the hand you might get in a second game.

   (b) Store the 52 cards in C1, and then use the SAMPLE command to sample 13 cards. Put the results in C2, name C2 as 'my hand' and print the results. Repeat this process to determine the hand you might get in a second game.

   (c) Compare the four hands you have generated. Are they different? Would you expect this result?

4. We can also simulate the experiment of tossing a fair coin. The possible outcomes resulting from tossing a coin are heads or tails. Assign the outcome heads the number 2 and the outcome tails the number 1. Use RANDOM with INTEGER subcommand to simulate the act of tossing a coin 10 times. Use TALLY with COUNTS and PERCENTS subcommands to tally the results. Repeat the experiment with 10 tosses. Do the percents of outcomes seem to change? Repeat the experiment again with 100 tosses.

COMMAND SUMMARY

Instead of using menu options and dialog boxes, you can type commands directly into the Session window. Notice that you can enter data via the session window with the commands READ and SET rather than through the data window. The following commands will enable you to open worksheets, enter data, manipulate data, save worksheets, etc. Note: Switch to the Session window; the menu choice ➤**Editor**➤**Enable Command** allows you to enter commands directly into the Session window and also shows the commands corresponding the to menu choices.

HELP   gives general information about MINITAB

   WINDOWS menu: **Help**

INFO   gives the status of the worksheet

STOP   ends MINITAB session

   WINDOWS menu: ➤**File**➤**Exit**

To Enter Data

READ C...C   puts data into designated columns

READ C...C

  File "filename"   reads data from file into columns

SET C   puts data into single designated column

SET C

  File "filename"   reads data from file into column

NAME C "name" names column C

     WINDOWS menu: You can enter data in rows or columns and name the column

in the DATA window. To access the Data window, select ➤**Window**➤**Worksheet**

RETRIEVE 'filename'

     WINDOWS menu: ➤**File**➤**Open Worksheet**

To Edit Data

LET C(K) = K   changes the value in row K of column C

INSERT K K C C   inserts data between rows K and K of C...C

DELETE K K C C   deletes data between rows K and K from column C to C

     WINDOWS menu: You can edit data in rows or columns in the Data window.

                 To access the Data window, select ➤**Window**➤**Worksheet.**

OMIT[C] K...K   subcommand to omit designated rows

     WINDOWS menu: ➤**Data**➤**Copy** ➤**Columns to Columns**

ERASE E...E   erases designated columns or constants

     WINDOWS menu: ➤**Data**➤**Erase Variables**

To Output Data

PRINT E...E   print data on your screen

     WINDOWS menu: ➤**Data**➤**Display Data**

SAVE 'filename' saves current worksheet or project.

PORTABLE   subcommand to make worksheet portable

     WINDOWS menu: ➤**File**➤**Save Project**

     WINDOWS menu: ➤**File**➤**Save Project as**

     WINDOWS menu: ➤**File**➤**Save Current Worksheet**

     WINDOWS menu: ➤**File**➤**Save Current Worksheet As...** you may select
portable

WRITE C...C

  File "filename" saves data in ASCII file

Miscellaneous

OUTFILE "filename": put all input and output in "filename"

NOOUTFILE : ends outfile

To Generate a Random Sample

RANDOM K C...C selects a random sample from the distribution described in the subcommand.

>   WINDOWS menu: **➤Calc➤Random data**

INTEGER K K Specifies distribution to sample, with discrete uniform on integers from minimum value =

>   K to maximum value = K

Other distributions that may be used with the RANDOM command. We will study many of these in later chapters.

>   **BERNOULLI K**
>
>   **BINOMIAL K K**
>
>   **CHISQUARE K**
>
>   **DISCRETE C C**
>
>   **F K K**
>
>   **NORMAL [K [k]]**
>
>   **POISSON K**
>
>   **T K**
>
>   **UNIFORM [K K]**

SAMPLE K C...C   generate k rows of random data from specified input columns, C...C and stores in specified storage columns, C...C.

>   **REPLACE** causes the sample to be taken with replacement.
>
>   **NOREPLACE** causes the sample to be taken with replacement.

To Organize Data

SORT C[C...C] C[C...C] Sorts C, carrying [C..C], and places results into C[C...C]

>   WINDOWS menu: **➤Data➤Sort**
>
>   **DESCENDING C...C** subcommand to sort in descending order

TALLY C...C tallies data in columns with integers.

>   **COUNTS**
>
>   **PERCENTS**
>
>   **CUMCOUNTS**

CUMPERCENTS

ALL gives all four values.

  WINDOWS menu: ➤**Stats**➤**Tables**➤**Tally Individual Variables**

CHAPTER 2 ORGANIZING DATA

HISTOGRAMS (SECTION 2.2 OF *UNDERSTANDABLE STATISTICS*)

MINITAB has graphics in two modes. The default mode is high resolution or professional graphics. There is also an option to use character graphics, which appear in text mode. We will use the high resolution mode.

➤Graph➤Histogram

Dialogue Box Responses

> Choose graph type: you may choose "simple"

> Graph variables: Column containing data

> Click on Data Options and you may select the type for the graph.

> After the histogram is displayed on screen, double click anywhere on the histogram. A dialogue box will show up. Click on Binning. This will allow you to choose type of interval as well as definition of interval. For example, you may choose:

> > "Cutpoints" for type of interval

> > "Midpoint/cutpoint positions" for definition of intervals:

> > > List the class boundaries (as computed in *Understandable Statistics*)

Note: If you do not use Binning selections, the computer sets the number of classes automatically. It uses the convention that data falling on a boundary are counted in the class above the boundary.

Example

Let's make a histogram of the data we stored in the worksheet Ads (created in Chapter 1). We'll use C1, the column with the number of ads per hour on prime time TV. Use four classes.

First we need to retrieve the worksheet. Use **➤File➤Open Worksheet.** Scroll to the drive containing the worksheet. We used 3½ disk drive A. Click on the file.

The number of ads per hour of TV is in column C1. Use **➤Graph➤Histogram.** The dialogue boxes follow.

Choose "simple" and click on OK. The following dialogue box will show up. Choose C1 as the graph variable.

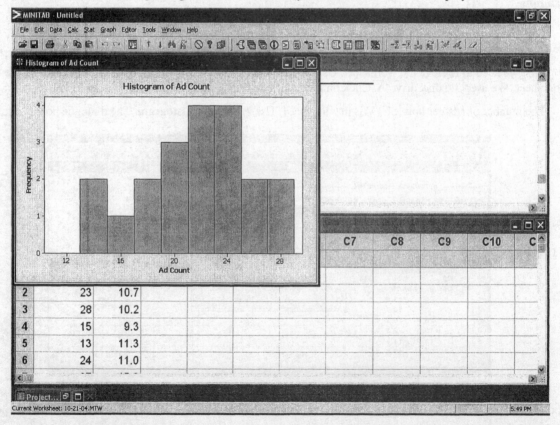

Click on OK the following histogram with automatically selected classes will be displayed.

Now, double click anywhere on the histogram. A dialogue box will show up. Click on Binning. You will see another dialogue box. Choose "Cutpoint" for Interval Type. Note that the low data value is 13 and the high value is 28. Using techniques shown in the text *Understandable Statistics*, we see that the class boundaries for 4 classes are 12.5; 16.5; 20.5; 24.5; 28.5. List these values under Interval Definition as "Midpoint/Cutpoint positions", as shown below.

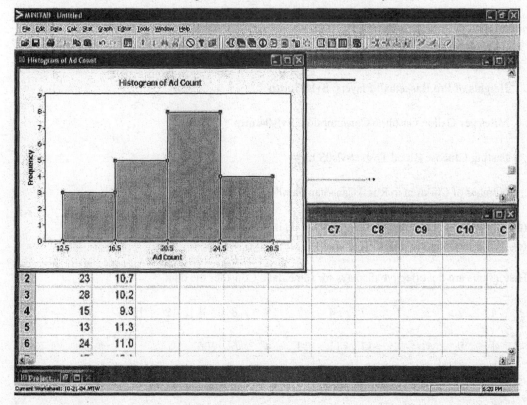

Click OK. You will see the new histogram with the four classes defined by us.

LAB ACTIVITIES FOR HISTOGRAMS

1. The Ads worksheet contains a second column of data that records the number of minutes per hour consumed by ads during prime time TV. Retrieve the Ads worksheet again and use Column C2 to

   (a) make a histogram, letting the computer scale it.

   (b) sort the data and find the smallest data value.

   (c) make a histogram using the smallest data value as the starting value and an increment of 4 minutes. Do this by using cutpoints, with the smallest value as the first cutpoint and incrementing cutpoints by 4 units.

2. As a project for her finance class, Linda gathered data about the number of cash requests made at an automatic teller machine located in the student center between the hours of 6 P.M. and 11 P.M. She recorded the data every day for four weeks. The data values follow.

25	17	33	47	22	32	18	21	12	26	43	25
19	27	26	29	39	12	19	27	10	15	21	20
34	24	17	18								

   (a) Enter the data.

   (b) Use the command HISTOGRAM to make a histogram.

   (c) Use the SORT command to order the data and identify the low and high values. Use the low value as the start value and an increment of 10 to make another histogram.

3. Choose one of the following files from the CD-ROM.

   Disney Stock Volume: **Svls01.mtp**

   Weights of Pro Football Players: **Svls02.mtp**

   Heights of Pro Basketball Players: **Svls03.mtp**

   Miles per Gallon Gasoline Consumption: **Svls04.mtp**

   Fasting Glucose Blood Tests: **Svls05.mtp**

   Number of Children in Rural Canadian Families: **Svls06.mtp**

   (a) Make a histogram, letting MINITAB scale it.

   (c) Make a histogram using five classes.

4. Histograms are not effective displays for some data. Consider the data

1	2	3	6	7	4	7	9	8	4	12	10
1	9	1	12	12	11	13	4	6	206		

Enter the data and make a histogram, letting MINITAB do the scaling. Next scale the histogram with starting value 1 and increment 20. Where do most of the data values fall? Now drop the high value 206 from the data. Do you get more refined information from the histogram by eliminating the high and unusual data value?

STEM-AND-LEAF DISPLAYS (SECTION 2.3 OF *UNDERSTANDABLE STATISTICS*)

MINITAB supports many of the exploratory data analysis methods. You can create a stem-and-leaf display with the following menu choices.

➤Graph➤Stem-and-Leaf

Dialogue Box Responses

Graph variables: Column numbers C# containing the data

Increment: Difference in value between smallest possible data in any adjacent lines.

For example, if the stem unit is ten, then choose increment 10 for 1 line per stem, or 5 for 2 lines per stem.

Example

Let's take the data in the worksheet Ads and make a stem-and-leaf display of C1. Recall that C1 contains the number of ads occurring in an hour of prime time TV.

Use the menu **➤Graph➤Stem-and-Leaf.**

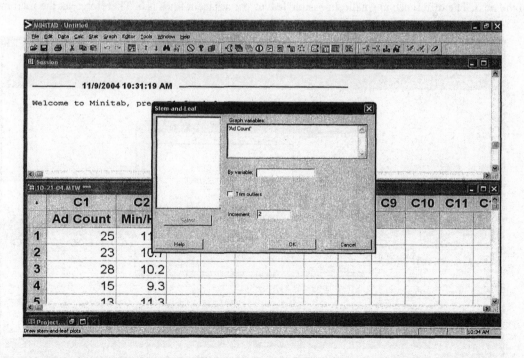

The increment defaulted to 2, so leaf units 0 and 1 are on one line, 2 and 3 on the next, 4 and 5 on the next, etc. The results follow.

```
MINITAB - Untitled
File  Edit  Data  Calc  Stat  Graph  Editor  Tools  Window  Help

Session

Stem-and-Leaf Display: Ad Count

Stem-and-leaf of Ad Count   N  = 20
Leaf Unit = 1.0

     2    1   33
     3    1   5
     4    1   7
     7    1   899
     9    2   01
    (6)   2   222333
     5    2   455
     2    2   7
     1    2   8
```

·	C1	C2	C3	C4	C5	C6	C7	C8	C9	C10	C11	C·
	Ad Count	**Min/Hr**										
1	25	11.5										

Welcome to Minitab, press F1 for help.   Editable   10:37 AM

The first column gives the depth of the data. The line containing the middle value is indicated by (number of data in this line), which is (6) in this example. The remaining numbers in the first column are divided into two parts: the part above (6) indicates the number of data accumulated from the top, and the part below (6) is for that from the bottom. The second column gives the stem and the last gives the leaves.

Let's remake a stem leaf with 2 lines per stem. That means that leaves 0–4 are on one line and leaves 5–9 are on the next. The difference in smallest possible leaves per adjacent lines is 5. Therefore, set the increment as 5.

```
MINITAB - Untitled
File  Edit  Data  Calc  Stat  Graph  Editor  Tools  Window  Help

Session

               Stem-and-Leaf                              ×

                              Graph variables:
                              'Ad Count'


                              By variable:

                              ☐ Trim outliers

                  Select      Increment: 5

         Help              OK           Cancel
```

·	C1	C2	...	C9	C10	C11	C·
	Ad Count	**Min/...**					
1	25	11.5					
2	23	10.7					
3	28	10.2					
4	15	9.3					

Draw stem-and-leaf plots   Editable   10:46 AM

The results follow.

```
Stem-and-Leaf Display: Ad Count

Stem-and-leaf of Ad Count   N  = 20
Leaf Unit = 1.0


  2    1   33
  7    1   57899
 (9)   2   012223334
  4    2   5578
|
```

2	23	10.7
3	28	10.2
4	15	9.3

LAB ACTIVITIES FOR STEM-AND-LEAF DISPLAYS

1. Retrieve worksheet Ads again, and make a stem-and-leaf display of the data in C2. This data gives the number of minutes of ads per hour during prime time TV programs.

   (a) Use an increment of 2.

   (b) Use an increment of 5.

2. In a physical fitness class students ran 1 mile on the first day of class. These are their times in minutes.

12	11	14	8	8	15	12	13	12
10	8	9	11	14	7	14	12	9
13	10	9	12	12	13	10	10	9
12	11	13	10	10	9	8	15	17

   (a) Enter the data in a worksheet.

   (b) Make a stem-and-leaf display and let the computer set the increment.

   (c) Use the TRIM option and let the computer set the increment. How does this display differ from the one in part (b)?

   (d) Set your own increment and make a stem-and-leaf display.

COMMAND SUMMARY

To Organize Data

SORT C[C...C] C[C...C] sorts the data in the first column and carries the other columns along.

>   WINDOWS menu: **➤Data➤Sort**

>   **DESCENDING C...C** subcommand to sort in descending order

TALLY C...C displays a one-way table for each variable in C...C

>   **COUNTS**

>   **PERCENTS**

>   **CUMCOUNTS**

>   **CUMPERCENTS**

>   **ALL** gives all four values.

>   WINDOWS menu: **➤Stat➤Tables➤Tally Individual Variables**

HISTOGRAM C...C prints a separate histogram for data in each of the listed columns.

>   **MIDPOINT K...K** Places ticks at midpoints of the intervals K ... K

>   WINDOWS menu: (for numerical variables) **➤Graph➤Histogram (options for cutpoints)**

>   WINDOWS menu: (for categorical variables) **➤Graphs➤Bar Chart**

STEM-AND-LEAF C...C makes separate stem-and-leaf displays of data in each of the listed
                    columns.

>   **INCREMENT = K** sets the distance between two display lines.

>   **TRIM** trims all values beyond the inner fences

>   WINDOWS menu: **➤Graph ➤Stem-and-Leaf**

CHAPTER 3 AVERAGES AND VARIATION

AVERAGES AND STANDARD DEVIATION OF UNGROUPED DATA (SECTIONS 3.1 AND 3.2 OF *UNDERSTANDABLE STATISTICS*)

The command DESCRIBE of MINITAB gives many of the summary statistics described in *Understandable Statistics*.

➤**Stat**➤**Basic Statistics**➤**Display Descriptive Statistics** prints descriptive statistics for each column of data.

Dialogue Box Response

Variables: List the columns C1…CN that contain the data.

Graphs option: You may print histograms, etc. directly from this menu.

The labels for Display Descriptive Statistics are

N	number of data in C
N*	number of missing data in C
MEAN	arithmetic mean of C
SEMEAN	standard error of the mean, **STDEV/SQRT(N)** (we will use this value in Chapter 7)
STDEV	the sample standard deviation of C, *s*
MIN	minimum data value in C
Q1	1st quartile of distribution in C
MEDIAN	median or center of the data in C
Q3	3rd quartile of distribution in C
MAX	maximum data value in C

(**Q1** and **Q3** are similar to Q_1 and Q_3 as discussed in Section 3.4 of *Understandable Statistics*. However, the computation process is slightly different and gives values slightly different from those in the text.)

Example

Let's again consider the data about the number and duration of ads during prime time TV. We will retrieve worksheet Ads and use DESCRIBE on C2, the number of minutes per hour of ads during prime time TV.

First use ➤**File**➤**Open Worksheet** to open worksheet Ads.

Next use ➤**Stat**➤**Basic Statistics**➤**Display Descriptive Statics.**

Select **Min/Hr** and click on **OK.**

The results follow.

ARITHMETIC IN MINITAB

The standard deviation given in STDEV is the sample standard deviation

$$s = \sqrt{\frac{\sum(x - \overline{x})^2}{N - 1}}$$

We can compute the population standard deviation σ by multiplying s by the factor below:

$$\sigma = s\sqrt{\frac{N-1}{N}}$$

MINITAB allows us to do such arithmetic. Use the built-in calculator under menu selection ➤**Calc**➤**Calculator.** Note that * means multiply and ** means exponent.

Example

Let's use the arithmetic operations to evaluate the population standard deviation and population variance for the minutes per hour of TV ads. Notice that the sample standard deviation $s = 1.849$ and the sample size is 20.

Use the CALCULATOR as follows: Select ➤**Calc**➤**Calculator.** Then enter the expression for the population variance on the calculator.

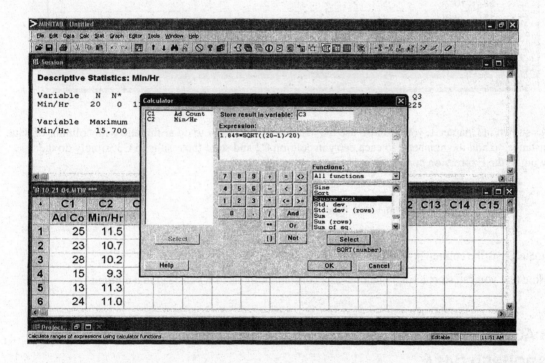

Note that we access the square root function in the function menu. The result is stored in C3. Go to the Data Window and name C3 as PopStDv. Then use ➤**Data**➤**Display Data.** The result follows.

As shown in Chapter 1, you can also use the MINITAB calculator to do arithmetic with columns of data. For instance, to add the number 5 to each entry in column C2 and store the results in C3, simply do the following in the Expression box:

Type C2 click on + key type 5

Designate C3 as the column in which to store results.

Note that you can store a single number as a constant designated K# instead in a column.

LAB ACTIVITIES FOR AVERAGES AND STANDARD DEVIATION OF UNGROUPED DATA

1. A random sample of 20 people were asked to dial 30 telephone numbers each. The incidences of numbers misdialed by these people follow:

3	2	0	0	1	5	7	8	2	6
0	1	2	7	2	5	1	4	5	3

   Enter the data and use the menu selections ➤**Basic Statistics**➤**Display Descriptive Statistics** to find the mean, median, minimum value, maximum value, and standard deviation.

2. Consider the test scores of 30 students in a political science class.

85	73	43	86	73	59	73	84	100	62

75	87	70	84	97	62	76	89	90	83
70	65	77	90	94	80	68	91	67	79

(a) Use the menu selections ➤**Basic Statistics**➤**Display Descriptive Statistics** to find the mean, median, minimum value, maximum value, and standard deviation.

(b) Greg was in political science class. Suppose he missed a number of classes because of illness, but took the exam anyway and made a score of 30 instead of 85 as listed in the data set. Change the 85 (first entry in the data set) to 30 and use the DESCRIBE command again. Compare the new mean, median and standard deviation with the ones in part (a). Which average was most affected: median or mean? What about the standard deviation?

3. Consider the 10 data values

     4   7   3   15   9   12   10   2   9   10

(a) Use the menu selections ➤**Basic Statistics**➤**Display Descriptive Statistics** to find the sample standard deviation of these data values. Then, following example 2 as a model, find the population standard deviation of these data. Compare the two values.

(b) Now consider these 50 data values in the same range.

7	9	10	6	11	15	17	9	8	2
2	8	11	15	14	12	13	7	6	9
3	9	8	17	8	12	14	4	3	9
2	15	7	8	7	13	15	2	5	6
2	14	9	7	3	15	12	10	9	10

Again use the menu selections ➤**Basic Statistics**➤**Display Descriptive Statistics** to find the sample standard deviation of these data values. Then, as above, find the population standard deviation of these data. Compare the two values.

(c) Compare the results of parts (a) and (b). As the sample size increases, does it appear that the difference between the population and sample standard deviations decreases? Why would you expect this result from the formulas?

4. In this problem we will explore the effects of changing data values by multiplying each data value by a constant, or by adding the same constant to each data value.

(a) Make sure you have a new worksheet. Then enter the following data into C1:

     1   8   3   5   7   2   10   9   4   6   32

Use the menu selections ➤**Basic Statistics**➤**Display Descriptive Statistics** to find the mean, median, minimum and maximum values and sample standard deviation.

(b) Now use the calculator box to create a new column of data C2 = 10*C1. Use menu selections again to find the mean, median, minimum and maximum values, and sample standard deviation of C2. Compare these results to those of C1. How do the means compare? How do the medians compare? How do the standard deviations compare? Referring to the formulas for these measures (see Sections 3.1 and 3.2 of *Understandable Statistics*), can you explain why these statistics behaved the way they did? Will these results generalize to the situation of multiplying each data entry by 12 instead of 10? Confirm your answer by creating a new C3 that has each datum of C1 multiplied by 12. Predict the corresponding statistics that would occur if we multiplied each datum of C1 by 1000. Again, create a new column C4 that does this, and use DESCRIBE to confirm your prediction.

(c) Now suppose we add 30 to each data value in C1. We can do this by using the calculator box to create a new column of data C6 = C1 + 30. Use menu selection ➤**Basic Statistics**➤**Display Descriptive Statistics** on C6 and compare the mean, median, and standard deviation to those shown for C1. Which are the same? Which are different? Of those that are different, did each change by being 30 more than the corresponding value of part (a)? Again look at the formula for the standard deviation. Can you predict the observed behavior from the formulas? Can you generalize these results? What if we added 50 to each datum of C1? Predict the values for the mean, median, and sample standard deviation. Confirm your predictions by creating a column C7 in which each datum is 50 more than that in the respective position of C1. Use menu selections ➤**Basic Statistic**➤**Display Descriptive Statistics** on C7.

(d) Name C1 as 'orig', C2 as 'T10', C3 as 'T12', C4 as 'T1000', C6 as 'P30', C7 as 'P50'. Now use the menu selections ➤**Basic Statistic**➤**Display Descriptive Statistics** C1-C4 C6 C7 and look at the display. Is it easier to compare the results this way?

BOX-AND-WHISKER PLOTS (SECTION 3.4 OF *UNDERSTANDABLE STATISTICS*)

The box-and-whisker plot is another of the explanatory data analysis techniques supported by MINITAB. With MINITAB version 14, unusually large or small values are displayed beyond the whisker and labeled as outliers by a *. The upper whisker extends to the highest data value within the upper limit, here the upper limit = Q3 + 1.5 (Q3 − Q1). Similarly, the lower whisker extends to the lowest value within the lower limit, and the lower limit = Q1− 1.5 (Q3 − Q1). By default, the top of the box is the third quartile (Q3) and the bottom of the box is the first quartile (Q1). The line in the box indicates the value of median.

The menu selections are

➤**Graph**➤**Boxplot**

Dialogue Box Responses: choose type of plot, such as "Simple"

Graph variables: enter the column number C# containing the data.

Labels: open box and you can title the graph.

Data view: IQ Range Box with Outliers shown.

There are other boxes available within this box. See Help to learn more about these options.

Example

Now let's make a box-and-whisker plot of the data stored in worksheet ADS. C1 contains the number of ads per hour of prime time TV, while C2 contains the duration per hour of the ads.

Use the menu selection ➤**Graph**➤**Boxplot**. Choose "simple" for the plot type, then choose C1 for graph variable. Click on OK.

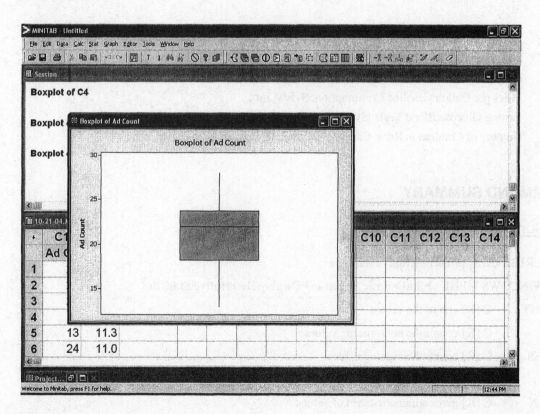

The results follow.

LAB ACTIVITIES FOR BOX-AND-WHISKER PLOTS

1. State-regulated nursing homes have a requirement that there be a minimum of 132 minutes of nursing care per resident per 8-hr shift. During an audit of Easy Life Nursing home, a random sample of 30 shifts showed the number of minutes of nursing care per resident per shift to be

200	150	190	150	175	90	195	115	170	100
140	270	150	195	110	145	80	130	125	115
90	135	140	125	120	130	170	125	135	110

   (a) Enter the data.

   (b) Make a box-and-whisker plot. Are there any unusual observations?

   (c) Make a stem-and-leaf plot. Compare the two ways of presenting the data.

   (d) Make a histogram. Compare the information in the histogram with that in the other two displays.

   (e) Use the ➤**Stat**➤**Basic Statistics**➤**Display Descriptive Statistics** menu selections.

   (f) Now remove any data beyond the outer fences. Do this by inserting an asterisk * in place of the number in the data cell. Use the menu selections ➤**Stat**➤**Basic Statistics**➤**Display Descriptive Statistics** on this data. How do the means compare?

   (h) Pretend you are writing a brief article for a newspaper. Describe the information about the time nurses spend with residents of a nursing home. Use non-technical terms. Be sure to make some comments about the "average" of the data measurements and some comments about the spread of the data.

2. Select one of these data files from the CD-ROM and repeat parts (b) and (h).

   Disney Stock Volume: **Svls01.mtp**

   Weights of Pro Football Players: **Svls02.mtp**

   Heights of Pro Basketball Players: **Svls03.mtp.**

   Miles per Gallon Gasoline Consumption: **Svls04.mtp**

   Fasting Glucose Blood Tests: **Svls05.mtp**

   Number of Children in Rural Canadian Families: **Svls06.mtp**

COMMAND SUMMARY

To Summarize Data by Column

DESCRIBE C...C prints descriptive statistics

   WINDOWS MENU: ➤**Stat**➤**Basic Statistics**➤**Display Descriptive Statistics**

COUNT	**C [K]** counts the values	
N	**C [K]** counts the non-missing values	
NMISS	**C [K]** counts the missing values	
SUM	**C [K]** sums the values	
MEAN	**C [K]** gives arithmetic mean of values	
STDEV	**C [K]** gives sample standard deviation	

MEDIAN   C [K] gives the median of the values

MINIMUM   C [K] gives the minimum of the values

MAXIMUM C [K] gives the maximum of the values

SSQ   C [K] gives the sum of squares of values

To Summarize Data by Row

RCOUNT   E...E C

RN   E...E C

RNMISS   E...E C

RSUM   E...E C

RMEAN   E...E C

RSTDEV   E...E C

RMEDIAN   E...E C

RMINIMUM E...E C

RMAXIMUM E...E C

RSSQ   E...E C

To Display Data

BOXPLOT C...C makes a separate box-and-whisker plot for each column C

    WINDOWS MENU: (professional graphics) ➤Graph➤Boxplot

To Do Artithmetic

LET E = expression

    evaluates the expression and stored the result in E, where E may be a column or a constant.

    ** raises to a power

    * multiplication

    / division

    + addition

    – subtraction

SQRT E E   takes the square root

ROUND(E E)   rounds numbers to the nearest integer

    Other arithmetic operations are possible.

    WINDOWS menu selections: ➤Calc➤Calculator

CHAPTER 4 ELEMENTARY PROBABILITY THEORY

RANDOM VARIABLES AND PROBABILITY

MINITAB supports random samples from a column of numbers or from many probability distributions. See the options under **≻Calc≻Random Data.** By using some of the same techniques shown in Chapter 2 of this guide for random samples, you can simulate a number of probability experiments.

Example

Simulate the experiment of tossing a fair coin 200 times. Look at the percent of heads and the percent of tails. How do these compare with the expected 50% of each?

Assign the outcome heads to digit 1 and tails to digit 2. We will draw a random sample of size 200 from the distributions of integers from a minimum of 1 to a maximum of 2.

Use the menu selections **≻Calc≻Random Data≻Integer.** In the dialog box, enter 200 for the number of rows, 1 for the minimum and 2 for the maximum. Put the data in column C1 and label the column Coin.

To tally the results use **≻Stat≻Tables≻Tally Individual Variables** and check the counts and percents options.

The results are shown. (Your results will vary.)

LAB ACTIVITIES FOR RANDOM VARIABLES AND PROBABILITY

1. Use the RANDOM command and INTEGER A = 0 to B = 1 subcommand to simulate 50 tosses of a fair coin. Use the TALLY command with COUNT and PERCENT subcommands to record the percent of each outcome. Compare the result with the theoretical expected percents (50% heads, 50% tails). Repeat the process for 1000 trials. Are these outcomes closer the results predicted by the theory?

2. We can use the RANDOM 50 C1 C2 command (that is, in the dialog box of ➤**Calc**➤**Random Data**➤**Integer,** enter C1 C2 for "store in colomns") with INTEGER A = 1 to B = 6 subcommand to simulate the experiment of rolling two dice 50 times and recording each sum. This command puts outcomes of die 1 into C1 and those of die 2 into C2. Put the sum of the dice into C3. Then use the TALLY command with COUNT and PERCENT subcommands to record the percent of each outcome. Repeat the process for 1000 rolls of the dice.

CHAPTER 5 THE BINOMIAL PROBABILITY DISTRIBUTION AND RELATED TOPICS

BINOMIAL PROBABILITY DISTRIBUTIONS (SECTIONS 5.2, 5.3 OF

UNDERSTANDABLE STATISTICS)

The binomial probability distribution is a discrete probability distribution controlled by the number of trials, n, and the probability of success on a single trial, p.

MINITAB has three main commands for studying probability distributions.

The PDF (probability density function) gives the probability of a specified value for a discrete distribution.

The CDF (cumulative distribution function) for a value X gives the probability a random variable with distribution specified in a subcommand is less than or equal to X.

The INVCDF gives the inverse of the CDF. In other words, for a probability P, INVCDF returns the value X such that $P \approx CDF(X)$. In this case of a binomial distribution. INVCDF often gives the two values of X for which P lies between the respective CDF(X).

The three commands PDF, CDF, and INVCDF apply to many probability distributions. To apply them to a binomial distribution, we need to use the menu selections.

Calc➤Probability distributions➤Binomial

Dialog Box Responses

        Select Probability for PDF; Cumulative probability for CDF;

        Inverse cumulative probability for INVCDF

        Number of trials: use the value of n in a binomial experiment,

        Probability of success: use the value of p, the probability of success on a single trial,

        Input column: Put the values of r, the number of successes in a binomial experiment in a

        column such as C1. Select an optional storage column.

        Note: MINITAB uses X instead of r to count the number of successes,

        Input constant: Instead of putting values of r in a column, you can type a specific value of

                r in the dialog box.

Example

A surgeon performs a difficult spinal column operation. The probability of success of the operation is $p = 0.73$. Ten such operations are scheduled. Find the probability of success for 0 through 10 successes out of these ten operations.

First enter the possible values of r, 0 through 10, in C1 and name the column r. We will put the probabilities in C2, so name the column $P(r)$.

Fill in the dialog box as shown below.

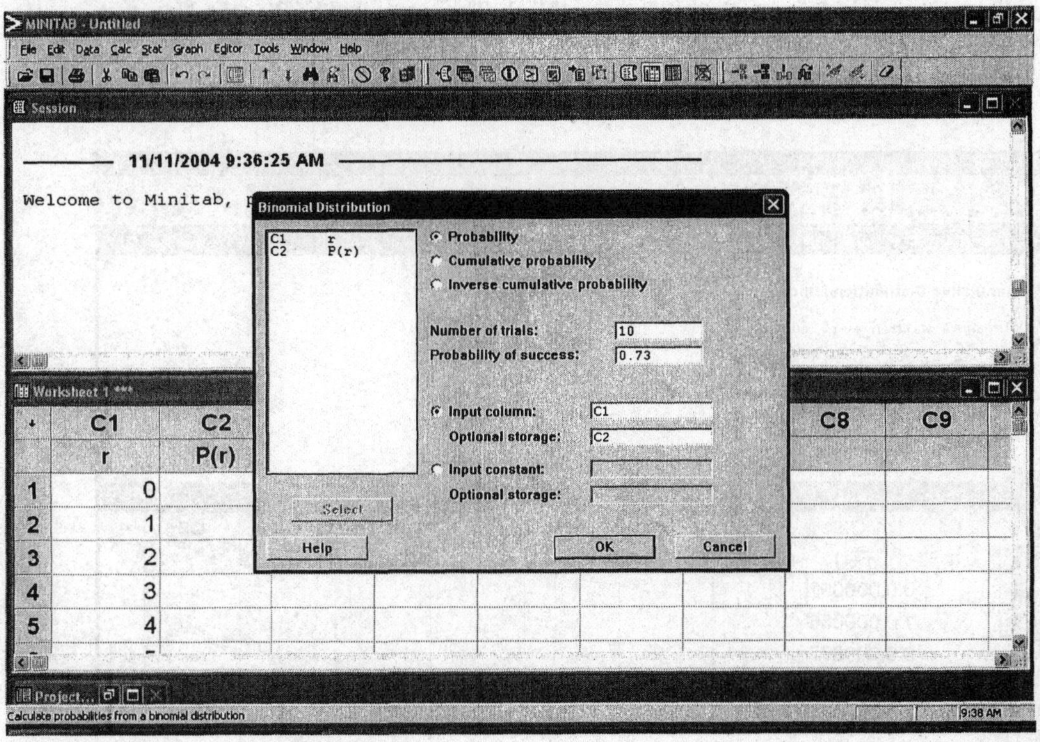

Then use the ➤**Data**➤**Display data** command.

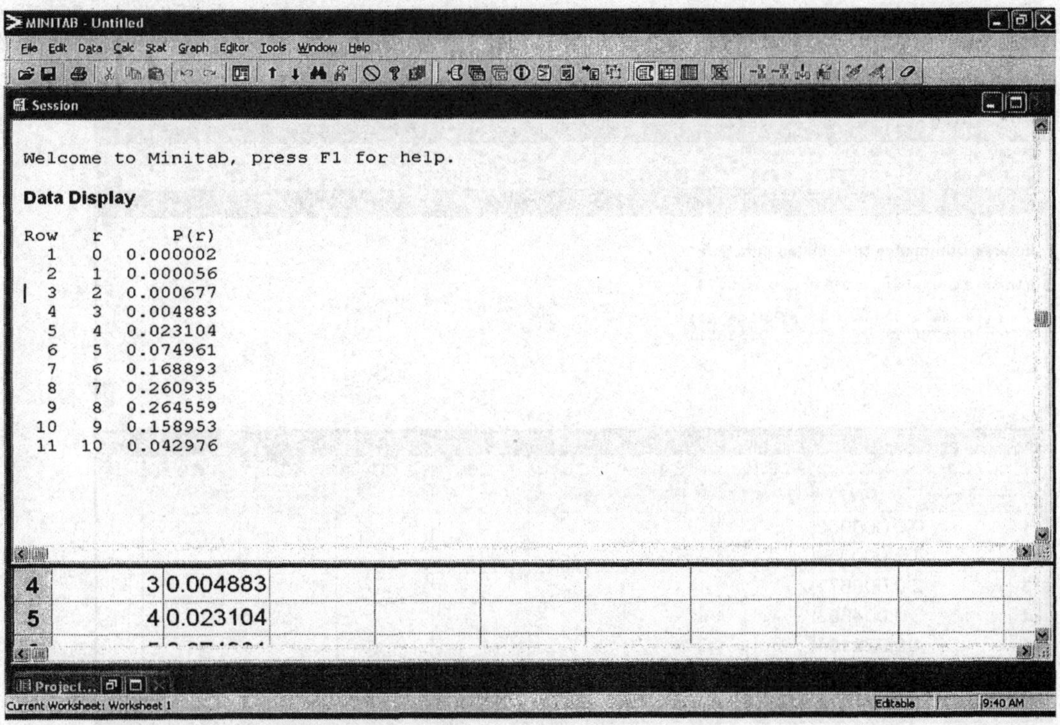

Next use the CDF command to find the probability of 5 or fewer successes. In this case use the option for an input constant of 5. Leave Optional storage blank. The output will be P(r ≤ 5). Note that MINITAB uses X in place of r.

The results follow.

Finally use INVCDF to determine how many operations should be performed in order for the probability of that many or fewer successes to be 0.5. We select Inverse cumulative probability. Use .5 as the input constant.

The results follow.

LAB ACTIVITIES FOR BINOMIAL PROBABILITY DISTRIBUTIONS

1. You toss a coin 8 times. Call heads success. If the coin is fair, the probability of success P is 0.5. What is the probability of getting exactly 5 heads out of 8 tosses? of exactly 20 heads out of 100 tosses?

2. A bank examiner's record shows that the probability of an error in a statement for a checking account at Trust Us Bank is 0.03. The bank statements are sent monthly. What is the probability that exactly two of the next 12 monthly statements for our account will be in error? Now use the CDF option to find the probability that <u>at least</u> two of the next 12 statements contain errors. Use this result with subtraction to find the probability that <u>more than</u> two of the next 12 statements contain errors. You can use the Calculator key to do the required subtraction.

3. Some tables for the binomial distribution give values only up to 0.5 for the probability of success p. There is symmetry to the values for p greater than 0.5 with those values of p less than 0.5.

   (a) Consider the binomial distribution with $n = 10$ and $p = .75$. Since there are 0–10 successes possible, put $0 - 10$ in C1. Use PDF option with C1 and store the distribution probabilities in C2. Name C2 = 'P = .75'. We will print the results in part (c).

   (b) Now consider the binomial distribution with $n = 10$ and $p = .25$. Use PDF option with C1 and store the distribution probabilities in C3. Name C3 = 'P = .25'.

   (c) Now display C1 C2 C3 and see if you can discover the symmetries of C2 with C3. How does $P(K = 4$ successes with $p = .75)$ compare to $P(K = 6$ successes with $p = .25)$?

   (d) Now consider a binomial distribution with $n = 20$ and $p = .035$. Use PDF on the number 5 to get $P(K = 5$ successes out of 20 trials with $p = .35)$. Predict how this result will compare to the probability $P(K = 15$ successes out of 20 trials with $p = .65)$. Check your prediction by using the PDF on 15 with the binomial distribution $n = 20$, $p = .65$.

The INVCDF command for a binomial distribution can be used in the solution of Quota problems as described in Section 5.3 of *Understandable Statistics*.

4. Consider a binomial distribution with $n = 15$ and $p = 0.64$. Use the INVCDF to find the smallest number of successes K for which $P(X \le K) = 0.98$. What is the smallest number of successes K for which $P(X \le K) = 0.09$?

COMMAND SUMMARY

To Find Probabilities

PDF E [E] calculates probabilities for the specified values of a discrete distribution and calculates the probability density function for a continuous distribution.

CDF E [E] gives the cumulative distribution. For any value X, CDF X gives the probability that a random variable with the specified distribution has a value less than or equal to X.

INVCDF E [E] gives the inverse of the CDF.

Each of these commands apply the following distributions (as well as some others). If no subcommand is used, the default distribution is the standard normal.

   BINOMIAL n = K p = K

   POISSON   μ = K (note that for the Poisson distribution $\mu = \lambda$)

   INTEGER   a = K b = K

   DISCRETE values in C, probabilities in C

   NORMAL   [μ = K [σ = K]]

   UNIFORM [a = k b = K]

T   d.f. = K

F   d.f. numerator = K d.f. denominator = K

CHISQUARE   d.f. = K

WINDOWS menu selection: ➤**Calc**➤**Probability Distribution**➤**Select distribution**

In the dialog box, select **Probability for PFD; Cumulative probability for CDF; Inverse cumulative for INV;** Enter the required information such as **E, n, p, or μ, d.f.** and so forth.

CHAPTER 6 NORMAL DISTIBUTIONS

GRAPHS OF NORMAL DISTRIBUTIONS (SECTION 6.1 OF *UNDERSTANDABLE STATISTICS*)

The normal distribution is a continuous probability distribution determined by the value of μ and σ. We can sketch the graph of a normal distribution by using the menu selection

> ➤**Cal**➤**Probability Distribution**➤**Normal**

Dialog Box Responses

> Select Probability density for **PDF,** Cumulative probability for **CDF,** or Inverse cumulative probability for **INVCDF.**
>
> Enter the mean.
>
> Enter the standard deviation.
>
> Select an input column: Put the value of x for which you want to compute P(x) in the designated column. Designate an optional storage column.
>
> Select an input constant: If you wish to compute P(x) for a single value x, enter value as the constant.

We will use the normal probability density option **PDF** to create a graph of a normal distribution with a specified mean and standard deviation.

Menu Options for Graphs

To graph functions in MINITAB, we use the menu ➤**Graph**➤**Scatterplot.**

Dialog Box Responses

> Choose "simple" as graph type, click on OK
>
> Specify which column contains data for the X values.
>
> Specify which column contains data for the Y values.
>
> Click on Data views, then select the Connect line option.

There are other options available. See the Help menu for more information.

Example

Graph the normal distribution with mean $\mu = 10$ and standard deviation $\sigma = 2$.

Since most of the normal curve occurs over the values $\mu - 3\sigma$ to $\mu + 3\sigma$, we will start the graph at $10 - 3(2) = 4$ and end it at $10 + 3(2) = 16$. We will let MINITAB set the scale on the vertical axis.

To graph a normal distribution, put X values into C1 and Y values (height of the graph at point X) into C2.

Use the menu option ➤**Calc**➤**Make Patterned Data**➤**Simple set of numbers.** Store values in C1. First value is 4, last value is 16, and increment is 0.25. Variables and sequences occur one time each. Name C1 as X.

Use ➤**Calc**➤**Probability Distributions**➤**Normal** to generate the Y values which we will store in C2. Name C2 as P(X).

Finally, use ➤**Graph**➤**Scatterplot.** Set the options as shown. Then click on Data View

The following dialog box will show up. Choose the Connect line option.

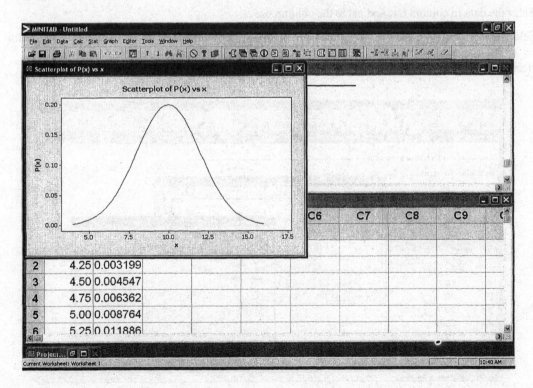

The graph then follows.

CONTROL CHARTS (SECTION 6.1 OF *UNDERSTANDABLE STATISTICS*)

MINITAB supports a variety of control charts. The type discussed in Section 6.1 of *Understandable Statistics* is called an individual chart. The menu selection is **➤Stat➤Control Chart➤Variables Charts for Individuals➤Individuals.**

Dialog Box Responses

Variable: Designate the column number C1 where the data is located.

Click on "I Chart options".

Enter values for the historical mean and standard deviation.

Tests option button lists out-of-control tests. Select numbers 1, 2 and 5 for signals discussed in *Understandable Statistics*.

For information about the other options, see the Help menu.

Example

In a packaging process, the weight of popcorn that is to go in a bag has a normal distribution with $\mu = 20.7$ oz and $\sigma = 0.7$ oz. During one session of packaging, eleven samples were taken. Use an individual control chart to show these observations. The weights were (in oz).

19.5	20.3	20.7	18.9	19.5	20.5
20.7	21.4	21.9	22.7	23.8	

Enter the data in column C1, and name the column oz.

Select **Stat➤Control Charts➤Variables Charts for Individuals ➤Individuals.** Fill in the values for

μ and σ.

The control chart is

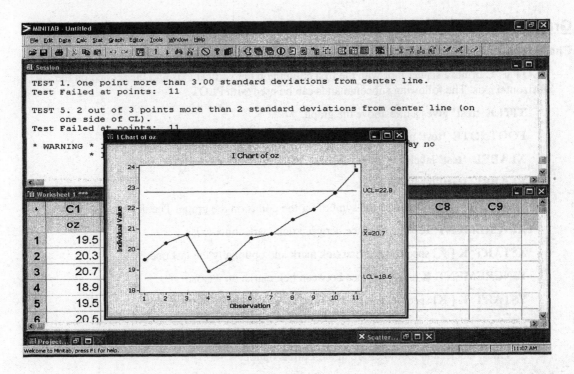

LAB ACTIVITIES FOR GRAPHS OF NORMAL DISTRIBUTIONS AND CONTROL CHARTS

1. **(a)** Sketch a graph of the standard normal distribution with $\mu = 0$ and $\sigma = 1$. Generate C1 using data from –3 to 3 in increments of 0.5 (generates data from –3 to 3 with .5 increments).

   (b) Sketch a graph of a normal distribution with $\mu = 10$ and $\sigma = 1$. Generate C1 using data from –7 to 7 in increments of 0.5. Compare this graph to that of part (a). Do the height and spread of the graphs appear to be the same? What is different? Why would you expect this difference?

   (c) Sketch a graph of a normal distribution with $\mu = 0$ and $\sigma = 2$. Generate C1 using data from –6 to 6 in increments of 0.5. Compare this graph to that of part (a). Do the height and spread of the graphs appear to be the same? What is different? Why would you expect this difference? Note, to really compare the graphs, it is best to graph them using the same scales. Redo the graph of part (a) using X from –6 to 6. Then redo the graph in this part using the same X values as in part (a) and Y values ranging from 0 to the high value of part (a).

2. Use one of the following MINITAB portable worksheets found on the CD-ROM. In each of the files the target value for the mean μ is stored in the C2(1) position and the target value for the standard deviation is stored in the C3(1) position. Use the targeted MU and SIGMA values.

   Yield of Wheat: Tscc01.mtp

   PepsiCo Stock Closing Prices: Tscc02.mtp

   PepsiCo Stock Volume of Sales: Tscc03.mtp

   Futures Quotes for the Price of Coffee Beans: Tscc04.mtp

   Incidence of Melanoma Tumors: Tscc05.mtp

   Percent Change in Consumer Price Index: Tscc06.mtp

COMMAND SUMMARY

Graphing Commands

Character Graphics Commands

PLOT C C prints a scatter plot with the first column on the vertical axis and the second on the horizontal axis. The following subcommands can be used with PLOT.

> **TITLE 'test'** gives a title above the graph.

> **FOOTNOTE 'test'** places a line of text below the graph.

> **XLABEL 'test'** labels the *x*-axis.

> **YLABEL 'test'** labels the *y*-axis.

> **SYMBOL 'symbol'** selects the symbol for the points on the graph. The default is *.

> **XINCREMENT K** is the distance between tick marks on *x*-axis.

> **XSTART K [K]** specifies the first tick mark and optionally the last one.

> **YINCREMENT K** is the distance between tick marks on *y*-axis.

> **YSTART K [K]** specifies the first tick mark and optionally the last one.

> WINDOWS menu selection: **➤Graph➤Scatterplot**

> > Titles, labels, and footnotes are in the **Labels** option.

Professional Graphics

Plot C * C prints a scatter plot with the first column on the vertical axis and the second on the horizontal axis. Note that the columns must be separated by an asterisk *.

Connect connects the points with a line.

Other subcommands may be used to title the graph and set the tick marks on the axes. See your MINITAB software manual for details.

> WINDOWS menu selection: **➤Graph➤Scatterplot**

> > Use the dialog boxes to title the graph, label the axes, set the tick marks, and so forth.

> > See your MINITAB software manual for details.

Control Charts

ICHART C...C draws an individuals control chart
MU K specifies the historical means

SIGMA K specifies the standard deviation.

WINDOWS menu selection: **➤Stat➤Control Charts➤Variables Charts for Individuals ➤Individuals**

> Enter choices for MU and SIGMA in the dialog box.

CHAPTER 7

INTRODUCTION TO SAMPLING DISTRIBUTIONS

Note: This section uses session window commands instead of menu choices

CENTRAL LIMIT THEOREM (SECTION 7.2 OF *UNDERSTANDABLE STATISTICS*)

The Central Limit Theorem says that if x is a random variable with <u>any</u> distribution having mean μ and standard deviation σ, then the distribution of sample means $\bar{x}$ based on random samples of size n is such that for sufficiently large n:

(a) The mean of the $\bar{x}$ distribution is approximately the same as the mean of the x distribution.

(b) The standard deviation of the x distribution is approximately $\sigma/\sqrt{n}$.

(c) The $\bar{x}$ distribution is approximately a normal distribution.

Furthermore, as the sample size n becomes larger and larger, the approximations mentions in (a), (b) and (c) become better.

We can use MINITAB to demonstrate the Central Limit Theorem. The computer does not prove the theorem. A proof of the Central Limit Theorem requires advanced mathematics and is beyond the scope of an introductory course. However, we can use the computer to gain a better understanding of the theorem.

To demonstrate the Central Limit Theorem, we need a specific x distribution. One of the simplest is the <u>uniform probability distribution</u>.

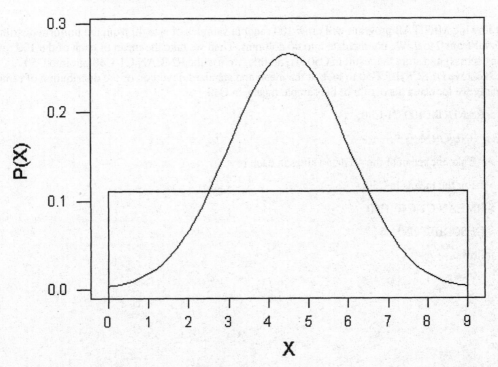

Uniform Distribution and a Normal Distribution

The normal distribution is the usual bell-shaped curve, but the uniform distribution is the rectangular or box-shaped graph. The two distributions are very different.

The uniform distribution has the property that all subintervals of the same length inside the interval 0 to 9 have the same probability of occurrence no matter where they are located. This means that the uniform distribution on the interval from 0 to 9 could be represented on the computer by selecting random numbers from 0 to 9. Since all numbers from 0 to 9 would be equally likely to be chosen, we say we are dealing with a uniform (equally likely) probability distribution. Note that when we say we are selecting random numbers from 0 to 9, we do not just mean whole numbers or integers; we mean real numbers in decimal form such as 2.413912, and so forth.

Because the interval from 0 to 9 is 9 units long and because the total area under the probability graph must by 1 (why?), the height of the uniform probability graph must be 1/9. The mean of the uniform distribution on the interval from 0 to 9 is the balance point. Looking at the Figure, it is fairly clear that the mean is 4.5. Using advanced methods of statistics, it can be shown that for the uniform probability distribution x between 0 and 9,

$$\mu = 4.5 \text{ and } \sigma = 3\sqrt{3}/2 \approx 2.598$$

The figure shows us that the uniform x distribution and the normal distribution are quite different. However, using the computer we will construct one hundred sample means $\bar{x}$ from the x distribution using a sample size of $n = 40$. We can vary the sample size n according to how many columns we use in RANDOM command.

> **RANDOM 100 C1-C40**
>
> **UNIFORM 0 9** (that is, from a = 0 to b = 9)

We will see that even though the uniform distribution is very different from the normal distribution, the histogram of the sample means is somewhat bell shaped. Looking at the DESCRIBE command, we will also see that the mean or the $\bar{x}$ distribution is close to the predicted mean of 4.5 and that the standard deviation is close to $\sigma/\sqrt{n}$ or $2.598/\sqrt{40}$ or 0.411.

Example

The following MINITAB program will draw 100 random samples of size 40 from the uniform distribution on the interval from 0 to 9. We put the data into 40 columns. Then we take the mean of each of the 100 rows (40 columns across) and store the result in C50. To do this, we use the RMEAN C1-C40 put into C50 command. Next we DESCRIBE C50 to look at the mean and standard deviation of the distribution of sample means. Finally we look at a histogram of the sample means in C50.

```
MTB > RANDOM 100 C1-C40;

SUBC > UNIFORM 0 9.

MTB > # Take the mean of the 40 data values in each row.

MTB > # Put the means in C50.

MTB > RMEAN C1-C40 C50

MTB > DESCRIBE C50
```

The result follows.

```
MINITAB - Untitled

File  Edit  Data  Calc  Graph  Editor  Tools  Window  Help

Session

MTB > RANDOM 100 C1-C40;
SUBC> UNIFORM 0 9.
MTB > RMEAN C1-C40 C50
MTB > DESCRIBE C50

Descriptive Statistics: C50

Variable    N    N*     Mean   SE Mean    StDev   Minimum       Q1    Median       Q3
C50       100    0    4.4850    0.0399   0.3991    3.4920   4.2140    4.5083   4.7319

Variable   Maximum
C50         5.2908

MTB > GSTD
|
```

3						4.20570
4						4.29182
5						4.19137

Current Worksheet: Worksheet 1   Editable   12:21 PM

Note the MEAN and STDEV are very close to the values predicted by the Central Limit Theorem.

MTB > GSTD

MTB > HISTOGRAM C50

The result follows.

```
MINITAB - Untitled

File  Edit  Data  Calc  Stat  Graph  Editor  Tools  Window  Help

Session

Histogram

Histogram of C50   N = 100

Midpoint       Count
   3.4            1    *
   3.6            3    ***
   3.8            6    ******
   4.0            5    *****
   4.2           16    ****************
   4.4           17    *****************
   4.6           25    *************************
   4.8           13    *************
   5.0            6    ******
   5.2            8    ********

MTB >
```

3						4.20570
4						4.29182
5						4.19137

Current Worksheet: Worksheet 1   Editable   12:17 PM

The histogram for this sample does not appear very similar to a normal distribution. Let's try another sample. The following are the results.

```
MINITAB - Untitled                                                    _ ⊞ ☒
File  Edit  Data  Calc  Stat  Graph  Editor  Tools  Window  Help

[toolbar icons]

Session                                                               _ ⊡ ☒

Descriptive Statistics: C50

Variable      N   N*     Mean   SE Mean   StDev   Minimum      Q1   Median       Q3
C50         100   0   4.5959    0.0418  0.4184    3.5303  4.2818   4.5785   4.8897

Variable    Maximum
C50          5.5551

MTB >
```

```
Worksheet 1                                                           _ ⊡ ☒

  ·    C42      C43      C44      C45      C46      C47      C48      C49      C50      C
  1                                                                            4.37644
  2                                                                            4.89299
  3                                                                            4.37989
  4                                                                            4.87991
  5                                                                            4.84541

Project...  ⊞ ☐ ☒
Current Worksheet: Worksheet 1                                   Editable      12:01 PM
```

```
MINITAB - Untitled                                                    _ ⊡ ☒
File  Edit  Data  Calc  Stat  Graph  Editor  Tools  Window  Help

[toolbar icons]

Session                                                               _ ⊡ ☒

  Histogram of C50    N = 100

  Midpoint         Count
       3.6             1    *
       3.8             3    ***
       4.0             6    ******
       4.2            16    ****************
       4.4            17    *****************
       4.6            16    ****************
       4.8            18    ******************
       5.0             9    *********
       5.2            10    **********
       5.4             2    **
       5.6             2    **

  MTB >
```

```
  3                                                                            4.37989
  4                                                                            4.87991
  5                                                                            4.84541

Welcome to Minitab, press F1 for help.                          Editable      12:05 PM
```

This histogram looks more like a normal distribution. You will get slightly different results each time you draw 100 samples.

The number of samples used is determined by K, and the size of the samples is determined by the number of columns in the command.

> **RANDOM K C1-Cn**
>
> **UNIFORM a b**

Be sure that when you take the RMEAN of the rows, you use the same number of columns as you used in the random command

> **RMEAN C1-Cn C**

Then use the DESCRIBE and HISTOGRAM commands on the column C where you put the means.

You can sample from a variety of distributions, some of which were listed under the RANDOM command in the Chapter 1 Command Summary.

LAB ACTIVITIES FOR CENTRAL LIMIT THEOREM

1. Repeat the experiment of Example 1. That is, draw 100 random samples of size 40 each from the uniform probability distribution between 0 and 9. Then take the means of each of these samples and put the results in C50. Use the commands

> **RANDOM 100 C1-C40**
>
> **UNIFORM 0 9**
>
> **RMEAN C1-C40 C50**

Next use DESCRIBE on C50. How does the mean and standard deviation of the distribution of sample means compare to those predicted by the Central Limit Theorem? Use HISTOGRAM C50 to draw a histogram of the distribution of sample means. How does it compare to a normal curve?

2. Next take 100 random samples of size 20 from the uniform probability distribution between 0 and 9. To do this, use only 20 columns in the RANDOM and RMEAN commands. Again put the means in C50, use DESCRIBE and HISTOGRAM on C50 and comment on the results. How do these results compare to those in problem 1? How do the standard deviations compare?

CHAPTER 8 ESTIMATION

CONFIDENCE INTERVALS FOR A MEAN OR FOR A PROPORTION (SECTIONS 8.1–8.3 OF *UNDERSTANDABLE STATISTICS*)

Student's *t* Distribution

In Section 8.1 of *Understandable Statistics*, confidence intervals for μ when σ is known are presented. In Section 8.2, Student's *t* distribution is introduced and confidence intervals for μ when σ is unknown are discussed. If the value of σ is unknown then the $\bar{x}$ distribution follows the Student's *t* distribution with degrees of freedom $(n - 1)$.

$$t = \frac{\bar{x} - \mu}{s/\sqrt{n}}$$

There is a different Student's *t* distribution for every degree of freedom. MINITAB includes Student's *t* distribution in its library of probability distributions. You may use the RANDOM, PDF, CDF, INVCDF commands with Student's *t* distribution as the specified distribution.

Menu selection: **Calc➤Probability Distributions➤t**

Dialog Box Responses

Select from Probability Density (PDF), Cumulative Probability (CDF), Inverse Cumulative Probability (INVCDF)

Degrees of Freedom: enter value

Input Column:

Column containing values for which you wish to compute the probability and optional storage column

Input Constant:

If you want the probability of just one value, use a constant rather than an entire column. Designate optional storage constant or column.

For CDF and INVCDF, set the value of "Noncentrality parameter" to be 0.

You can graph different *t*-distributions by using **➤Graph➤Scatterplot.** Follow steps similar to those given in Chapter 6 for graphing a normal distribution. Student's *t* distribution is symmetric and centered at 0. Select X values from about –4 to 4 in increments of 0.10 and place the values in a column, say C1. Then use **Calc➤Probability Distributions➤t** with Probability Density to generate a column, say C2, of Y values. The graph shown represents 10 degrees of freedom.

Student's Distribution with 10 Degrees of Freedom

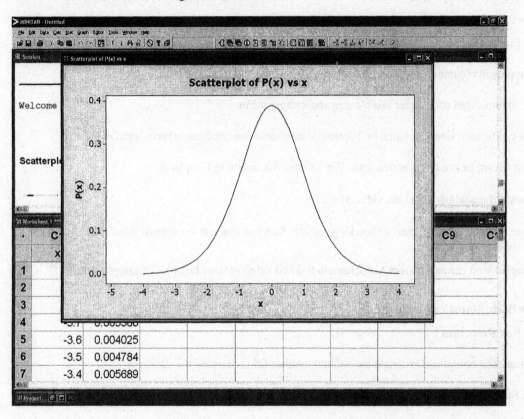

Confidence Intervals for Means

Confidence intervals for μ depend on the sample size n, and knowledge about the standard deviation σ. For small samples we assume the x distribution is approximately normal (mound-shaped and symmetric). When sample size is large, we do not need to make assumptions on the x distribution. The relationship between confidence intervals for μ, sample size, and knowledge about σ are shown in the table below.

	Confidence Interval
Large samples	
Small samples from an approximately normal distribution	
σ is known	$\bar{x} - z\left(\dfrac{\sigma}{\sqrt{n}}\right)$ to $\bar{x} + z\left(\dfrac{\sigma}{\sqrt{n}}\right)$
σ is unknown	$\bar{x} - t\left(\dfrac{s}{\sqrt{n}}\right)$ to $\bar{x} + t\left(\dfrac{s}{\sqrt{n}}\right)$

In MINITAB we can generate confidence intervals for μ by using the menu selections.

➤Stat➤Basic Statistics➤1-sample z

Dialog Box Responses

Samples in columns: Designate the column number C# containing the data.

or, Summarized data: Enter sample size and sample mean.

For confidence Interval: Click on [Options], then enter the confidence level, such as 90%.

Test Mean: Leave blank at this time. We will use the option in Chapter 9.

Standard deviation: Enter the value of σ.

Note that MINITAB requires knowledge of σ before you can use the normal distribution for confidence intervals.

Graphs: You can select from histogram, individual value plot, or box plot of sample data.

➤Stat➤Basic Statistics➤1-sample t

Dialog Box Responses

Samples in columns: Designate the column number C# containing the data.

or, Summarized data: Enter sample size, sample mean, and sample standard deviation.

For confidence Interval: Click on [Options], then enter the confidence level, such as 90%.

Test Mean: Leave blank at this time. We will use this option in Chapter 9.

Graphs: You can select from histogram, individual value plot, or box plot of sample data.

Example

The manager of First National Bank wishes to know the average waiting times for student loan application action. A random sample of 20 applications showed the waiting times from application submission (in days) to be

3	7	8	24	6	9	12	25	18	17
4	32	15	16	21	14	12	5	18	16

Find a 90% confidence interval for the population mean of waiting times.

In this example, the value of σ is not known. We need to use Student's t distribution. Enter the data into column C1 and name the column Days. Use the menu selection **➤Stat➤Basic Statistics➤1-sample t**.

The results are

```
One-Sample T: Days

Variable    N     Mean    StDev  SE Mean      90% CI
Days       20  14.1000   7.7044   1.7228  (11.1211, 17.0789)
```

Confidence Intervals for Proportions

This option is in more recent versions of MINITAB including Version 14.

Menu selection

➤Stat➤Basic Statistics➤1 Proportion

Dialog Box Responses

Select the option of Summarized Data.

> Number of Trials: Enter value (*n* in *Understandable Statistics*)
>
> Number of Events: Enter value of successes (*r* in *Understandable Statistics*)

Click on [Options]; enter confidence level and click on Use test and interval based on normal distribution.

> In Chapter 9 we will see how to interpret the results from this test.

Example

The public television station BPBS wants to find the percent of its viewing population who give donations to the station. A random sample of 300 viewers was surveyed and it was found that 123 made contributions to the station. Find a 95% confidence interval for the probability that a viewer of BPBS selected at random contributions to the station.

Use the menu selection ➤Stat➤Basic Statistics➤1 Proportion. Click on Summarized Data. Use 300 for number of trials and 123 for number of events. Click on [Options]. Enter 95 for the confidence level.

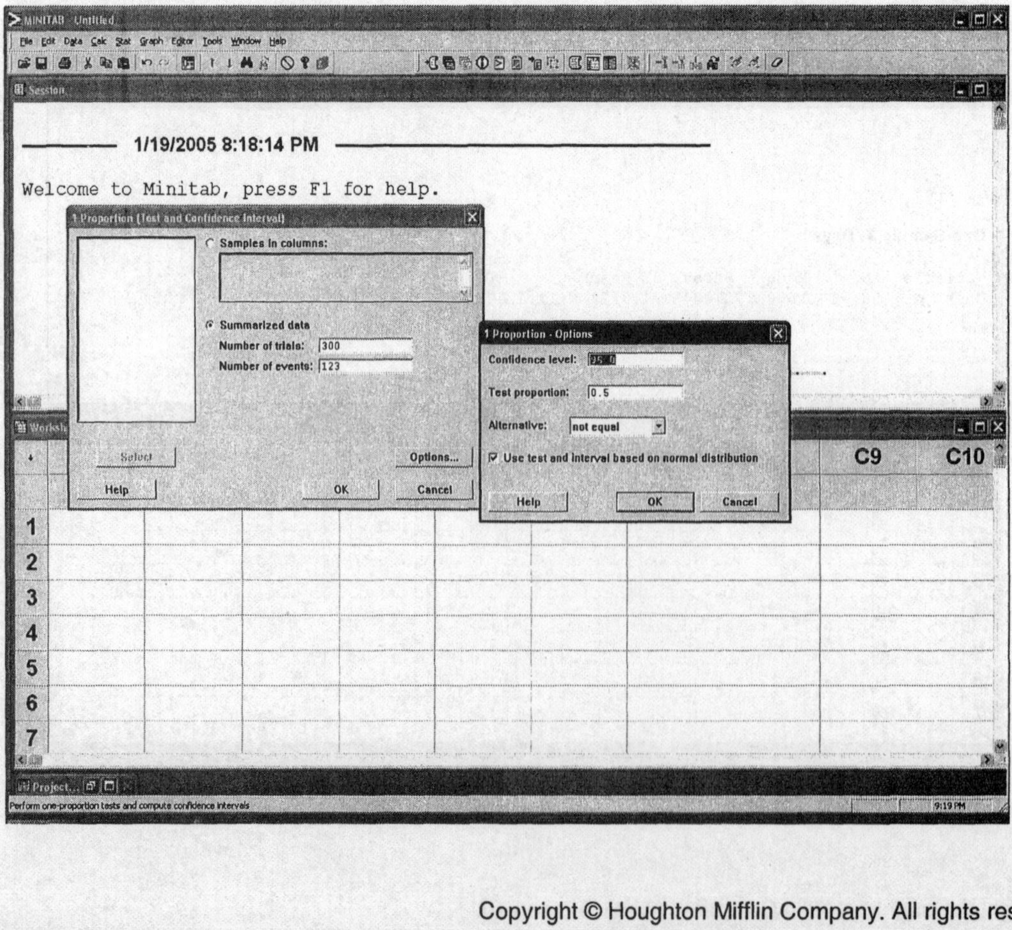

The results follow.

```
MINITAB - Untitled
File Edit Data Calc Stat Graph Editor Tools Window Help

Session

Test and CI for One Proportion

Test of p = 0.5 vs p not = 0.5

Sample    X     N   Sample p          95% CI         Z-Value  P-Value
1        123   300  0.410000  (0.354345, 0.465655)   -3.12    0.002
```

The output regarding test of p, Z-value, and P-value will be discussed in Chapter 9.

Confidence Intervals for Difference of Means or Difference of Proportions

In MINITAB, confidence intervals for difference of means and difference of proportions are included in the menu selection for tests of hypothesis for difference of means and tests of hypothesis for difference of proportions respectively. These menu selections with their dialog boxes will be discussed in Chapter 9.

LAB ACTIVITIES FOR CONFIDENCE INTERVALS FOR A MEAN OR FOR A PROPORTION

1. Snow King Ski resorts is considering opening a downhill ski slope in Montana. To determine if there would be an adequate snow base in November in the particular region under consideration, they studied snowfall records for the area over the last 100 years. They took a random sample of 15 years. The snowfall during November for the sample years was (in inches):

26   35   42   18   29   42   28   35
47   29   38   27   21   35   30

(a) To find a confidence interval for μ, do we use a normal distribution or Student's t distribution?

(b) Find a 90% confidence interval for the mean snowfall.

(c) Find a 95% confidence interval for the mean snowfall.

(d) Compare the intervals of parts (b) and (c). Which one is narrower? Why would you expect this?

2. Consider the snowfall data of problem 1. Suppose you knew that the snowfall in the region under consideration for the ski area in Montana (see problem 1) had a population standard deviation of 8 inches.

   (a) Since you know σ, (and the distribution of snowfall is assumed to be approximately normal) do you use the normal distribution or Student's t for confidence intervals?

   (b) Find a 90% confidence interval for the mean snowfall.

   (c) Find a 95% confidence interval for the mean snowfall.

   (d) Compare the respective confidence intervals created in problem 1 and in this problem. Of the 95% intervals, which is longer, the one using the t distribution or the one using the normal distribution? Why would you expect this result?

3. Retrieve the worksheet Svls01.mpt from the CD-ROM. This worksheet contains the number of shares of Disney Stock (in hundreds of shares) sold for a random sample of 60 trading days in 1993 and 1994. The data is in column C1.

   Use the sample standard deviation computed with menu options ➤**Stat**➤**Basic Statistics**➤**Display Descriptive Statistics** as the value of σ. You will need to compute this value first, and then enter it as a number in the dialog box for 1-sample z.

   (a) Find a 99% confidence interval for the population mean volume.

   (b) Find a 95% confidence interval for the population mean volume.

   (c) Find a 90% confidence interval for the population mean volume.

   (d) Find an 85% confidence interval for the population mean volume.

   (d) What do you notice about the lengths of the intervals as the confidence level decreases?

4. There are many types of errors that will cause a computer program to terminate or give incorrect results. One type of error is punctuation. For instance, if a comma is inserted in the wrong place, the program might not run. A study of programs written by students in a beginning programming course showed that 75 out of 300 errors selected at random were punctuation errors. Find a 99% confidence interval for the proportion of errors made by beginning programming students that are punctuation errors. Next, find a 90% confidence interval. Is this interval longer or shorter?

5. Sam decided to do a statistics project to determine a 90% confidence interval for the probability that a student at West Plains College eats lunch in the school cafeteria. He surveyed a random sample of 12 students and found that 9 ate lunch in the cafeteria. Can Sam use the program to find a confidence interval for the population proportion of students eating in the cafeteria? Why or why not? Try the program with $N = 12$ and $R = 9$. What happens? What should Sam do to complete his project?

COMMAND SUMMARY

Probability Distribution Subcommand

T K is the subcommand that calls up Student's *t* distribution with specified degrees of freedom K. This subcommand may be used with RANDOM, PDF, CDF, INVCDF.

>WINDOWS menu selection: **➤Calc➤Probability Distributions➤t**

>>In the dialog box select PDF, CDF, or Inverse, then enter the degrees of freedom.

To Generate Confidence Intervals

ZINTERVAL K σ**=K C...C**

>generates a confidence interval for μ using the normal distribution with confidence level K%.. You must enter a value for σ, either actual or estimated. A separate interval is given for data in each column. If K is not specified, a 95% confidence interval will be given.

>WINDOWS menu selection: **➤Stat➤Basic Statistics➤1-sample z**

>>In the dialog box click [Options] and enter the confidence level.

TINTERVAL K C...C

>generates a confidence interval for μ using Student's *t* distribution with confidence level K%. A separate interval is given for data in each column. If K is not specified, a 95% confidence interval is given.

>WINDOWS menu selection: **➤Stat➤Basic Statistics➤1-sample t**

>>In the dialog box click [Options] and enter the confidence level.

PONE K K with **subcommand Confidence K** generates a confidence interval for one proportion.

>WINDOWS menu selection: **➤Stat➤Basic Statistics➤1 Proportion**

CHAPTER 9 HYPOTHESIS TESTING

TESTING A SINGLE POPULATION MEAN OR PROPORTION (SECTIONS 9.1–9.3 OF *UNDERSTANDABLE STATISTICS*)

Tests involving a single mean are found in Sections 9.2. In MINITAB, the user concludes the test by comparing the P value of the test statistic to the level of significance α. The method of using P values to conclude tests of hypotheses is explained in Section 9.2. Section 9.3 discusses tests of a single proportion.

For tests of the mean when σ is known, use ➤**Stat**➤**Basic Statistics**➤**1-sample z.**

Dialog Box Responses

> Samples in columns: Enter column number where data is located;
>
> or, Summarized data: Enter sample size and sample mean.
>
> Select Test Mean. Enter the value of k for the null hypothesis.
>
> > $H_0: \mu = k$
>
> Click on [Options] and then select Alternative: Scroll to the appropriate alternate hypothesis:
>
> > $H_1: \mu \neq k$ (not equal)
> >
> > $H_1: \mu > k$ (greater than)
> >
> > $H_1: \mu < k$ (less than)
>
> Standard deviation: Enter the value of σ.

For tests of the mean when σ is unknown, use ➤**Stat**➤**Basic Statistics**➤**1-sample t.**

Dialog Box Responses

> Samples in columns: Enter column number where data is located;
>
> or, Summarized data: Enter sample size and sample mean.
>
> Select Test Mean. Enter the value of k for the null hypothesis
>
> > $H_0: \mu = k$
>
> Click on [Options] and then select Alternative: Scroll to the appropriate alternate hypothesis:
>
> > $H_1: \mu \neq k$ (not equal)
> >
> > $H_1: \mu > k$ (greater than)
> >
> > $H_1: \mu < k$ (less than)

For tests of a single proportion use ➤**Stat**➤**Basic Statistics**➤**1 proportion.**

Dialog Box Responses

> Select the option of Summarized Data.
>
> Number of Trials: Enter value (*n* in *Understandable Statistics*)
>
> Number of Events: Enter value of successes (*r* in *Understandable Statistics*)
>
> Click on [Options].

Confidence Level: Enter a value such as 95.

Test proportion: Enter the value of k, where

$$H_0: p = k$$

Alternative: Scroll to the appropriate alternate hypothesis:

$H_1: p \neq k$ (not equal)

$H_1: p > k$ (greater than)

$H_1: p < k$ (less than)

Both the Z-sample and the T-sample operate on data in a column. They each compute the sample mean $\bar{x}$. The Z-sample converts the sample mean $\bar{x}$ to a z value, while the T-sample converts $\bar{x}$ to t using the respective formulas

$$z = \frac{\bar{x} - \mu}{\sigma/\sqrt{n}} \qquad t = \frac{\bar{x} - \mu}{s/\sqrt{n}}$$

The test of 1 proportion converts the sample proportion $\hat{p} = r/n$ to a z value using the formula

$$z = \frac{\hat{p} - p}{\sqrt{p(1-p)/n}}$$

The tests also give the P value of the sample statistic $\bar{x}$. The user can then compare the P value to α, the level of significance of the test. If

P value $\leq \alpha$ we reject the null hypothesis.

P value $> \alpha$ we do not reject the null hypothesis.

Example

Many times patients visit a health clinic because they are ill. A random sample of 12 patients visiting a health clinic had temperatures (in °F) as follows:

97.4   99.3   99.0   100.0   98.6

97.1   100.2   98.9   100.2   98.5

98.8   97.3

Dr. Tafoya believes that patients visiting a health clinic have a higher temperature than normal. The normal temperature is 98.6 degrees. Test the claim at the $\alpha = 0.01$ level of significance.

In this case, we do not know the value of σ. We need a t-test. Enter the data in C1 and name the column Temp. Then select ➤**Stat**➤**Basic Statistics**➤**1-sample t**.

Use a 98.6 as the value for Test mean. Click on [Options] and then select 'greater than' in the drop down menu next to Alternative.

The results follow.

Recall that SE Mean is the value of $\dfrac{s}{\sqrt{n}}$.

LAB ACTIVITIES FOR TESTING A SINGLE POPULATION MEAN OR PROPORTION

1. A new catch-and-release policy was established for a river in Pennsylvania. Prior to the new policy, the average number of fish caught per fisherman hour was 2.8. Two years after the policy went into effect, a random sample of 12 fisherman hours showed the following catches per hour.

3.2	1.1	4.6	3.2	2.3	2.5
1.6	2.2	3.7	2.6	3.1	3.4

   Test the claim that the per hour catch has increased, at the 0.05 level of significance.

   (a) Decide whether to use the Z-sample or T-sample menu choices. What is the value of μ in the null hypothesis?

   (b) What is the choice for ALTERNATIVE?

   (c) Compare the P value of the test statistic to the level of significance α. Do we reject the null hypothesis or not?

2. Open or retrieve the worksheet **Svls04.mtp** from the CD-ROM. The data in column C1 of this worksheet represent the miles per gallon gasoline consumption (highway) for a random sample of 55 makes and models of passenger cars (source: Environmental Protection Agency).

30	27	22	25	24	25	24	15
35	35	33	52	49	10	27	18
20	23	24	25	30	24	24	24
18	20	25	27	24	32	29	27
24	27	26	25	24	28	33	30
13	13	21	28	37	35	32	33
29	31	28	28	25	29	31	

   Test the hypothesis that the population mean miles per gallon gasoline consumption for such cars is greater than 25 mpg.

   (a) Do we know σ for the mpg consumption? Can we estimate σ by s, the sample standard deviation? Should we use the Z-sample or T-sample menu choice? What is the value of μ in the null hypothesis?

   (b) If we estimate σ by s, we need to instruct MINITAB to find the stdev, or s, of the data before we use Z-sample. Use ➤**Stat**➤**Basic Statistics**➤**Display Descriptive Statistics** to find s.

   (c) What is the alternative hypothesis?

   (d) Look at the P value in the output. Compare it to α. Do we reject the null hypothesis or not?

   (e) Using the same data, test the claim that the average mpg for these cars is not equal to 25. How has the P value changed? Compare the new P value to α. Do we reject the null hypothesis or not?

3. Open or retrieve the worksheet **Svss01.mtp** from the CD-ROM. The data in column C1 of this worksheet represent the number of wolf pups per den from a sample of 16 wolf dens (source: *The Wolf in the Southwest: The Making of an Endangered Species* by D.E. Brown, University of Arizona Press).

5	8	7	5	3	4	3	9
5	8	5	6	5	6	4	7

   Test the claim that the population mean number of wolf pups in a den is greater than 5.4.

4. Jones Computer Security is testing a new security device that is believed to decrease the incidence of computer break-ins. Without this device, the computer security test team can break security 47% of the time. With the device in place, the test team made 400 attempts and was successful 82 times. Select an appropriate program from the HYPOTHESIS TESTING menu and test the claim that the device reduces the proportion of successful break-ins. Use alpha = 0.05 and note the *P* value. Does the test conclusion change for alpha = 0.01?

TESTS INVOLVING PAIRED DIFFERENCES (DEPENDENT SAMPLES)
(SECTION 9.4 OF *UNDERSTANDABLE STATISTICS*)

To perform a paired difference test, we put our paired data into two columns. Now in later versions of MINITAB, including the Student Edition 12 for Windows, is a menu item for testing data pairs. Select

➤Stat➤Basic Statistics➤Paired t

Dialog Box Responses

First Sample: column number where before data is located

Second Sample: column number where after data is located

Click [Options]

Confidence Level: Enter a value such as 95.

Test mean: Leave as default 0.0.

Alternative: Scroll to not equal, greater than, or less than as appropriate.

H_1: $\mu \neq 0$ (not equal)

H_1: $\mu > 0$ (greater than)

H_1: $\mu < 0$ (less than)

Example

Promoters of a state lottery decided to advertise the lottery heavily on television for one week during the middle of one of the lottery games. To see if the advertising improved ticket sales, the promoters surveyed a random sample of 8 ticket outlets and recorded weekly sales for one week before the television campaign and for one week after the campaign. The results follow (in ticket sales) where B stands for "before" and A for "after" the advertising campaign.

B:	3201	4529	1425	1272	1784	1733	2563	3129
A:	3762	4851	1202	1131	2172	1802	2492	3151

We want to test to see if D = B − A is less than zero, since we are testing the claim that the lottery ticket sales are greater after the television campaign. We will put the before data in C1, the after data in C2. Select **➤Stat➤Basic Statistics➤Paired t.** Use less than for Alternative, and use a Confidence level of 95.0.

The results follow.

Since the *P* value 0.139 is larger than the level of significance of 0.05, we do not reject the null hypothesis.

LAB ACTIVITIES FOR TESTS INVOLVING PAIRED DIFFERENCES (DEPENDENT SAMPLES)

1. Open or retrieve the worksheet **Tvds01.mtp** from the CD-ROM. The data are pairs of values, where the entry in C1 represents the average salary (in thousands of dollars/year) for male faculty members at an institution and C2 represents the average salary for female faculty members (in thousands of dollars/year) at the same institution. A random sample of 22 U.S. colleges and universities was used (source: *Academe, Bulletin of the American Association of University Professors*).

(34.5, 33.9)	(30.5, 31.2)	(35.1, 35.0)	(35.7, 34.2)	(31.5, 32.4)
(34.4, 34.1)	(32.1, 32.7)	(30.7, 29.9)	(33.7, 31.2)	(35.3, 35.5)
(30.7, 30.2)	(34.2, 34.8)	(39.6, 38.7)	(30.5, 30.0)	(33.8, 33.8)
(31.7, 32.4)	(32.8, 31.7)	(38.5, 38.9)	(40.5, 41.2)	(25.3, 25.5)
(28.6, 28.0)	(35.8, 35.1)			

   (a) The data is in C1 and C2.

   (b) Use the **>Stat>Basic Statistics>Paired t** menu to test the hypothesis that there is a difference in salaries. What is the *P* value of the sample test statistic? Do we reject or fail to reject the null hypothesis at the 5% level of significance? What about at the 1% level of significance?

   (c) Use the **>Stat>Basic Statistics>Paired t** menu to test the hypothesis that female faculty members have a lower average salary than male faculty members. What is the test conclusion at the 5% level of significance? At the 1% level of significance?

2. An audiologist is conducting a study on noise and stress. Twelve subjects selected at random were given a stress test in a room that was quiet. Then the same subjects were given another stress test, this time in a room with high-pitched background noise. The results of the stress tests were scores 1 through 20, with 20 indicating the greatest stress. The results follow, where B represents the score of the test administered in the quiet room and A represents the scores of the test administered in the room with the high-pitched background noise.

Subject		1	2	4	5	6	7	8	9	10	11	12
	B	13	12	16	19	7	13	9	15	17	6	14
	A	18	15	14	18	10	12	11	14	17	8	16

Test the hypothesis that the stress level was greater during exposure to noise. Look at the *P* value. Should you reject the null hypothesis at the 1% level of significance? At the 5% level?

TESTS OF DIFFERENCE OF MEANS (INDEPENDENT SAMPLES) (SECTION 9.5 OF *UNDERSTANDABLE STATISTICS*)

We consider the $\bar{x}_1 - \bar{x}_2$ distribution. The null hypothesis is that there is no difference between means, so $H_0: \mu_1 = \mu_2$, or $H_0: \mu_1 - \mu_2 = 0$.

Large Samples

MINITAB has a slightly different approach to testing difference of means with large samples (each sample size 30 or more, whether σ_1 and σ_2 are known does not matter) than that shown in *Understandable Statistics*. In MINITAB Student's *t* distribution is used instead of the normal distribution. The degrees of freedom used by MINITAB for this application of the *t* distribution are at least as large as the smaller sample. Therefore, we have degrees of freedom at 30 or more. In such cases, the normal and Student's t distribution give reasonably similar results. However, the results will not be exactly the same.

The menu choice MINITAB uses to test the difference of means is ➤**Stat**➤**Basic Statistics**➤**2-sample t**. The null hypothesis is always H_0: $\mu_1 = \mu_2$. The alternate hypothesis H_1: $\mu_1 \neq \mu_2$, corresponds to the choice "not equal." To do a left-tailed or right-tailed test, you need to use the choice of "less than" for ALTERNATIVE on a left-tailed test and "greater than" for ALTERNATIVE on a right-tailed test.

WINDOWS menu selection: ➤**Stat**➤**Basic Statistics**➤**2-sample t**

Dialog Box Responses

Select Samples in Different Columns and enter the C# for the columns containing the data.

Assume equal variances: Do **not** select for large samples.

Click on [Options] then:

Alternative: Scroll to the appropriate choice.

Confidence Level: Enter a value such as 95.

Small Samples

To do a test of difference of sample means with small samples with the assumption that the samples come from populations with the same standard deviation, we use the ➤**Stat**➤**Basic Statistics**➤**2-sample t** menu selection. If we believe that the two populations have unequal variances and leave the box **Assume equal variances** unchecked, MINITAB will produce a test using Satterthwaite's approximation for the degree of freedom. When we check that box equal variances are assumed, MINITAB automatically pools the standard deviations.

➤**Stat**➤**Basic Statistics**➤**2-sample t**

Dialog Box Responses

Select Samples in Different Columns and enter the C# for the columns containing the data.

Assume equal variances: If checked, the pooled standard deviation is used.

Click on [Options] then;

Alternative: Scroll to the appropriate choice.

Confidence Level: Enter a value such as 95.

Example

Sellers of microwave French fry cookers claim that their process saves cooking time. McDougle Fast Food Chain is considering the purchase of these new cookers, but wants to test the claim. Six batches of French fries were cooked in the traditional way. Cooking times (in minutes) are

15   17   14   15   16   13

Six batches of French fries of the same weight were cooked using the new microwave cooker. These cooking times (in minutes) are

11   14   12   10   11   15

Test the claim that the microwave process takes less time. Use $\alpha = 0.05$.

Under the assumption that the distributions of cooking times for both methods are approximately normal and that $\sigma_1 = \sigma_2$, we use the **≻Stat≻Basic Statistics≻2-sample t** menu choices with the assumption of equal variances checked. We are testing the claim that the mean cooking time of the second sample is less than that of the first sample, so our alternate hypothesis will be H_1: $\mu_1 > \mu_2$. We will use a right-tailed test and scroll to "greater than" for ALTERNATIVE.

The results follow.

```
MINITAB - Untitled
File Edit Data Calc Stat Graph Editor Tools Window Help

Session

  Two-Sample T-Test and CI: Traditional, New

  Two-sample T for Traditional vs New

                 N    Mean   StDev  SE Mean
  Traditional    6   15.00    1.41     0.58
  New            6   12.17    1.94     0.79


  Difference = mu (Traditional) - mu (New)
  Estimate for difference:  2.83333
  95% lower bound for difference:  1.05646
  T-Test of difference = 0 (vs >): T-Value = 2.89  P-Value = 0.008  DF = 10
  Both use Pooled StDev = 1.6980
```

3	14	12
4	15	10
5	16	11
6	13	15
7		
8		

```
Project...

Current Worksheet: Worksheet 1                          Editable    8:03 PM
```

We see that the P value of the test is 0.008. Since the P value is less than $\alpha = 0.05$, we reject the null hypothesis and conclude that the microwave method takes less time to cook French fries.

LAB ACTIVITIES USING DIFFERENCE OF MEANS (INDEPENDENT SAMPLES)

1. Calm Cough Medicine is testing a new ingredient to see if its addition will lengthen the effective cough relief time of a single dose. A random sample of 15 doses of the standard medicine were tested, and the effective relief times were (in minutes):

42	35	40	32	30	26	51	39	33	28
37	22	36	33	41					

   A random sample of 20 doses was tested when the new ingredient was added. The effective relief times were (in minutes):

43	51	35	49	32	29	42	38	45	74
31	31	46	36	33	45	30	32	41	25

   Assume that the standard deviations of the relief times are equal for the two populations. Test the claim that the effective relief time is longer when the new ingredient is added. Use $\alpha = 0.01$.

2. Open or retrieve the worksheet **Tvis06.mtp** from the CD-ROM. The data represent the number of cases of red fox rabies for a random sample of 16 areas in each of two different regions of southern Germany.

   NUMBER OF CASES IN REGION 1

10   2   2   5   3   4   3   3   4   0   2   6   4   8   7   4

NUMBER OF CASES IN REGION 2

1   1   2   1   3   9   2   2   4   5   4   2   2   0   0   2

Test the hypothesis that the average number of cases in Region 1 is greater than the average number of cases in Region 2. Use a 1% level of significance.

3. Open or retrieve the MINITAB worksheet **Tvis02.mtp** from the CD-ROM. The data represent the petal length (cm) for a random sample of 35 *Iris Virginica* and for a random sample of 38 *Iris Setosa* (source: Anderson, E., Bulletin of American Iris Society).

PETAL LENGTH (CM) IRIS VIRGINICA

5.1   5.8   6.3   6.1   5.1   5.5   5.3   5.5   6.9   5.0   4.9   6.0   4.8   6.1   5.6   5.1
5.6   4.8   5.4   5.1   5.1   5.9   5.2   5.7   5.4   4.5   6.1   5.3   5.5   6.7   5.7   4.9
4.8   5.8   5.1

PETAL LENGTH (CM) IRIS SETOSA

1.5   1.7   1.4   1.5   1.5   1.6   1.4   1.1   1.2   1.4   1.7   1.0   1.7   1.9   1.6   1.4
1.5   1.4   1.2   1.3   1.5   1.3   1.6   1.9   1.4   1.6   1.5   1.4   1.6   1.2   1.9   1.5
1.6   1.4   1.3   1.7   1.5   1.7

Test the hypothesis that the average petal length for the *Iris Setosa* is shorter than the average petal length for the *Iris Virginica*. Assume that the two populations have unequal variances.

COMMAND SUMMARY

To Test a Single Mean

ZTEST [K] K C...C performs a z-test on the data in each column. The first **K** is μ and the second **K** is σ. If you do not specify μ, it is assumed to be 0. You need to supply a value for σ. If the ALTERNATIVE subcommand is not used, a two-tailed test is conducted.

> WINDOWS menu selection: **➤Stat➤Basic Statistics➤1-sample z**

In dialog box select alternate hypothesis, specify the mean for H_0, specify the standard deviation.

TTEST [K] C...C performs a separate t-test on the data of each column. The value **K** is μ. If you do not specify μ, it is assumed to be 0. The computer evaluates s, the sample standard deviation for each column, and uses the computed s value to conduct the test. If the ALTERNATIVE subcommand is not used, a two-tailed test is conducted.

> WINDOWS menu selection: **➤Stat➤Basic Statistics➤1-sample t**

> In dialog box select alternate hypothesis, specify the mean for H_0.

> **ALTERNATIVE K** is the subcommand required to conduct a one-tailed test.

> If K = –1, then a left-tailed test is done. If K = 1, then a right-tailed test is done.

To Test a Difference of Means (Independent Samples)

TWOSAMPLE [K] C ...C does a two (independent) sample t test and (optionally) confidence interval for data in the two columns listed. **K** is optional and for K% confidence. The first data set is put into the first column, and the second data set into the second column. Unless the ALTERNATIVE subcommand is used, the alternate hypothesis is assumed to be $H_1: \mu_1 \neq \mu_2$. Samples are assumed to be independent.

ALTERNATIVE K is the subcommand to change the alternate hypothesis to a left-tailed test with K = –1 or right-tailed test with K = 1.

POOLED is the subcommand to be used only when the two samples come from populations with equal standard deviations.

> WINDOWS menu selection: **➤Stat➤Basic Statistics➤2-sample t**

> In dialog box select alternate hypothesis, specify the mean for H_0. For large samples do not check assume equal variances. For small samples check assume equal variances.

To Do a Paired Difference Test

(Available on newer versions of MINITAB, such as release 12)

PAIRED C...C tests for a difference of means in paired (dependent) data and gives a confidence interval if requested.

TEST 0.0 is a subcommand to set the null hypothesis to 0

ALTERNATIVE K is the subcommand to change the alternate hypothesis to a left-tailed test with K = –1 or right-tailed test with K = 1.

> WINDOWS menu selection: **➤Stat➤Basic Statistics➤paired t**

> In dialog box select alternate hypothesis, specify the mean for H_0.

CHAPTER 10 REGRESSION AND CORRELATION

SIMPLE LINEAR REGRESSION: TWO VARIABLES (SECTIONS 10.1–10.3 OF *UNDERSTANDABLE STATISTICS*)

Chapter 10 of *Understandable Statistics* introduces linear regression. The formula for the correlation coefficient r is given in Section 10.1. Formulas to find the equation of the least squares line,

$$y = a + bx$$

are given in Section 10.2. This section also contains the formula for the coefficient of determination r^2. The equation for the standard error of estimate as well as the procedure to find a confidence interval for the predicted value of y are given in Section 10.3.

The menu selection ➤**Stat**➤**Regression**➤**Regression** gives the equation of the least-squares line, the value of the standard error of estimate (s = standard error of estimate), the value of the coefficient of determination r^2 (R – sq), as well as several other values such as R – sq adjusted (an unbiased estimate of the population r^2). For simple regression with a response variable and one explanatory variable, we can get the value of the Pearson product moment correlation coefficient r by simply taking the square root of R – sq.

The standard deviation, t-ratio and P values of the coefficients are also given. The P value is useful for testing the coefficients to see that the population coefficient is not zero (see Section 10.4 of *Understandable Statistics* for a discussion about testing the coefficients). For the time being we will not use these values.

Depending on the amount of output requested (controlled by the options selected under the **[Results]** button) you will also see an analysis of variance chart, as well as a table of x and y values with the fitted values y_p and residuals $(y - y_r)$. We will not use the analysis of variance chart in our introduction to regression. However, in more advanced treatments of regression, you will find it useful.

To find the equation of the least-squares line and the value of the correlation coefficient, use the menu options

➤**Stat**➤**Regression**➤**Regression**

Dialog Box Responses

Response: Enter the column number C# of the column containing the responses (that is Y values).

Predictor: Enter the column number C# of the column containing the explanatory variables (that is, X values).

[Graphs]: Do not click on at this time.

[Results]: Click on and select the second option, Regression equation, etc.

[Options]: We will click on this option when we wish to do predictions for new variables.

[Storage]: Click on and select the fits and residual options if you wish.

To graph the scatter plot and show the least-squares line on the graph, use the menu options

➤**Stat**➤**Regression**➤**Fitted Line Plot**

Dialog Box Responses

Response: List the column number C# of the column containing the Y values.

Predictor: List the column number C# of the column containing the X values.

Type of Regression model: Select Linear.

[Options]: Click on and select Display Prediction Interval for a specified confidence level of

prediction band. Do not use if you do not want the prediction band.

[Storage]: This button gives you the same storage options as found under regression.

To find the value of the correlation coefficient directly and to find its corresponding P value, use the menu selection

> **Stat➤Basic Statistics➤Correlation**

Dialog Box Responses

Variables: List the column number C# of the column containing the X variable and the column

number C# of the column containing the Y variable.

Select Display P- values option.

Example

Merchandise loss due to shoplifting, damage, and other causes is called shrinkage. Shrinkage is a major concern to retailers. The managers of H.R. Merchandise think there is a relationship between shrinkage and number of clerks on duty. To explore this relationship, a random sample of 7 weeks was selected. During each week the staffing level of sales clerks was kept constant and the dollar value (in hundreds of dollars) of the shrinkage was recorded.

X	10	12	11	15	9	13	8	
Y	19	15	20	9	25	12	31	(in hundreds)

Store the value of X in C1 and name C1 as X. Sore the values of Y in C2 and name C2 as Y.

Use menu choices to give descriptive statistics regarding the values of X and Y. Use commands to draw an (X, Y) scatter plot and then to find the equation of the regression line. Find the value of the correlation coefficient, and test to see if it is significant.

(a) First we will use ➤**Stat➤Basic Statistics➤Display Descriptive Statistics** and each of the columns X and Y. Note that we select both C1 and C2 in the variables box.

(b) Next we will use **>Stat>Regression>Fitted Line Plot** to graph the scatter plot and show the least-squares line on the graph. We will not use prediction bands.

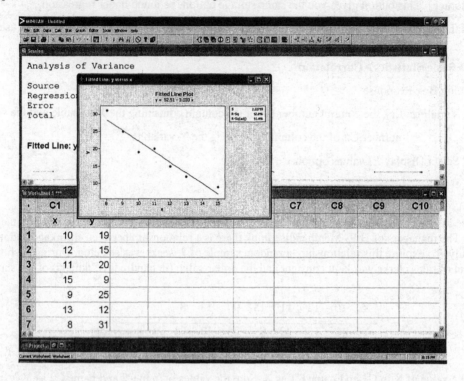

Notice that the equation of the regression line is given on the figure, as well as the value of r^2.

(c) However, to find out more information about the linear regression model, we use the menu selection **>Stat>Regression>Regression.** Enter C2 for Response and C1 for Predictor.

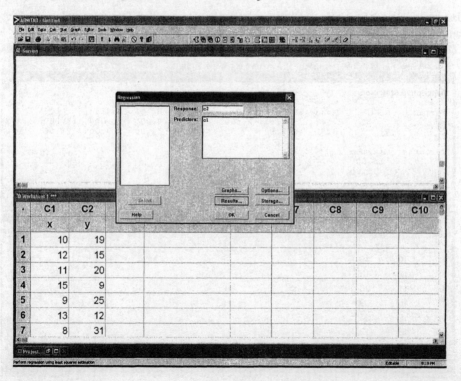

The results follow.

```
MINITAB - Untitled
File Edit Data Calc Stat Graph Editor Tools Window Help

Session

Regression Analysis: y versus x

The regression equation is
y = 52.5 - 3.03 x


Predictor       Coef   SE Coef      T      P
Constant      52.508     4.288  12.24  0.000
x             -3.0328    0.3774  -8.04  0.000


S = 2.22799   R-Sq = 92.8%   R-Sq(adj) = 91.4%


Analysis of Variance

Source          DF       SS       MS      F      P
Regression       1   320.61   320.61  64.59  0.000
Residual Error   5    24.82     4.96
Total            6   345.43
```

| 6 | 13 | 12 |
| 7 | 8 | 31 |

```
Current Worksheet: Worksheet 1                               Editable    8:22 PM
```

Notice that the regression equation is given as

$$y = 52.5 - 3.03x$$

The value of the standard error of estimate S_e is given as S = 2.28. We have the value of r^2, R-square = 92.8%. Find the value of r by taking the square root. It is 0.963 or 96.3%.

(d) Next, let's use the prediction option to find the shrinkage when 14 clerks are available.

Use➤**Stat**➤**Regression**➤**Regression.** Your previous selections should still be listed. Now press [Options]. Enter 14 in the prediction window.

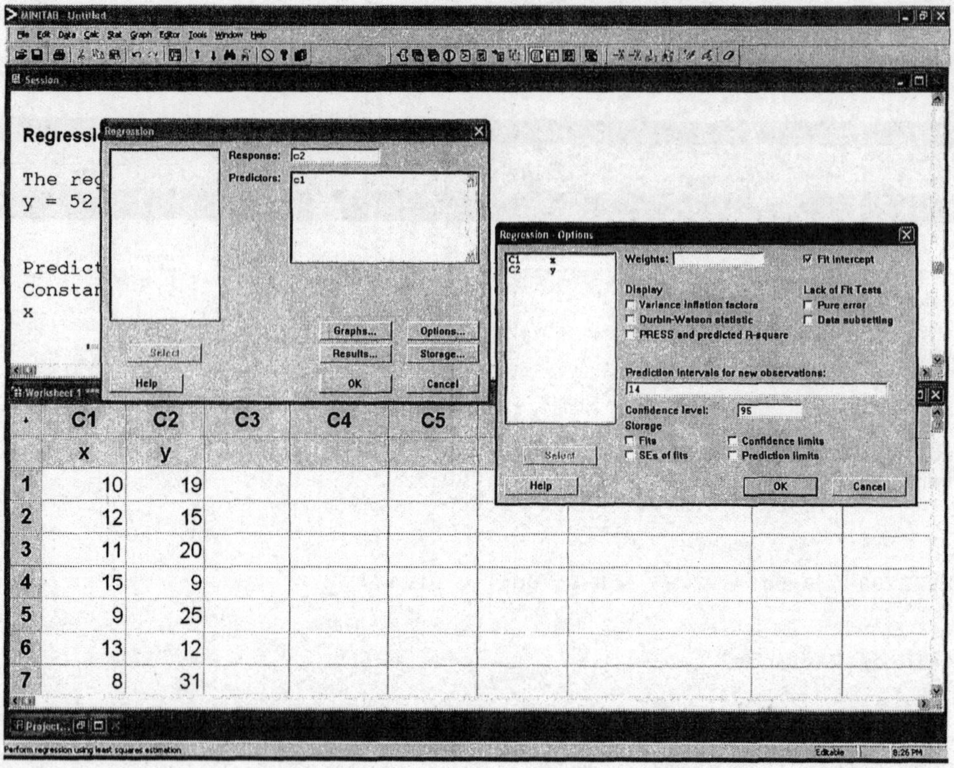

The results follow.

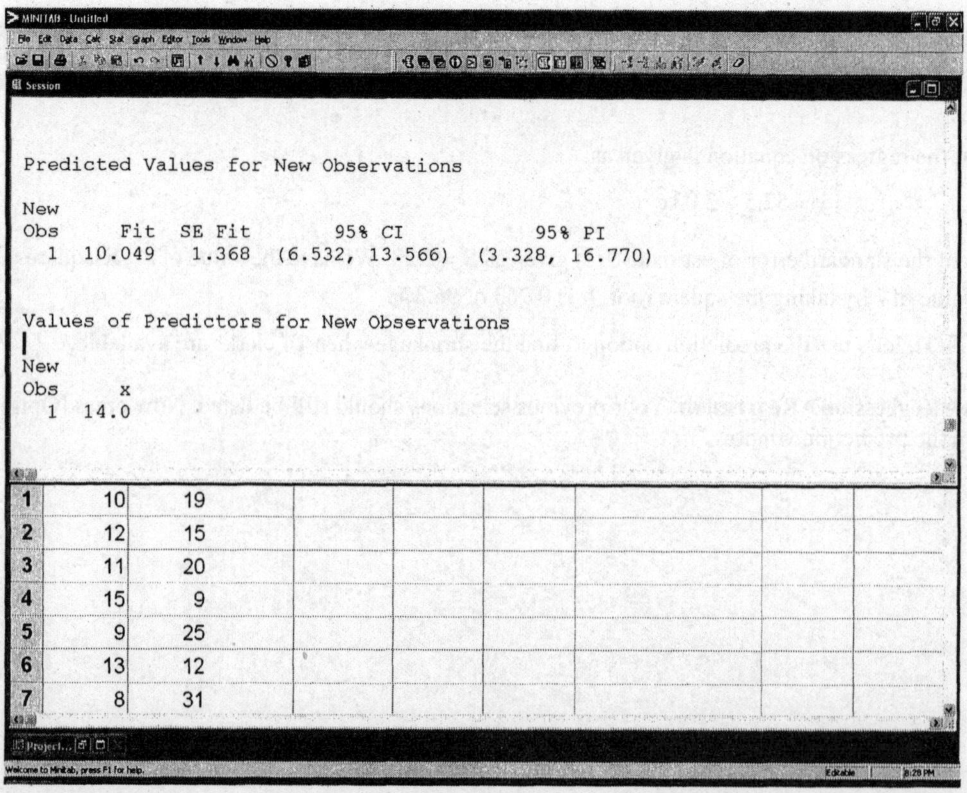

The predicted value of the shrinkage when 14 clerks are on duty is 10.05 hundred dollars, or $1005. A 95% prediction interval goes from 3.33 hundred dollars to 16.77 hundred dollars—that is, from $333 to $1677.

(e) Graph a prediction band for predicted values.

Now we use **>Stat>Regression>Fitted Line Plot** with the [Option] Display Prediction Interval selected.

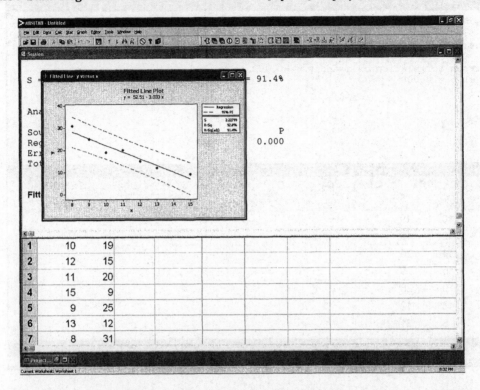

(f) Find the correlation coefficient and test it against the hypothesis that there is no correlation. We use the menu options **>Stat>Basic Statistics>Correlation.**

The results are

Notice $r = -0.963$ and the P value is so small as to be 0. We reject the null hypothesis and conclude that there is a linear correlation between the number of clerks on duty and the amount of shrinkage.

LAB ACTIVITIES FOR SIMPLE LINEAR REGRESSION: TWO VARIABLES

1. Open or retrieve the worksheet **Slr01.mtp** from the CD-ROM. This worksheet contains the following data, with the list price in column C1 and the best price in the column C2. The best price is the best price negotiated by a team from the magazine.

List Price versus Best Price for a New GMC Pickup Truck

In the following data pairs (X, Y)

X = List Price (in $1000) for a GMC Pickup Truck

Y = Best Price (in $1000) for a GMC Pickup Truck

SOURCE: CONSUMERS DIGEST, FEBRUARY 1994

(12.4, 11.2)	(14.3, 12.5)	(14.5, 12.7)
(14.9, 13.1)	(16.1, 14.1)	(16.9, 14.8)
(16.5, 14.4)	(15.4, 13.4)	(17.0, 14.9)
(17.9, 15.6)	(18.8, 16.4)	(20.3, 17.7)
(22.4, 19.6)	(19.4, 16.9)	(15.5, 14.0)

$$(16.7, 14.6) \qquad (17.3, 15.1) \qquad (18.4, 16.1)$$
$$(19.2, 16.8) \qquad (17.4, 15.2) \qquad (19.5, 17.0)$$
$$(19.7, 17.2) \qquad (21.2, (18.6)$$

(a) Use MINITAB to find the least-squares regression line using the best price as the response variable and list price as the explanatory variable.

(b) Use MINITAB to draw a scatter plot of the data.

(c) What is the value of the standard error of estimate?

(d) What is the value of the coefficient of determination r^2? of the correlation coefficient r?

(e) Use the least-squares model to predict the best price for a truck with a list price of $20,000. Note: Enter this value as 20 since X is assumed to be in thousands of dollars. Find a 95% confidence interval for the prediction.

2. Other MINITAB worksheets appropriate to use for simple linear regression are

Cricket Chirps Versus Temperature: **Slr02.mtp**

Source: *The Song of Insects* by Dr. G.W. Pierce, Harvard College Press

The chirps per second for the striped grouped cricket are stored in C1; the corresponding temperature in degrees Fahrenheit is stored in C2.

Diameter of Sand Granules Versus Slope on a Beach:

Slr03.mtp; source *Physical Geography* by A.M. King, Oxford press

The median diameter (mm) of granules of sand in stored in C1; the corresponding gradient of beach slope in degrees is stored in C2.

National Unemployment Rate Male Versus Female: **Slr04.mtp**

Source: *Statistical Abstract of the United States*

The national unemployment rate for adult males is stored in C1; the corresponding unemployment rate for adult females for the same period of time is stored in C2.

The data in these worksheets are described in the Appendix of this *Guide*. Select these worksheets and repeat parts (a)–(d) of problem 1, using C1 as the explanatory variable and C2 as the response variable.

3. A psychologist interested in job stress is studying the possible correlation between interruptions and job stress. A clerical worker who is expected to type, answer the phone, and do reception work has many interruptions. A store manager who has to help out in various departments as customers make demands also has interruptions. An accountant who is given tasks to accomplish each day and who is not expected to interact with other colleagues or customers except during specified meeting times has few interruptions. The psychologist rated a group of jobs for interruption level. The results follow, with X being interruption level of the job on a scale of 1 to 20, with 20 having the most interruptions, and Y the stress level on a scale of 1 to 50, with 50 the most stressed.

Person	1	2	3	4	5	6	7	8	9	10	11	12
X	9	15	12	18	20	9	5	3	17	12	17	6
Y	20	37	45	42	35	40	20	10	15	39	32	25

(a) Enter the X values into C1 and the Y values into C2. Use the menu selections **>Stat>Basic Statistics>Display Descriptive Statistics** on the two columns. What is the mean of the Y-values? Of the X-values? What are the respective standard deviations?

(b) Make a scatter plot of the data using the **>Stat>Regression>Fitted Line** menu selection. From the diagram do you expect a positive or negative correlation?

(c) Use the ➤**Stat**➤**Basic Statistics**➤**Correlation** menu choices to get the value of *r*. Is this value consistent with your response in part (b)?

(d) Use the ➤**Stat**➤**Regression**➤**Regression** menu choices with Y as the response variable and X as the explanatory variable. Use the [Option] button with predictions 5, 10, 15, 20 to get the predicted stress level of jobs with interruption levels of 5, 10, 15, and 20. Look at the 95% P.I. intervals. Which are the longest? Why would you expect these results? Find the standard error of estimate. Is R – sq equal to the square of *r* as you found in part (c)? What is the equation of the least-squares line?

(e) Redo the ➤**Stat**➤**Regression**➤**Regression** menu option, this time using X as the response variable and Y as the explanatory variable. Is the equation different than that of part (d)? What about the value of the standard error of estimate (s on your output)? Did it change? Did R – sq change?

4. The researcher of problem 3 was able to add to her data. Another random sample of 11 people had their jobs rated for interruption level and were then evaluated for stress level.

Person	13	14	15	16	17	18	19	20	21	22	23
X	4	15	19	13	10	9	3	11	12	15	4
Y	20	35	42	37	40	23	15	32	28	38	12

Add this data to the data in problem 3, and repeat parts (a) through (e). Compare the values of s, the standard error of estimate in parts (d). Did more data tend to reduce the value of s? Look at the 95% P.I. intervals. How do they compare to the corresponding ones of problem 3? Are they shorter or longer? Why would you expect this result?

MULTIPLE REGRESSION (SECTION 10.4 OF *UNDERSTANDABLE STATISTICS*)

The ➤**Stat**➤**Regression**➤**Regression** menu choices also do multiple regression.

➤**Stat**➤**Regression**➤**Regression**

Dialog Box Responses

Response: Enter the column number C# of the column containing the responses (that is Y values).

Predictor: Enter the column number C# of the columns containing the explanatory variables.

[Graphs]: Do not click on at this time.

[Results]: Click on and select the second option, Regression equation, etc.

[Options]: We will click on this option when we wish to do predictions for new variables.

[Storage]: Click on and select the fits and residual options if you wish.

Example

Bowman Brothers is a large sporting goods store in Denver that has a giant ski sale every year during the month of October. The chief executive officer at Bowman Brothers is studying the following variables regarding the ski sale:

X_1 = Total dollar receipts from October ski sale

X_2 = Total dollar amount spent advertising ski sale on local TV

X_3 = Total dollar amount spent advertising ski sale on local radio

X_4 = Total dollar amount spent advertising ski sale in Denver newspapers

Data for the past eight years is shown below (in thousands of dollars):

Year	1	2	3	4	5	6	7	8
X1	751	768	801	832	775	718	739	780
X2	19	23	27	32	25	18	20	24
X3	14	17	20	24	19	9	10	19
X4	11	15	16	18	12	5	7	14

(a) Enter the data in C1–C4. Name C1 = 'X1', C2 = 'X2', C3 = 'X3', C4 = 'X4'. Use ➤**Stat**➤**Basic Statistics**➤**Display Description Statistics** to study the data.

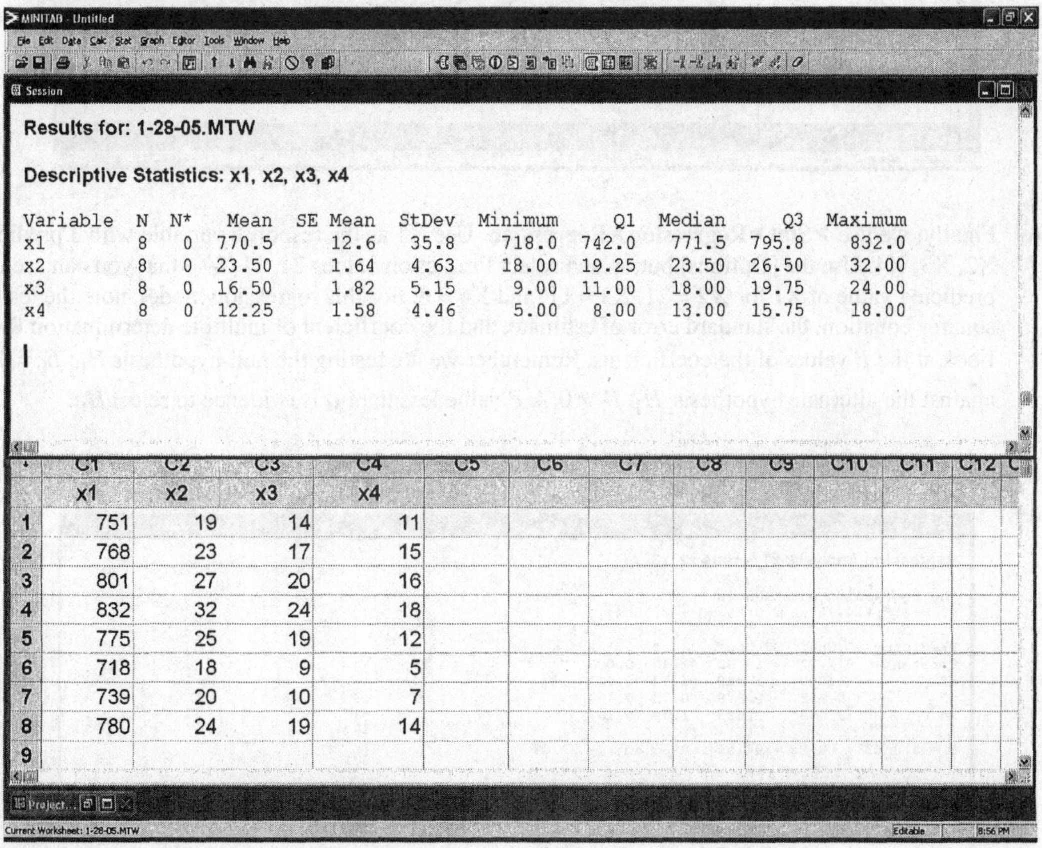

(b) Next use ➤**Stat**➤**Basic Statistics**➤**Correlation** menu option to see the correlation between each pair of columns of data.

```
>MINITAB - Untitled
File Edit Data Calc Stat Graph Editor Tools Window Help

Session

   Correlations: x1, x2, x3, x4

           x1       x2       x3
   x2    0.974
   x3    0.968    0.934
   x4    0.937    0.864    0.950

   Cell Contents: Pearson correlation
```

	C1	C2	C3	C4	C5	C6	C7	C8	C9	C10	C11	C12	C
	x1	x2	x3	x4									
1	751	19	14	11									
2	768	23	17	15									
3	801	27	20	16									
4	832	32	24	18									
5	775	25	19	12									
6	718	18	9	5									
7	739	20	10	7									
8	780	24	19	14									
9													

```
Current Worksheet: 1-28-05.MTW                           Editable   8:59 PM
```

(c) Finally, we use **>Stat>Regression>Regression.** Use X1 as the response variable with 3 predictors X2, X3, X4. Use the [Options] button and select Prediction values 21, 11, 8 so that you can see the predicted value of X1 for X2 = 21, X3 = 11, and X4 = 8. For this regression model, note the least-squares equation, the standard error of estimate, and the coefficient of multiple determination R – sq. Look at the P values of the coefficients. Remember we are testing the null hypothesis H_0: $\beta_1 = 0$. against the alternate hypothesis H_1: $\beta_1 \neq 0$. A P value less than α is evidence to reject H_0.

```
>MINITAB - Untitled
File Edit Data Calc Stat Graph Editor Tools Window Help

Session

   Regression Analysis: x1 versus x2, x3, x4

   The regression equation is
   x1 = 618 + 4.70 x2 + 0.65 x3 + 2.58 x4

   Predictor    Coef   SE Coef      T      P
   Constant   617.72    14.92    41.40  0.000
   x2           4.698    1.369     3.43  0.027
   x3           0.652    1.979     0.33  0.758
   x4           2.580    1.623     1.59  0.187

   S = 5.86631   R-Sq = 98.5%   R-Sq(adj) = 97.3%

   Analysis of Variance
   Source           DF       SS       MS      F      P
   Regression        3   8820.3   2940.1  85.43  0.000
   Residual Error    4    137.7     34.4
   Total             7   8958.0

   Source   DF   Seq SS
   x2        1   8497.6
   x3        1    235.7
   x4        1     87.0

   Predicted Values for New Observations
   New
   Obs    Fit   SE Fit        95% CI             95% PI
     1  744.20    4.36   (732.10, 756.30)   (723.91, 764.49)

   Values of Predictors for New Observations
   New
   Obs   x2     x3     x4
     1  21.0   11.0   8.00
```

```
Welcome to Minitab, press F1 for help                    Editable   9:04 PM
```

Note that we will not use the results of the Analysis of Variance.

LAB ACTIVITIES FOR MULTIPLE REGRESSION

Use the Section 10.4 problems 3–6. Each of these problems has MINITAB worksheets stored on the CD-ROM.

Section 10.5 problem #3 (Systolic Blood Pressure Data)

MINITAB worksheet: **Mlr02.mtp**

Section 10.5 problem #4 (Test Scores for General Psychology)

MINITAB worksheet: **Mlr03.mtp**

Section 10.5 problem #5 (Hollywood Movies data)

MINITAB worksheet: **Mlr04.mtp**

Section 10.5 problem #6 (All Greens Franchise Data)

MINITAB worksheet: **Mlr05.mtp**

Two additional case studies are available on the CD-ROM. The data are listed in the Appendix. For each of these studies, explore the relationships among the variables.

MINITAB WORKSHEET Mlr07.mtp

This is a case study of public health, income, and population density for small cities in eight midwestern states: Ohio, Indiana, Illinois, Iowa, Missouri, Nebraska, Kansas, and Oklahoma. The data is for a sample of 53 small cities in these states.

$X1$ = Death Rate per 1000 Residents

$X2$ = Doctor Availability per 100,000 Residents

$X3$ = Hospital Availability per 100,000 Residents

$X4$ = Annual per Capita Income in Thousands of Dollars

$X5$ = Population Density People per Square Mile

MINITAB WORKSHEET Mlr06.mtp

This is a case study of education, crime, and police funding for small cities in ten eastern and southeastern states. The states are New Hampshire, Connecticut, Rhode Island, Maine, New York, Virginia, North Carolina, South Carolina, Georgia, and Florida. The data is for a sample of 50 small cities in these states.

$X1$ = Total Overall Reported Crime Rate per 1 Million Residents

$X2$ = Reported Violent Crime Rate per 100,000 Residents

$X3$ = Annual Police Funding in Dollars per Resident

$X4$ = Percent of People 25 Years and Older that have had 4 years

of High School

$X5$ = Percent of 16 to 19 Year-Olds Not in High School and Not High School

Graduates

$X6$ = Percent of 18 to 24 Year-Olds Enrolled in College

$X7$ = Percent of People 25 Years and Older with at Least 4 Years of College

COMMAND SUMMARY

To Perform Simple or Multiple Regression

REGRESS C K C...C

does regression with the first column containing the response variable, K explanatory variables in the remaining columns. Following are some of the subcommands.

PREDICT E...E predicts the response variable for the given values of the explanatory variable(s).

RESIDUALS C stores the residuals in column C.

WINDOWS menu selection: ➤**Stat**➤**Regression**➤**Regression**

Use the dialog box to list the response and explanatory (prediction) variables. Mark the residuals box. In the Options dialog box list the values of the explanatory variable(s) for which you wish to make a prediction. Select the P.I. confidence interval.

BRIEF K controls the amount of output for K = 0, 1, 2, 3 with 3 giving the most output. Default selection is K=2. This command is not available from a menu.

There are other subcommands for REGRESS. See the MINITAB Help for your release of MINITAB for a list of the subcommands and their descriptions.

To Find the Pearson Product Moment Correlation Coefficient

CORRELATION C...C calculates the correlation coefficient for all pairs of columns.

WINDOWS menu selection: ➤**Stat**➤**Basic Statistics**➤**Correlation**

To Graph the Scatter Plot for Simple Regression

With **GSTD** use the **PLOT C C** command.

WINDOWS menu selection:

➤**Stat**➤**Regression**➤**Fitted Line Plot**

CHAPTER 11 CHI-SQUARE AND *F* DISTRIBUTIONS

CHI-SQUARE TESTS OF INDEPENDENCE (SECTION 11.1 OF *UNDERSTANDABLE STATISTICS*)

In chi-square tests of independence we use the hypotheses.

H_0: The variables are independent

H_1: The variables are not independent

To use MINITAB for tests of independence, we enter the values of a contingency table row by row. The command CHISQUARE then prints a contingency table showing both the observed and expected counts. It computes the sample chi-square value using the following formula, in which E stands for the expected count in a cell and O stands for the observed count in that same cell. The sum is taken over all cells.

$$\chi^2 = \sum \frac{(O-E)^2}{E}$$

Then MINITAB gives the number of degrees of the chi-square distribution. To conclude the test, use the P value of the sample chi-square statistic if your version of MINITAB provides it. Otherwise, compare the calculated chi-square value to a table of the chi-square distribution with the indicated degrees of freedom. We may use Table 8 of Appendix II of *Understandable Statistics*. If the calculated sample chi-square value is larger than the value in Table 8 for a specified level of significance, we reject H_0.

Use the menu selection

➤**Stat➤Tables➤Chi-square Test**

Dialog Box Responses

List the columns containing the data from the contingency table. Each column must contain integer values.

Example

A computer programming aptitude test has been developed for high school seniors. The test designers claim that scores on the test are independent of the type of school the student attends: rural, suburban, urban. A study involving a random sample of students from these types of institutions yielded the following contingency table. Use the CHISQUARE command to compute the sample chi-square value, and to determine the degrees of freedom of the chi-square distribution. Then determine if type or school and test score are independent at the $\alpha = 0.05$ level of significance.

	School Type		
Score	**Rural**	**Suburban**	**Urban**
200–299	33	65	83
300–399	45	79	95
400–500	21	47	63

To use the menu selection ➤**Stat➤Tables➤Chi-square Test** with C1 containing test scores for rural schools, C2 corresponding test scores for suburban schools, and C3 corresponding test scores for urban schools.

```
> MINITAB - Untitled                                                    _ 8 X
File Edit Data Calc Stat Graph Editor Tools Window Help
[toolbar]                              [toolbar]

Session                                                              _ □ X

Chi-Square Test: Rural, Subruban, Urban

Expected counts are printed below observed counts
Chi-Square contributions are printed below expected counts

        Rural  Subruban  Urban  Total
  1       33        65     83     181
        33.75     65.11  82.15
        0.016     0.000  0.009

  2       45        79     95     219
        40.83     78.77  99.40
        0.426     0.001  0.194

  3       21        47     63     131
        24.42     47.12  59.46
        0.480     0.000  0.211

Total     99       191    241     531

Chi-Sq = 1.338, DF = 4, P-Value = 0.855
```

	C1	C2	C3	C4	C5	C6	C7	C8	C9	C10
	Rural	Subruban	Urban							
1	33	65	83							
2	45	79	95							
3	21	47	63							

Since the P value, 0.855, is greater than $\alpha = 0.05$, we do not reject the null hypothesis.

LAB ACTIVITIES FOR CHI-SQUARE TESTS OF INDEPENDENCE

Use MINITAB to produce a contingency table and compute the sample chi square value. If your version of MINITAB produces the P value of the sample chi-square statistic, conclude the test using P values. Otherwise, use Table 9 of *Understandable Statistics* to find the chi-square value for the given α and degrees of freedom. Compare the sample chi-square value to the value found in Table 8 to conclude the test.

1. We Care Auto Insurance had its staff of actuaries conduct a study to see if vehicle type and loss claim are independent. A random sample of auto claims over six months gives the information in the contingency table.

Total Loss Claims per Year per Vehicle

Type of vehicle	$0–999	$1000–2999	$3000–5999	$6000+
Sports car	20	10	16	8
Truck	16	25	33	9
Family Sedan	40	68	17	7
Compact	52	73	48	12

Test the claim that car type and loss claim are independent. Use $\alpha = 0.05$.

2. An educational specialist is interested in comparing three methods of instruction.

 SL–standard lecture with discussion

 TV–video taped lectures with no discussion

 IM–individualized method with reading assignments

  and tutoring, but no lectures.

The specialist conducted a study of these methods to see if they are independent. A course was taught using each of the three methods and a standard final exam was given at the end. Students were put into the different method sections at random. The course type and test results are shown in the next contingency table.

Final Exam Score

Course Type	< 60	60–69	70–79	80–89	90–100
SL	10	4	70	31	25
TV	8	3	62	27	23
IM	7	2	58	25	22

Test the claim that the instruction method and final exam test scores are independent, using $\alpha = 0.01$.

ANALYSIS OF VARIANCE (ANOVA) (SECTION 11.5 OF *UNDERSTANDABLE STATISTICS*)

Section 11.5 of *Understandable Statistics* introduces single factor analysis of variance (also called one-way ANOVA). We consider several populations that are each assumed to follow a normal distribution. The standard deviations of the populations are assumed to be approximately equal. ANOVA provides a method to compare several different populations to see if the means are the same. Let population 1 have mean μ_1, population 2 have mean μ_2, and so forth. The hypotheses of ANOVA are

$H_0: u_1 = u_2 = \ldots = u_n$

H_1: not all the means are equal.

In MINITAB we use the menu selection **>Stat>ANOVA>Oneway (Unstacked)** to perform one-way ANOVA. We put the data from each population in a separate column. The different populations are called levels in the output of ANOVAONEWAY. An analysis of variance table is printed, as well a confidence interval for the mean of each level. If there are only two populations, ANOVAONEWAY is equivalent to using the 2-Sample Test choice with the equal variances option checked.

➤**Stat➤ANOVA➤Oneway (Unstacked)**

Dialog Box Responses

Responses: Enter the columns containing the data.

Select a confidence level such as 95%.

Check [Store Residuals] and /or [Store fits] only when you want to store these results.

Example

A psychologist has developed a series of tests to measure a person's level of depression. The composite scores range from 50 to 100 with 100 representing the most severe depression level. A random sample of 12 patients with approximately the same depression level, as measured by the tests, was divided into 3 different treatment groups. Then, one month after treatment was completed, the depression level of each patient was again evaluated. The after-treatment depression levels are given below.

Treatment 1	70	65	82	83	71
Treatment 2	75	62	81
Treatment 3	77	60	80	75

Put treatment 1 responses in column C1, treatment 2 responses in C2, treatment 3 responses in C3.

Use the ➤**Stat➤ANOVA➤Oneway (Unstacked)** menu selections.

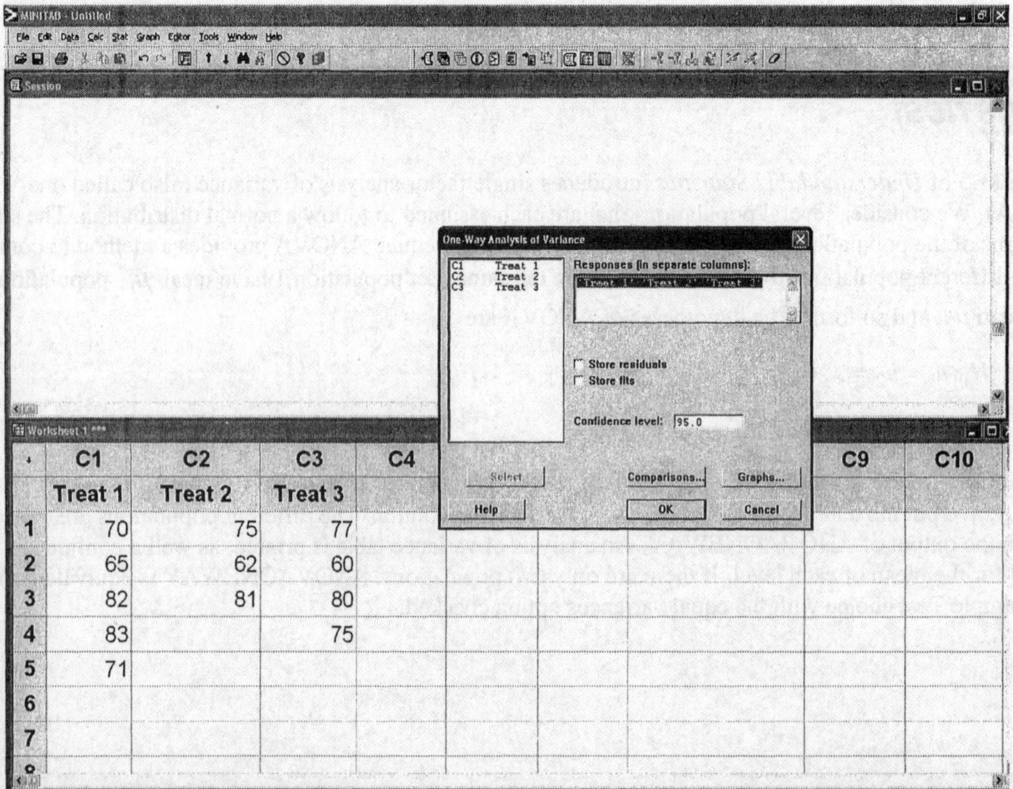

The results are

```
> MINITAB - Untitled
File Edit Data Calc Stat Graph Editor Tools Window Help

 Session

   One-way ANOVA: Treat 1, Treat 2, Treat 3

   Source  DF     SS     MS     F      P
   Factor   2    5.5    2.7   0.04  0.965
   Error    9  677.5   75.3
   Total   11  682.9

   S = 8.676   R-Sq = 0.80%   R-Sq(adj) = 0.00%


                               Individual 95% CIs For Mean Based on
                               Pooled StDev
   Level    N    Mean  StDev  --------+---------+---------+---------+-
   Treat 1  5  74.200  7.918             (---------------*--------------)
   Treat 2  3  72.667  9.713  (-------------------*-------------------)
   Treat 3  4  73.000  8.907     (-----------------*----------------)
                               --------+---------+---------+---------+-
                                     66.0      72.0      78.0      84.0

   Pooled StDev = 8.676
```

3	82	81	80
4	83		75
5	71		
6			
7			

Since the level of significance $\alpha = 0.05$ is less than the P value of 0.965, we do not reject H_0.

LAB ACTIVITIES FOR ANALYSIS OF VARIANCE

1. A random sample of 20 overweight adults was randomly divided into 4 groups. Each group was given a different diet plan, and the weight loss for each individual after 3 months follows:

Plan 1	18	10	20	25	17
Plan 2	28	12	22	17	16
Plan 3	16	20	24	8	17
Plan 4	14	17	18	5	16

   Test the claim that the population mean weight loss is the same for the four diet plans, at the 5% level of significance.

2. A psychologist is studying the time it takes rats to respond to stimuli after being given doses of different tranquilizing drugs. A random sample of 18 rats was divided into 3 groups. Each group was given a different drug. The response time to stimuli was measured (in seconds). The results follow.

Drug A	3.1	2.5	2.2	1.5	0.7	2.4
Drug B	4.2	2.5	1.7	3.5	1.2	3.1
Drug C	3.3	2.6	1.7	3.9	2.8	3.5

Test the claim that the population mean response times for the three drugs is the same, at the 5% level of significance.

3. A research group is testing various chemical combinations designed to neutralize and buffer the effects of acid rain on lakes. Eighteen lakes of similar size in the same region have all been affected in the same way by acid rain. The lakes are divided into four groups and each group of lakes is sprayed with a different chemical combination. An acidity index is then take after treatment. The index ranges from 60 to 100, with 100 indicating the greatest acid rain pollution. The results follow.

Combination I	63	55	72	81	75
Combination II	78	56	75	73	82
Combination III	59	72	77	60	
	72	81	66	71	

Test the claim that the population mean acidity index after each of the four treatments is the same at the 0.01 level of significance.

COMMAND SUMMARY

CHISQUARE C...C

produced a contingency table and computes the sample chi-square value

WINDOWS menu select: ➤**Stat**➤**Tables**➤**Chi-square test**

In the dialog box, specify the columns that contain the chi-square table.

AOVONEWAY C...C

performs a one-way analysis of variance. Each column contains data from a different population

WINDOWS menu select: ➤**Stat**➤**ANOVA**➤**Oneway (Unstacked)**

In the dialog box specify the columns to be included.

COMMAND REFERENCE

This appendix summarizes all the MINITAB commands used in this Guide. A complete list of commands may be found in the MINITAB Reference Manual that comes with the MINITAB software.

C denotes a column

E denotes either a column or constant

K denotes a constant

[] denotes optional parts of the command

GENERAL INFORMATION

HELP   gives general information about MINITAB.

WINDOWS menu: **Help**

STOP   ends MINITAB session

WINDOWS menu: **➤File➤Exit**

TO ENTER DATA

READ C...C   puts data into designated columns.

READ C...C

File "filename"   reads data from file into columns.

SET C   put data into single designated column.

SET C

File "filename"   reads data from file into column.

NAME C = 'name' names column C.

WINDOWS menu selection: You can enter data in rows or columns and name the column in the

DATA window. To access the data window select **➤Window➤Worksheet.**

RETREIEVE 'filename' retrieves worksheet.

WINDOWS menu selection: **➤File➤Open Worksheet.**

TO EDIT DATA

LET C(K) = K   changes the value in row K of column C.

INSERT K K C C inserts data between rows K and K into columns C to C.

DELETE K K C C deletes data between row K and K from columns C to C.

WINDOWS menu selection: You can edit data in rows or columns in the DATA window.

To access the data window select **➤Window➤Worksheet.**

COPY C C   copies column C into column C.

USE K...K   subcommand to copy designated rows

OMIT [C] K...K subcommand to omit designated rows

    WINDOWS menu selection: **➤Data➤Copy ➤Columns to Columns**

ERASE E...E   erases designated columns or constants.

    WINDOWS menu selection: **➤Data➤Erase Variables**

TO OUTPUT DATA

PRINT E...E   prints designated columns or constant.

    WINDOWS menu selection: **➤Data➤Display Data**

SAVE 'filename' saves current worksheet or project.

PORTABLE   subcommand to make worksheet portable

    WINDOWS menu: **➤File➤Save Project**

    WINDOWS menu: **➤File➤Save Project as**

    WINDOWS menu selection: **➤File➤Save Current Worksheet**

    WINDOWS menu selection: **➤File➤Save Current Worksheet As...** you may select portable.

WRITE C...C

    File "filename" saves data in ASCII file

MISCELLANEOUS

OUTFILE = 'filename' : put all input and output in "filename".

NOOUTFILE: ends OUTFILE.

TO DO ARITHMETIC

LET E = expression evaluates the expression and stores the result in E, where E may be a column or a constant.

    ** raise to a power

    * multiplication

    / division

    + addition

    – subtraction

SQRT E E   takes the square root.

ROUND(E E) rounds numbers to the nearest integer.

    There are other arithmetic operations possible.

    WINDOW menu selection: **➤Calc➤Calculator**

TO GENERATE A RANDOM SAMPLE

RANDOM K C…C selects a random sample from the distribution described in the subcommand.

    INTEGER K K   distribution of integers from K to K

    BERNOULLI K

    BINOMIAL .K K

    CHISQUARE K

    DISCRETE C C

    F K K

    NORMAL [K [K]]

    POISSON K

    T K

    UNIFORM [K K]

        WINDOWS menu selection: **➤Calc➤Random data**

SAMPLE K C…C generate k rows of random data from specified input columns, C…C and stores in specified storage columns, C…C.

REPLACE causes the sample to be taken with replacement.

NOREPLACE causes the sample to be taken with replacement.

TO ORGANIZE DATA

SORT C [C…C] C[C…C] sorts C, carrying [C..C], and places results into C[C…C]

DESCENDING C…C   subcommand to sort in descending order

    WINDOWS menu selection: **➤Data➤Sort**

TALLY C…C tallies data in columns. The data must be integer values.

    COUNTS

    PERCENTS

    CUMCOUNTS

    CUMPERCENTS

    ALL   gives all four values.

        WINDOWS menu selection: **➤Stats➤Tables➤Tally Individual Variables**

HISTOGRAM C…C prints a separate histogram for data in each of the listed columns.

    MIDPOINT K…K Places ticks at midpoints of the intervals K … K

        WINDOWS menu: (for numerical variables) **➤Graph➤Histogram (options for cutpoints)**

        WINDOWS menu: (for categorical variables) **➤Graphs➤Bar Chart**

STEM-AND-LEAF C…C

        makes separate stem-and-leaf displays of data in each of the listed columns.

    INCREMENT = K sets the distance between two display lines.

TRIM trims all values beyond the inner fences

      WINDOWS menu: ➤**Graph** ➤**Stem-and-Leaf**

BOXPLOT C...C makes a separate box-and-whisker plot for each column C

      WINDOWS MENU: (professional graphics) ➤**Graph**➤**Boxplot**

TO SUMMARIZE DATA BY COLUMN

DESCRIBE C...C prints descriptive statistics

      WINDOWS menu selection: ➤**Stat**➤**Descriptive Statistics**➤**Display Descriptive Statistics**

COUNT   **C [K]** counts the values.

N   **C [K]** counts the non-missing values.

NMIS   **C [K]** counts the missing values.

SUM   **C [K]** sums the values.

MEAN   **C [K]** gives arithmetic mean of values.

STDEV   **C [K]** gives standard deviation.

MEDIAN   **C [K]** gives the median of the values.

MINIMUM **C [K]** gives the minimum of the values.

MAXIMUM C [K] gives the maximum of the values.

SSQ   **C [K]** gives the sum of squares of values.

TO SUMMARIZE DATA BY ROW

RCOUNT **E...E C**

RN   **E...E C**

RNMIS   **E...E C**

RSUM   **E...E C**

RMEAN   **E...E C**

RSTDEV **E...E C**

RMEDIAN E...E C

RMIN   **E...E C**

RMAX   **E...E C**

RSSQ   **E...E C**

TO FIND PROBABLITIES

PDF for values in E [put into E] calculates probabilities for the specified values of a discrete distribution and calculates the probability density function for a continuous distribution.

CDF for values in E...E [put into E...E] gives the cumulative distribution. For any value X, CDF X gives the probability that a random variable with the specified distribution has a value less than or equal to X.

INVCDF for values in E [put into E] gives the inverse of the CDF.

Each of these commands apply the following distributions (as well as some others). If no subcommand is used, the default distribution is the standard normal.

BINOMIAL   **n = K, p = K**

POISSION μ **= K** (note that for the Poisson distribution, $\mu = \lambda$)

INTEGER   **a = K, b = K**

DISCRETE values in C, probabilities in C

NORMAL μ **= K,** σ **= K**

UNIFORM   **a = Km b = K**

T   **d.f. = K**

F   **d.f numerator = K, d.f. denominator = K**

CHISQUARE d.f. = K

WINDOWS menu selection: **➤Calc➤Probability Distribution➤Select distribution**

In the dialog box, select **Probability for PDF; Cumulative probability for CDF; Inverse cumulative for INV;** enter the required information such as **E, n, p, or** μ**, d.f.** and so forth.

GRAPHING COMMANDS

Character Graphics Commands

Note: In some versions of MINITAB, you must use the command **GSTD** before you use the following graphics commands.

PLOT C versus C prints a scatter plot with the first column on the vertical axis and the second on the horizontal axis. The following subcommands can be used with **PLOT.**

TITLE = 'text' gives a title above the graph.

FOOTNOTE = 'text' places a line of text below the graph.

XLABEL = 'text' labels the *x*-axis.

YLABEL = 'text' labels the *y*-axis.

SYMBOL = 'symbol' selects the symbol for the points on the graph. The default is *.

XINCREMENT = K is distance between tick marks on *x*-axis.

XSTART = K [end = k' specifies the first tick mark and optionally the last one.

YINCREMENT = K is distance between tick marks on *y*-axis.

YSTART = K [end = K] specifies the first tick mark and optionally the last one.

WINDOWS menu selection: **➤Graph➤Character Graphs➤Scatter Plot**

Titles, labels, and footnotes are in the **Annotate...** option.

Increment and start are in the **Scale** option.

Professional Graphics

Note: In some versions of MINITAB, you must use the command **GPRO** before you use the following graphics commands.

Plot C * C prints a scatter plot with the first column on the vertical axis and the second on the horizontal axis. Note that the columns must be separated by an asterisk *.

Connect connects the points with a line

Other subcommands may be used to title the graph and set the tick marks on the axes. See your MINITAB software manual for details.

>   WINDOWS menu selection: **➤Graph➤Plot**

>   Use the dialog boxes to title the graph, label the axes, set the tick marks, and so forth.

>   See your MINITAB software manual for details.

CONTROL CHARTS

Character Graphics Commands

Note: In some versions of Minitab, you must use the command **GSTD** before you use the following graphics commands.

CHART C...C produces a control chart under the assumption that the data come from a normal distribution with mean and standard deviation specified by the subcommands.

MU = K   gives the mean of the normal distribution.

SIGMA = K gives the standard deviation.

>   WINDOWS menu selection: none for character graphics. Use the commands in the session window.

Professional Graphics

Note: This may be the default mode for versions of MINITAB supporting professional graphics. If necessary, use the command **GPRO** before using the following commands.

CHART C...C produces a control chart under the assumption that the data come from a normal distribution with mean and standard deviation specified by the subcommands

MU K gives the mean of the normal distribution.

SIGMA K gives the standard deviation..

>   WINDOWS menu selection: **➤Stat➤Control Chart➤Individual**

>   Enter choices for MU and Sigma in the dialog box.

TO GENERATE CONFIDENCE INTERVALS

ZINTERVAL [K% confidence] σ = K on C...C generates a confidence interval for μ using the normal distribution. You must enter a value for σ, either actual or estimated. A separate interval is given for data in each column. If K is not specified, a 95% confidence interval will be given.

>   WINDOWS menu selection: **➤Stat➤Basic Statistics➤1-sample z**

>   In the dialog box select confidence interval and enter the confidence level.

TINTERVAL [K% confidence] for C...C generates a confidence interval for μ using the Student's t distribution. It automatically computes stdev s from the data as well as the number of degrees of freedom. If K is not specified, a 95% confidence interval is given.

TO TEST A SINGLE MEAN

ZTEST [μ = K] σ = K, for C...C performs a z-test on the data in each column. If you do not specify μ, it is assumed to be 0. You need to supply a value for σ (either actual, or estimated by the sample standard deviation *s* of a column in the case of large samples). If the ALTERNATIVE subcommand is not used, a two-tailed test is conducted.

>   WINDOWS menu selection: **➤Stat➤Basic Statistics➤1 sample z**

>   In dialog box select alternate hypothesis, specify the mean for Hü, specify the standard deviation.

TTEST [μ = K] on C…C performs a separate *t*-test on the data of each column. If you do not specify μ, it is assumed to be 0. The computer evaluates *s*, the sample standard deviation for each column, and uses the computed *s* value to conduct the test. If the ALTERNATIVE subcommand is not used, a two-tailed test is conducted.

>   WINDOWS menu selection: **➤Stat➤Basic Statistics➤1 sample t**

>   In dialog box select alternate hypothesis, specify the mean for H_0.

ALTERNATIVE = K is the subcommand required to conduct a one-tailed test. If K = –1, then a left-tailed test is done. If K = 1, then a right-tailed test is done.

TO TEST A DIFFERENCE OF MEANS (INDEPENDENT SAMPLES)

TWOSAMPLE [K% confidence] for C…C does a two (independent) sample *t* test and (optional confidence interval) for data in the two columns listed. The first data set is put into the first column, and the second data set into the second column. Unless the ALTERNATIVE subcommand is used, the alternate hypothesis is assumed to be $H_1: \mu_1 \neq \mu_2$. Samples are assumed to be independent.

ALTERNATIVE = K is the subcommand to change the alternate hypothesis to a left-tailed test with K = –1 or right-tailed test with K = 1.

POOLED is the subcommand to be used only when the two samples come from populations with equal standard deviations.

>   WINDOWS menu selection: **➤Stat➤Basic Statistics➤2 sample t**

>   In dialog box select alternate hypothesis, specify the mean for H_0, for small samples select equal variances.

TO PERFORM SIMPLE OR MULTIPLE REGRESSION

REGRESS C on K explanatory variables in C...C does regression with the first column containing the response variable, K explanatory variables in the remaining columns.

PREDICT E...E predicts the response variable for the given values of the explanatory variable(s).

> WINDOWS menu selection: ➤**Stat**➤**Regression**➤**Regression**

> Use the dialog box to list the response and explanatory (prediction) variables. Mark the residuals box. In the Options dialog box list the values of the explanatory variable(s) for which you wish to make a prediction. Select the P.I. confidence interval.

BRIEF K controls the amount of output for K = 1, 2, 3 with 3 giving the most output. This command is not available from a menu.

There are other subcommands for REGRESS. See the MINITAB reference manual for your release of MINITAB for a list of the subcommands and their descriptions.

TO FIND THE PEARSON PRODUCT MOMENT CORRELATION COEFFICIENT

CORRELATION for C...C calculates the correlation coefficient for all pairs of columns.

> WINDOWS menu selection: ➤**Stat**➤**Basic Statistics**➤**Correlation**

TO GRAPH THE SCATTER PLOT FOR SIMPLE REGRESSION

With **GSTD** use the **PLOT C vs C** command.

> WINDOWS menu selection (MINITAB SE for Windows or MINITAB 10 for Windows):
> ➤**Stat**➤**Regression**➤**Fitted Line Plot**

TO PERFORM CHI SQUARE TESTS AND ANOVA

CHISQUARE test on table stored in C...C produces a contingency table and computes the sample chi-square value

> WINDOWS menu selection: ➤**Stat**➤**Tables**➤**Chisquare Test**

> In the dialog box specify the columns which contain the chi-square table.

AOVONEWAY on C...C performs a one-way analysis of variance.

> Each column contains data from a different population.

> WINDOWS menu selection: ➤**Stat**➤**ANOVA**➤**Oneway (Unstacked)**

> In the dialog box specify the columns to be included.

Nonparametric commands

MANN-WHITNEY [confidence = K] on CC does a two-sample rank sum test for the difference of two population means. Data from each population is in each separate column. The test is a two-tailed test unless ALTERNATE subcommand is used.

> WINDOWS menu selection: ➤**Stat**➤**Nonprametrics**➤**Mann-Whitney**

PART IV

SPSS GUIDE

FOR

UNDERSTANDABLE STATISTICS

EIGHTH EDITION

CHAPTER 1 GETTING STARTED

GETTING STARTED WITH SPSS

In this chapter you will find

(a) general information about SPSS

(b) general directions for using the Windows style pull-down menus

(c) general instructions for choosing values for dialog boxes

(d) how to enter data

(e) other general commands.

General Information

SPSS is a powerful tool that can perform many statistical procedures. Data are entered in the data editor window. The data editor window offers to choices: data view screen and variable view screen. The variable view screen is for you to define variables, meaning to name variables, declare variable type, determine variable format, and declare measurement type. The choices for measurement type are scale, ordinal, or nominal. The data view screen is where you enter data. The data view screen has a spreadsheet format. Each column contains data for one variable. If the variable is not defined, then the default variable name "var00001" will be used for the first column, "var00002" for the second column, and so on.

Once data is entered, Windows style pull-down menus are used to select activities, graphs, or other statistical procedures.

Starting and Ending SPSS

The steps you use to start SPSS may differ according to the computer equipment you are using. You will need to get specific instructions for your installation from your professor or computer lab manager. Use this space to record the details of logging onto your system and accessing SPSS.

Once SPSS is activated, the first screen you see will look like this

Choose **Type in data,** you will have the data view screen of the data editor window.

Notice the main menu items:

<u>F</u>ile   <u>E</u>dit   <u>V</u>iew   D<u>a</u>ta   <u>T</u>ransform   <u>A</u>nalyze   <u>G</u>raphs   <u>U</u>tilities   <u>A</u>dd-ons   <u>W</u>indow   <u>H</u>elp

The toolbar contains icons for frequently used operations.

To end SPSS: Click on the <u>F</u>ile option. Select Exit and click on it or press ENTER.

Menu selection summary: **➤File➤Exit**

Entering Data

One of the first tasks you do when you begin an SPSS session is to enter data into the data editor window. To do that, you may choose to first define variables (columns) of the data. This is done in the variable view screen. By clicking on the **Variable View** button located at the bottom of the data editor window, you will get into the variable view screen, where you define variables' name, type, format, etc. After variables are defined, click on the **Data View** button, which is also located at the bottom of the data editor window, to get back to the data view screen, where you enter the data. Notice that the active cell is outlined by a heavier box.

To enter a number, type it in the active box and then press ENTER or TAB. The data value is entered and the next cell in the same column is activated. Arrow keys and mouse cursor may also be used to move around in the data view screen. Each column contains data for a specific variable. Notice that there is a cell for a column label above row number 1.

To change a data value in a cell, activate the cell by clicking on it, then correct the data in the entry bar above the data sheet, and press ENTER or TAB.

Example

Open a new data sheet by selecting **➤File➤New➤Data.**

Let's create a new data set that has data in it regarding ads on TV. A random sample of 20 hours of prime time viewing on TV gave information about the number of TV ads in each hour as well as the total time consumed in the hour by ads. We will enter the data into two variables (columns): one variable representing the number of ads and the other the time per hour devoted to ads.

First, let's get into the variable view screen to define the two variables. As shown below, we name the two variables as Ad_Count and Min_Per_Hr. They are both of the type of numeric. Width (number of digits) are 8 for both of them. Decimals (number of digits after the decimal point) are 0 for Ad_Count, and 2 for the other. We use "ad" and "mph" as the labels for the two variables, respectively. At this time data is not entered yet. Therefore, "values" (number of value entered in this column) and "missing" (number of missing values in this column) are None for both. "Column" stands for the column width, and we use 8 for both variables. They are both aligned to the right and are in scale measurement. The screen is displayed below.

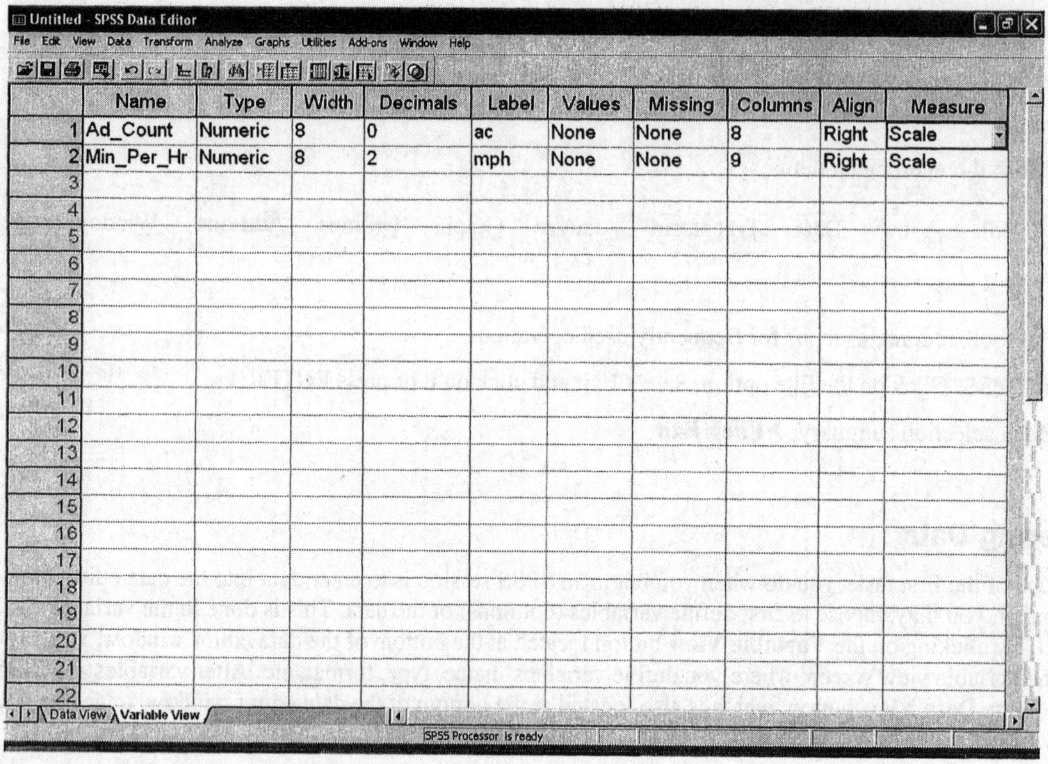

	Name	Type	Width	Decimals	Label	Values	Missing	Columns	Align	Measure
1	Ad_Count	Numeric	8	0	ac	None	None	8	Right	Scale
2	Min_Per_Hr	Numeric	8	2	mph	None	None	9	Right	Scale

New click on the **Data View** button to get into the data view screen, where we enter our data. The result is shown below.

	Ad_Count	Min_Per_Hr
1	25	11.50
2	23	10.70
3	28	10.20
4	15	9.30
5	13	11.30
6	24	11.00
7	27	15.00
8	22	12.00
9	17	10.00
10	19	10.50
11	20	14.30
12	22	11.70
13	18	14.90
14	19	10.70
15	23	12.30
16	13	10.10
17	23	11.20
18	21	10.80
19	22	10.30
20	25	15.70

Working with Data

In an SPSS data sheet, each column stands for one variable, and each row stands for a case (record). To delete a variable or a case, the Edit menu option may be used. To insert a variable or a case, the Data menu option may be used. To delete value in a certain cell, activate that cell and press the Delete key. However, the cell itself will not be deleted. Deleting value in a cell simply causes a "missing value" for the corresponding variable.

To insert a variable(column) to the left of column K: activate a cell in the column K, then select ➤**Data**➤**Insert Variable.**

To insert a case(row) above row K: activate a cell in the row K, then select ➤**Data**➤**Insert Cases.**

To delete a variable(column) or a case (row): select the column by clicking on the variable name (or the row by clicking on the row number), then use ➤**Edit**➤**Cut.**

Click on the Data menu item. You will see these cascading options in the pull down list.

Click on the Edit menu item. You will see these cascading options in the pull down list.

To print the data sheet, use ➤**File**➤**Print.**

Manipulating Data

You can also do calculations with entire columns. Click on the Transform menu item and select Compute
(➤**Transform**➤**Compute**). The dialog box appears:

Suppose you like to calculate a new variable x such that x = 3(Ad_Count)+4, and store the results in the third column. To do that, first type x in the "Target Variable" entry bar, then type 3, click on the multiply key * on the calculator, highlight [Ad_Count] and click ~ button to enter it into the "Numeric Expression" entry bar, click on the + key on the calculator, then type 4.

Now click on OK. The results of this arithmetic will appear in the third column (variable x) of the data sheet.

	Ad_Count	Min_Per_Hr	x	var	var	var	var	var	var
1	25	11.50	79.00						
2	23	10.70	73.00						
3	28	10.20	88.00						
4	15	9.30	49.00						
5	13	11.30	43.00						
6	24	11.00	76.00						
7	27	15.00	85.00						
8	22	12.00	70.00						
9	17	10.00	55.00						
10	19	10.50	61.00						
11	20	14.30	64.00						
12	22	11.70	70.00						
13	18	14.90	58.00						
14	19	10.70	61.00						
15	23	12.30	73.00						
16	13	10.10	43.00						
17	23	11.20	73.00						
18	21	10.80	67.00						
19	22	10.30	70.00						
20	25	15.70	79.00						
21									

Saving a Data

Click on the File menu and select Save As. A dialog box similar to the following appears. (Select
>**File**>**Save As.**)

For most computer labs, you will save your file on a 3½ inch floppy. Insert it in the appropriate drive.
Scroll down the Save in button until you find 3½ Floppy (A:). Then select a file name. In most cases you will
save the file as a SPSS file. If you change versions of SPSS or systems, you might select SPSS portable.

Example

Let's save the worksheet created in the previous example (information about ads on TV).

If you added the variable x as described under Manipulating the Data, clink on the variable name "x" to
highlight this column and press the Del key. Your data should have only two columns. Use >**File**>**Save As.**
Insert a diskette in drive A. Scroll down the Save in box and select 3½ Floppy (A:). Name the file ads. Click on
Save. The worksheet will be saved as ads.sav

LAB ACTIVITIES FOR GETTING STARTED WITH SPSS

1. Go to your computer lab (or use your own computer) and learn how to access SPSS.

2. (a) Use the data worksheet to enter the data:

   1   3.5   4   10   20   in column 1, name this variable as First.

   Enter the data

   3   7   9   8   12   in column 2, name this variable as Second.

   (b) Use ➤**Transform**➤**Compute** to generate a new variable Result, stored in column 3. The data in Result should be Result = 2*First + Second. Check to see that the first entry in column 3 for Result is 5. Do the other entries check?

   (c) Save the data as Prob 2 on a floppy diskette.

   (d) Retrieve the data by selecting ➤**File**➤**Open** ➤ **Data.**

   (e) Print the data. Use either the Print button or select ➤**File**➤**Print.**

RANDOM SAMPLES (SECTION 1.2 OF *UNDERSTANDABLE STATISTICS*)

In SPSS you can take random samples from a variety of distributions. We begin with one for the simplest: random samples from a range of consecutive integers under the assumption that each of the integers is equally likely to occur. To generate such a random sample, follow these steps:

1) In the data editor, enter the sample numbers in the first column. For example, to generate five random numbers, enter 1 through 5 in the first column.

2) Use the menu options **➤Transform➤Compute.** In the dialog box, first type in a variable name as the Target Variable, then select a function to generate random numbers from the desired distribution. The function **RV.UNIFORM(min, max)** under the group **Random Numbers** generate random numbers from the uniform distribution (min, max). Then function **TRUNC(k)** under the group **Arithmetic** truncates a real number to its integer part. Therefore, to generate random integer sample between two numbers, say between 1 and 100, the formula **TRUNC(RV.UNIFORM(1, 101))** should be used. Note that the real numbers between 100 and 101 will truncate to 100.

The random sample numbers are given in the order of occurrence. If you want them in ascending order (so you can quickly check to see if any values are repeated), use the menu options **➤Data➤Sort Cases.**

➤Data➤Sort Cases

Dialog Box Responses

Sort by: Enter the variable name that you wish to sort by.

Store order: Select either ascending or descending order.

Example

There are 175 students enrolled in a large section of introductory statistics. Draw a random sample of 15 of the students.

We number the students from 1 to 175, so we will be sampling from the integers 1 to 175. We don't want any student repeated, so if our initial sample has repeated values, we will continue to sample until we have 15 distinct students. We sort the data so that we can quickly see if any values are repeated.

First we follow the above two steps to generate the fifteen numbers. We use x as the variable containing the 15 random numbers. Note that we use 176 as the "max" for the **RV.UNIFORM** function. Displayed below is the responses in the **➤Transform➤Compute** dialog.

Now click on OK. Results are listed below. (Your results will vary.)

Next, sort the data.

Click on OK. Highlight the column for x to see the sorted data. The results are shown.

We see that no data are repeated. If you have repetitions, keep sampling until you get 15 distinct values.

Random numbers are also used to simulate activities or outcomes of a random experiment such as tossing a die. Since the six outcomes 1 through 6 are equally likely, we can use above procedure to simulate tossing a die any number of times. When outcomes are allowed to occur repeatedly, it is convenient to use frequency table to tally, count, and give percents of the outcomes. We do this with the menu options ➤**Analyze**➤**Descriptive Statistics**➤**Frequencies.**

➤Analyze➤Descriptive Statistics➤Frequencies

Dialog Box Responses

    Variables: variable name containing data

    Option to check: Display frequency tables.

Example

Use the above random number generating procedure with min = 1 and max = 7 (numbers between 6 and 7 will truncate to 6) to simulate 100 tosses of a fair die. Use the frequency table to give a count and percent of outcomes.

Generate the random sample using the function formula **TRUNC(RV.UNIFORM(1, 7)).** Use the name Outcome for the variable containing the outcomes. Then use ➤**Analyze**➤**Descriptive Statistics**➤**Frequencies** with "Display frequency tables" checked.

Click on OK. The results are shown below. (Your results will vary.)

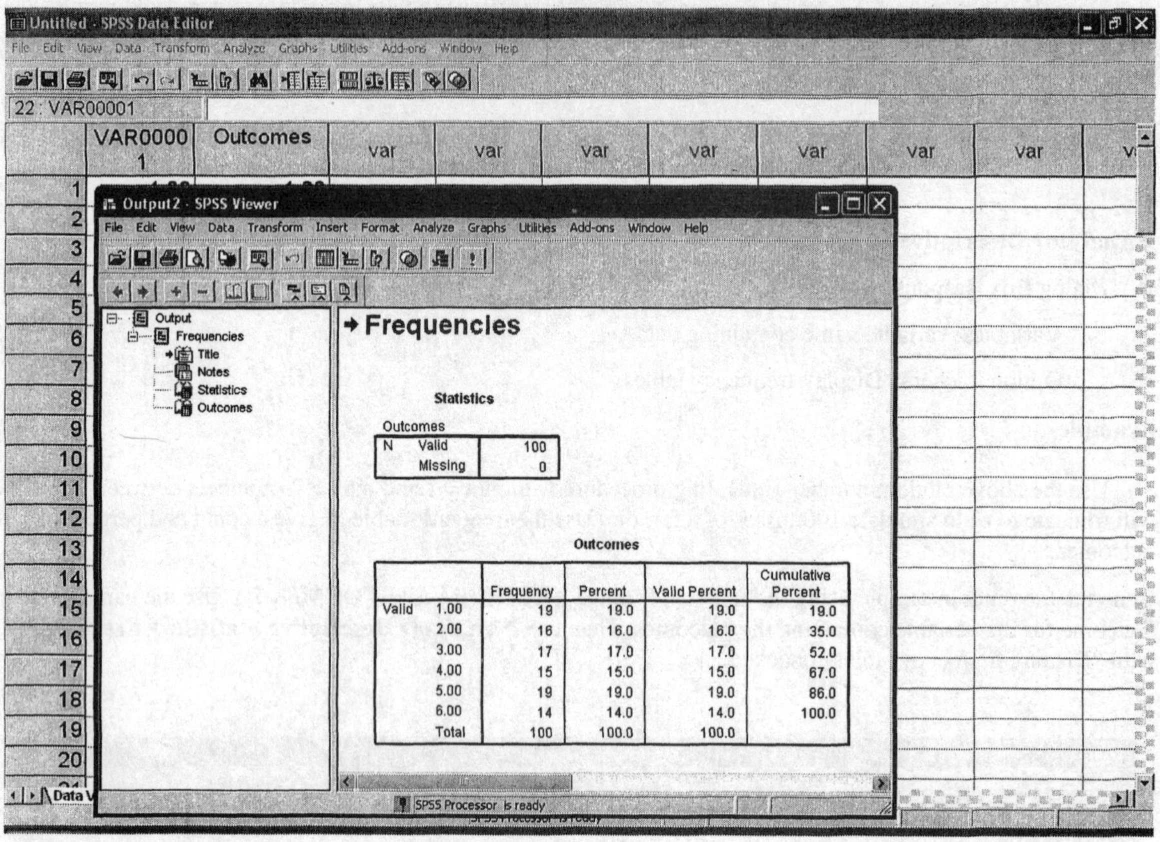

If you have a finite population, and wish to sample from it, you may use the menu options ➤**Data**➤**Select Cases** to accomplish that.

➤**Data**➤**Select Cases**

Dialog Box Responses

Select the variable to be sampled from.

Check: Random sample of cases

Check: either Filtered (the unselected cases will be marked with 0, selected cases with 1.) or Deleted (the unselected cases will be deleted).

Next, click on the Sample button. Another dialog will show up.

Dialog Box Responses

Check: Exactly

Enter number of cases to be selected from the first N (total number of cases) cases.

Example

Take a sample of size 10 without replacement from the population of numbers 1 through 100.

First, enter the numbers 1 through 100 in the first column, using x as the variable name. Then use ➤**Data**➤**Select Cases**. In the dialog box, select variable x, check Random sample of cases, and check Filtered. The dialog is shown below.

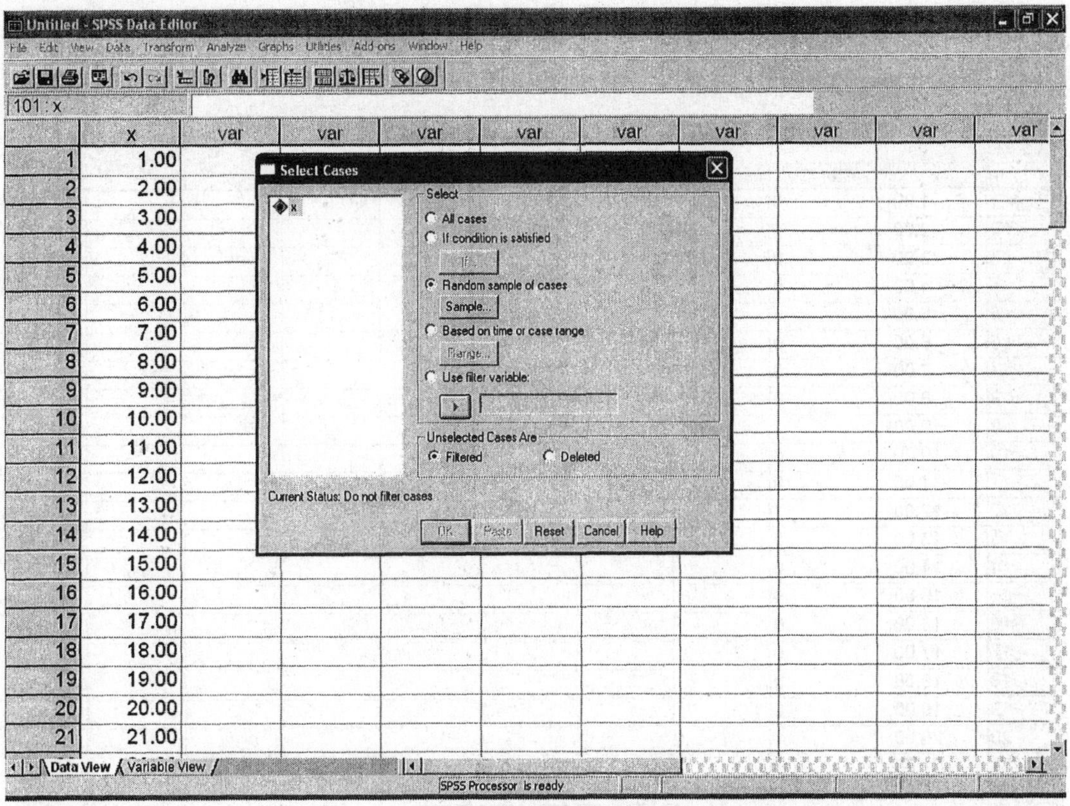

Next, click on the Sample button. In the Sample button dialog box, check "Exactly" and enter "10" cases from the first "100" cases. The dialog box is displayed below.

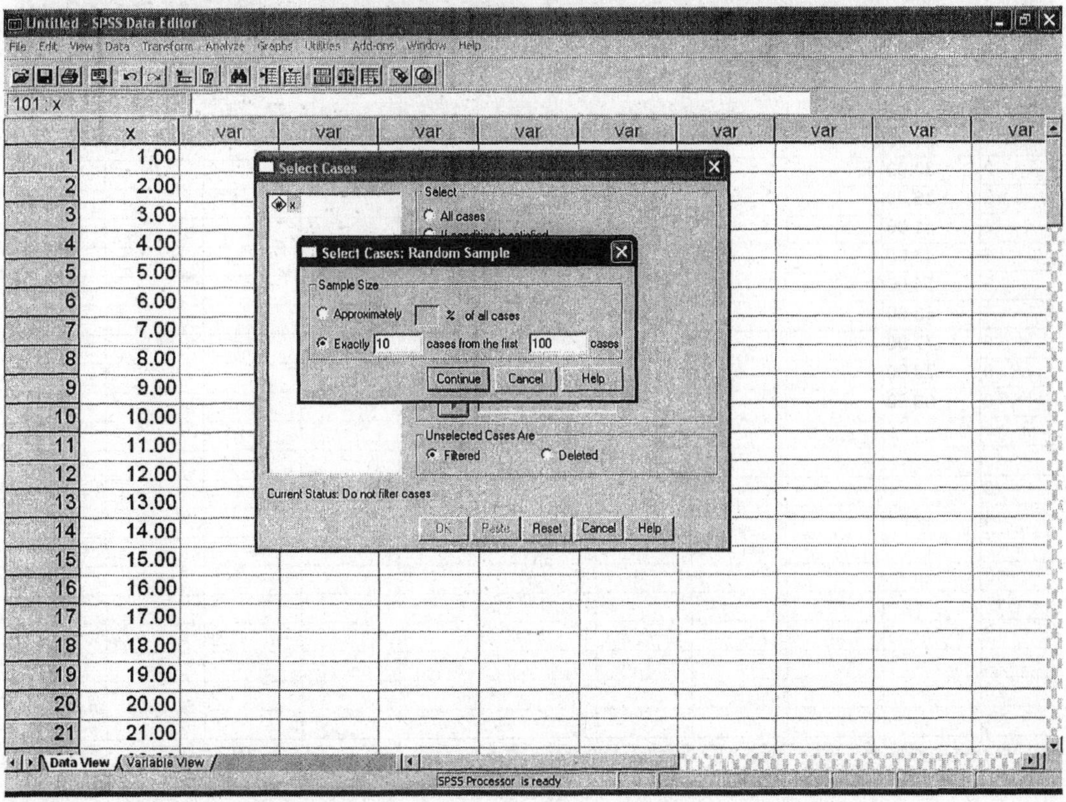

Now click on Continue. Then click on OK. The results follow:

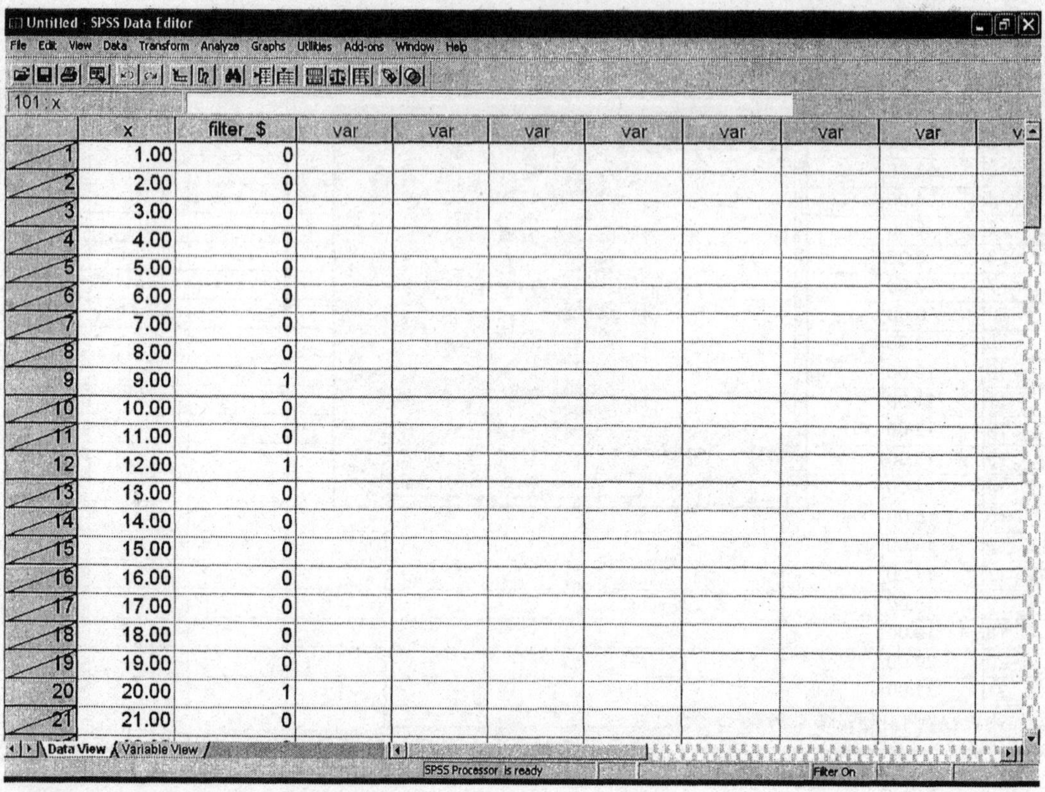

	x	filter_$	var	var	var	var	var	var	var	v
1	1.00	0								
2	2.00	0								
3	3.00	0								
4	4.00	0								
5	5.00	0								
6	6.00	0								
7	7.00	0								
8	8.00	0								
9	9.00	1								
10	10.00	0								
11	11.00	0								
12	12.00	1								
13	13.00	0								
14	14.00	0								
15	15.00	0								
16	16.00	0								
17	17.00	0								
18	18.00	0								
19	19.00	0								
20	20.00	1								
21	21.00	0								

Let us now check Deleted in the ➤**Data**➤**Select Cases** dialog box. This way only the selected cases will show up as shown below. Note that results varies from sampling to sampling.

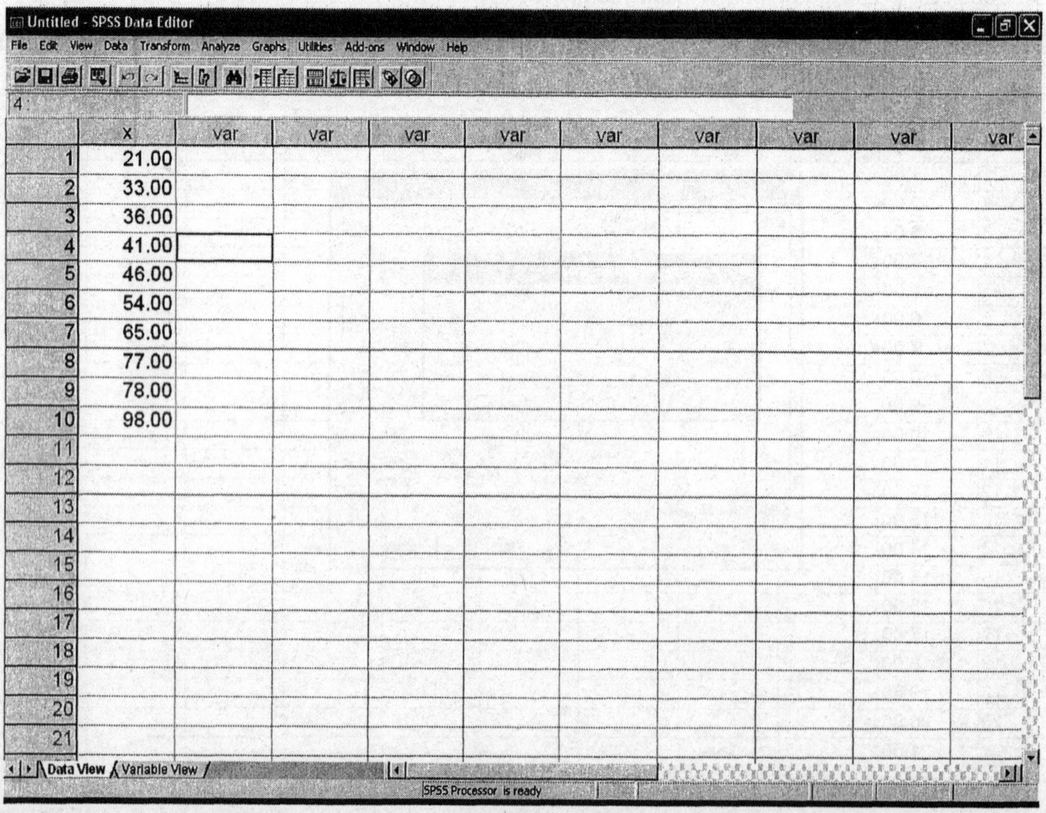

	x	var	var	var	var	var	var	var	var
1	21.00								
2	33.00								
3	36.00								
4	41.00								
5	46.00								
6	54.00								
7	65.00								
8	77.00								
9	78.00								
10	98.00								
11									
12									
13									
14									
15									
16									
17									
18									
19									
20									
21									

LAB ACTIVITIES FOR RANDOM SAMPLES

1. Out of a population of 8173 eligible count residents, select a random sample of 50 for prospective jury duty. Should you sample with or without replacement? Use the **TRUNC(RV.UNIFORM(min,max))** function to generate the sample. Use sorting procedure to sort the data so that you can check for repeated values. If necessary, repeat the procedure again to continue sampling until you have 50 different people.

2. Retrieve the SPSS data Svls02.sav on the CD-ROM. This file contains weights of a random sample of linebackers on professional football teams. The data is in Column 1. Use the menu options ➤**Data**➤**Select Cases** to take a random sample of 10 of these weights. Print the 10 weights included in the sample.

Simulating experiments in which outcomes are equally likely is another important use of random numbers.

3. We can simulate dealing bridge hands by numbering the cards in a bridge deck from 1 to 52. Then we draw a random sample of 13 numbers without replacement from the population of 52 numbers. A bridge deck has 4 suits: hearts, diamonds, clubs, and spades. Each suit contains 13 cards: those numbered 2 through 10, a jack, a queen, a king, and an ace. Decide how to assign the numbers 1 through 52 to the cards in the deck.

   (a) Use **TRUNC(RV.UNIFORM(min,max))** to get the numbers of the 13 cards in one hand. Translate the numbers into cards and tell what cards are in the hand. For a second game, the cards would be collected and reshuffled. Use the computer to determine the hand you might get in a second game.

   (b) Store the 52 cards in the first column, and then use ➤**Data**➤**Select Cases** to sample 13 cards. Print the results. Repeat this process to determine the hand you might get in a second game.

   (c) Compare the four hands you have generated. Are they different? Would you expect this result?

4. We can also simulate the experiment of tossing a fair coin. The possible outcomes resulting from tossing a coin are heads or tails. Assign the outcome heads the number 2 and the outcome tails the number 1. Use **TRUNC(RV.UNIFORM(min,max))** to simulate the act of tossing a coin 10 times. Use Frequency table to tally the results. Repeat the experiment with 10 tosses. Do the percents of outcomes seem to change? Repeat the experiment again with 100 tosses.

CHAPTER 2 ORGANIZING DATA

HISTOGRAMS (SECTION 2.2 OF *UNDERSTANDABLE STATISTICS*)

SPSS has two menu options for drawing histograms. The option ➤**Graphs**➤**Histogram** draws a histogram with default choice of boundaries (cutpoints), and the option ➤**Graphs**➤**Interactive** ➤**Histogram** allows user to define boundaries of the histogram. Histograms may also be drawn through ➤**Analyze**➤**Descriptive Statistics** ➤ **Frequencies**.

➤**Graphs**➤**Histogram**

Dialogue Box Responses

Select the variable to draw the histogram.

There are options for giving a title or displaying a normal curve with the histogram.

Click on OK to construct and display the histogram.

➤**Analyze**➤**Descriptive Statistics** ➤ **Frequencies**

Dialogue Box Responses

Select variable.

Click on Charts, another dialogue box will show up. Click Histogram in this dialogue box. Click on Continue.

Click on OK. The histogram will be displayed together with the frequency table.

➤**Graphs**➤**Interactive** ➤**Histogram** This menu option allows user to define the interval and cutpoints.

Dialogue Box Responses

Drag the variable into the domain target field.

Click on Options button to get into the options tab.

Under Scale Range choose the variable of the histogram (the same variable in the domain target field).

Uncheck Auto, this will allow you to enter Minimum (first cut point, that is, the lower bound of the first interval) and the Maximum (last cut point, that is, the upper bound of the last interval).

Click on Histogram button to get into the histogram tab.

Uncheck "Select interval size automatically", then either enter your choice of number of intervals or enter your choice of interval width.

Click on OK.

With SPSS, data falling on a boundary are counted in the class below the boundary.

Example

Let's make a histogram of the data we stored in the data file ads (created in Chapter 1). We'll use the variable Ad_Count (in the first column). This column contains the number of ads per hour on prime time TV.

First we need to retrieve the data file. Use ➤**File➤Open ➤ Data.** Scroll to the drive containing the worksheet. We used 3½ disk drive A. Click on the file.

The number of ads per hour of TV is in the first column under variable Ad_Count. Let us first use ➤**Graphs➤Histogram.** The dialogue box follows.

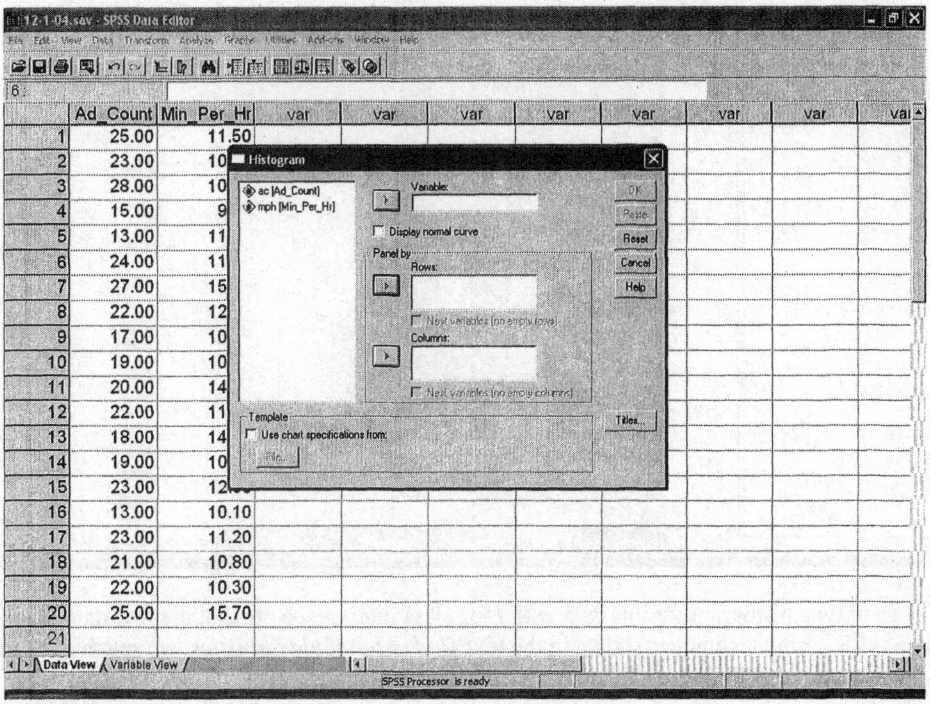

Select Ad_Count and enter it under Variable, as shown below.

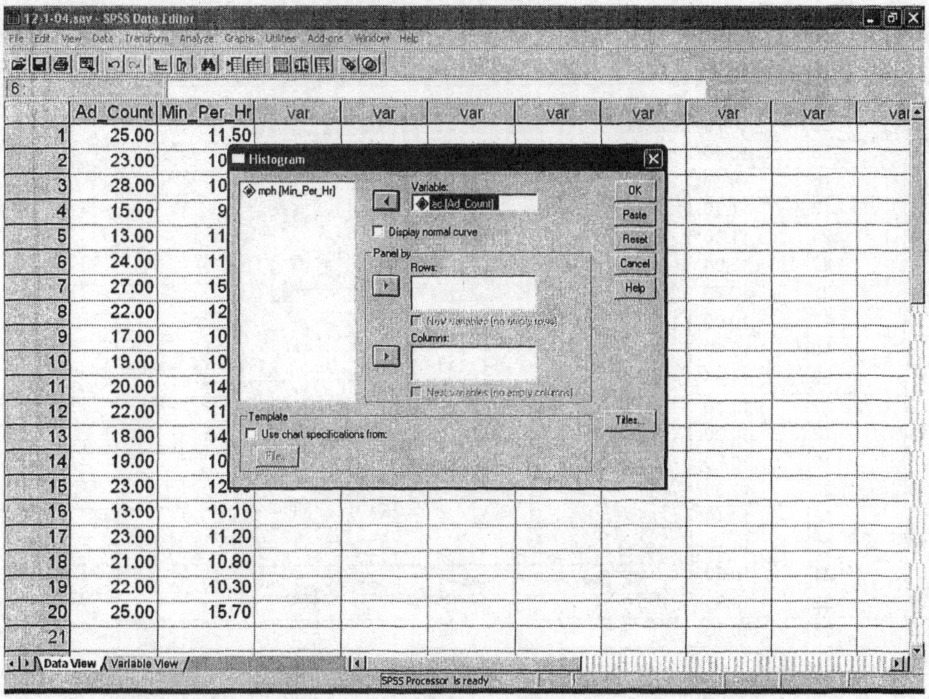

Click on OK. The following histogram with automatically selected classes will be displayed.

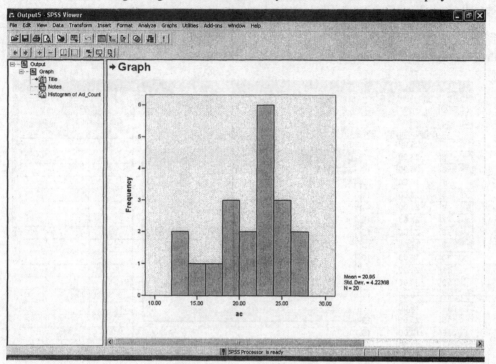

Now, let us draw a histogram for this data with four class intervals. Note that the low data value is 13 and the high value is 28. Using techniques shown in the text *Understandable Statistics*, we see that the class boundaries for 4 classes are 12.5; 16.5; 20.5; 24.5; 28.5. Use **➤Graphs➤Interactive ➤Histogram.** The dialog boxes follows.

Now drag the variable Ad_Count into the domain target field as shown below.

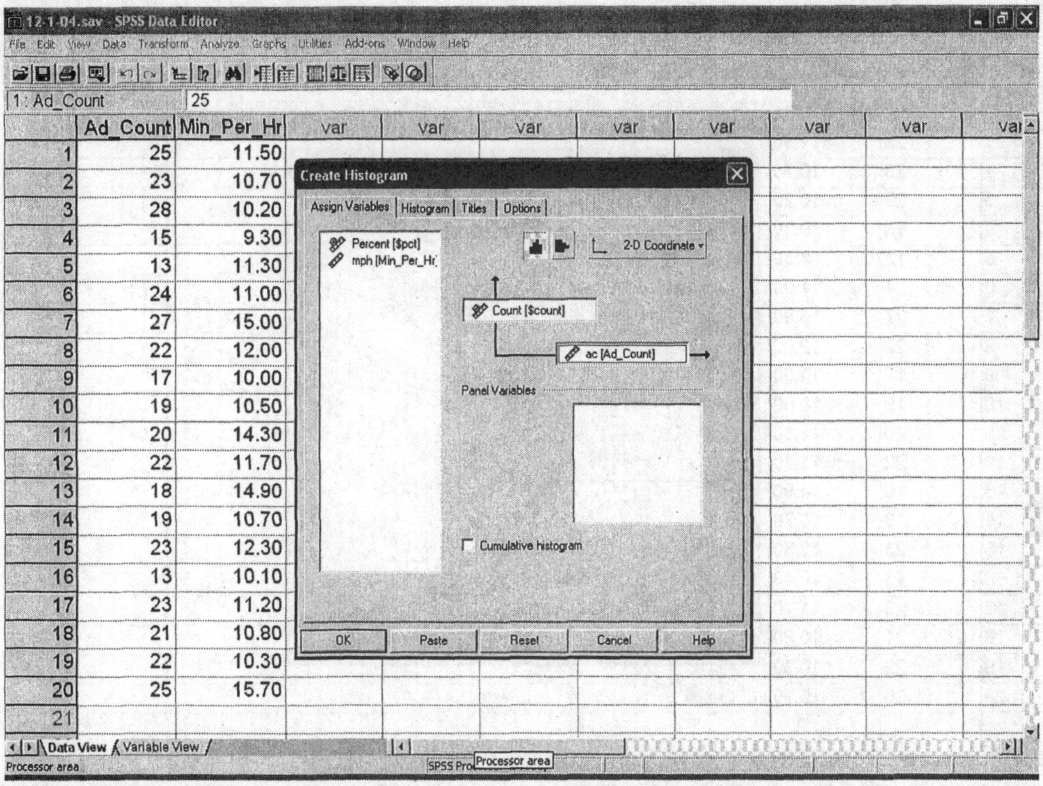

	Ad_Count	Min_Per_Hr	var	var	var	var	var	var	var	var
1	25	11.50								
2	23	10.70								
3	28	10.20								
4	15	9.30								
5	13	11.30								
6	24	11.00								
7	27	15.00								
8	22	12.00								
9	17	10.00								
10	19	10.50								
11	20	14.30								
12	22	11.70								
13	18	14.90								
14	19	10.70								
15	23	12.30								
16	13	10.10								
17	23	11.20								
18	21	10.80								
19	22	10.30								
20	25	15.70								
21										

Click on Options button to get into the options tab. Under Scale Range choose the variable ac (which is the label for Ad_Count). Uncheck Auto. Then enter 12.5 as Minimum and 28.5 as Maximum.

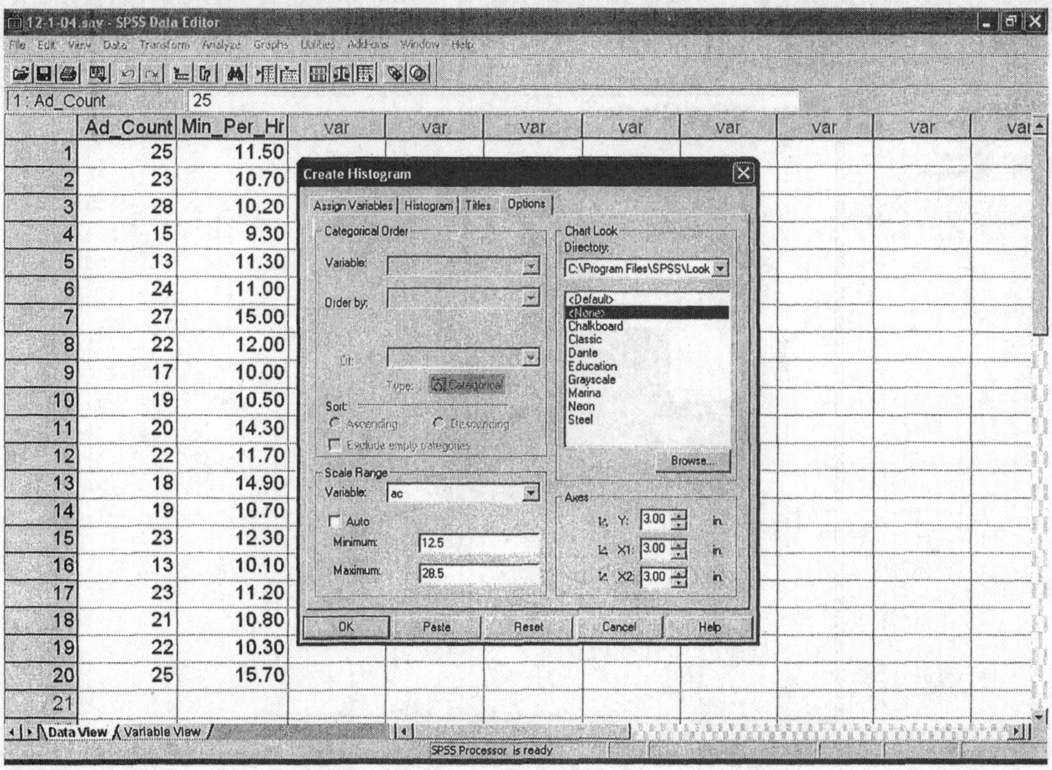

	Ad_Count	Min_Per_Hr	var	var	var	var	var	var	var	var
1	25	11.50								
2	23	10.70								
3	28	10.20								
4	15	9.30								
5	13	11.30								
6	24	11.00								
7	27	15.00								
8	22	12.00								
9	17	10.00								
10	19	10.50								
11	20	14.30								
12	22	11.70								
13	18	14.90								
14	19	10.70								
15	23	12.30								
16	13	10.10								
17	23	11.20								
18	21	10.80								
19	22	10.30								
20	25	15.70								
21										

Now, click on Histogram button to get into the histogram tab. Uncheck "Select interval size automatically", select Number of intervals, enter 4.

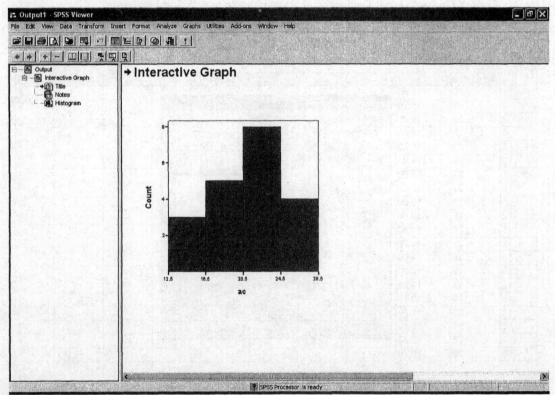

Click on OK. The histogram follows.

LAB ACTIVITIES FOR HISTOGRAMS

1. The ads data file contains a second column of data that records the number of minutes per hour consumed by ads during prime time TV. Retrieve the ads data file again and use the second column (under variable Min_Per_Hr) to

   (a) make a histogram, letting the computer scale it.

   (b) sort the data and find the smallest data value.

   (c) make a histogram using the smallest data value as the starting value and an increment of 4 minutes. Do this by setting the smallest value as the Minimum in the options tab and choosing 4 as the Width of interval in the histogram tab. You also need to choose an appropriate number as the Maximum in the options tab (Maximum = Minimum + Width*(number of intervals)).

2. As a project for her finance class, Linda gathered data about the number of cash requests made at an automatic teller machine located in the student center between the hours of 6 P.M. and 11 P.M. She recorded the data every day for four weeks. The data values follow.

25	17	33	47	22	32	18	21	12	26	43	25
19	27	26	29	39	12	19	27	10	15	21	20
34	24	17	18								

   (a) Enter the data.

   (b) Make a histogram for this data, letting the computer scale it.

   (c) Sort the data and identify the low and high values. Use the low value as the start value and an increment of 10 to make another histogram.

3. Choose one of the following files from the CD-ROM.

   Disney Stock Volume: **Svls01.sav**

   Weights of Pro Football Players: **Svls02.sav**

   Heights of Pro Basketball Players: **Svls03.sav**

   Miles per Gallon Gasoline Consumption: **Svls04.sav**

   Fasting Glucose Blood Tests: **Svls05.sav**

   Number of Children in Rural Canadian Families: **Svls06.sav**

   (a) Make a histogram, letting SPSS scale it.

   (b) Make a histogram using five classes.

4. Histograms are not effective displays for some data. Consider the data

1	2	3	6	7	4	7	9	8	4	12	10

1	9	1	12	12	11	13	4	6	206

Enter the data and make a histogram, letting SPSS do the scaling. Next scale the histogram with starting value 1 and increment 20. Where do most of the data values fall? Now drop the high value 206 from the data. Do you get more refined information from the histogram by eliminating the high and unusual data value?

STEM-AND-LEAF DISPLAYS (SECTION 2.3 OF *UNDERSTANDABLE STATISTICS*)

SPSS supports many of the exploratory data analysis methods. You can create a stem-and-leaf display with the following menu choices.

➤Analyze➤Descriptive Statistics ➤ Explore

Dialogue Box Responses

Dependent List: Enter variables for making the display.

Display: Choose Plots under Display.

Then, click on the button Plots. Another dialog box will show up.

Select Stem-and –leaf. Click on Continue.

Click on OK.

Example

Let's take the data in the data file Ads and make a stem-and-leaf display of Ad_Count. Recall that this variable contains the number of ads occurring in an hour of prime time TV.

Use the menu **➤Analyze➤Descriptive Statistics ➤ Explore.** In the dialog box, enter Ad_Count into the dependent list. Choose Plots under Display.

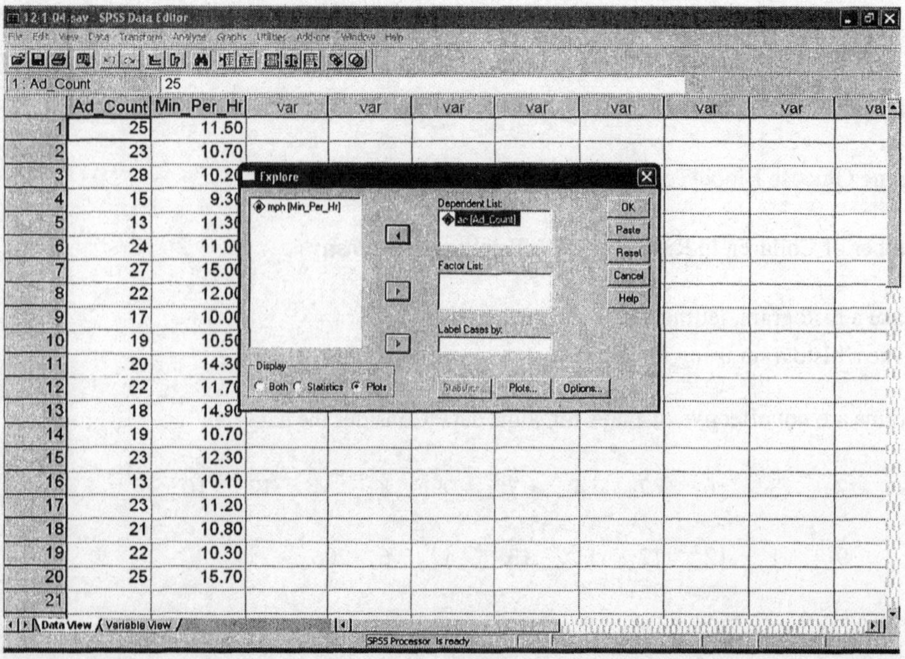

Now click on the button Plots. Another dialog box shows up. Select Stem-and-leaf.

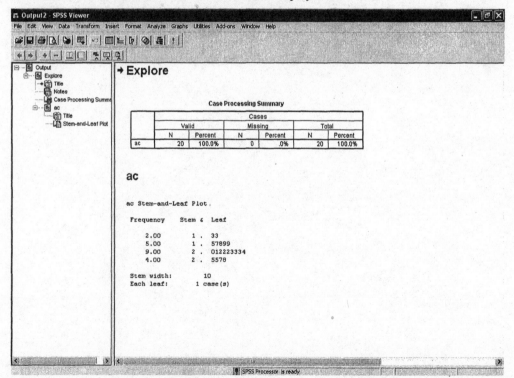

Click on Continue. Then click on OK. The stem-and-leaf display follows.

The first column gives the depth of the data, that is, the number of data in this line. The second column gives the stem and the last gives the leaves. The display has 2 lines per stem. That means that leaves 0–4 are on one line and leaves 5–9 are on the next.

LAB ACTIVITIES FOR STEM-AND-LEAF DISPLAYS

1. Retrieve data file Ads again, and make a stem-and-leaf display of the data in the second column. This data gives the number of minutes of ads per hour during prime time TV programs.

2. In a physical fitness class students ran 1 mile on the first day of class. These are their times in minutes.

12	11	14	8	8	15	12	13	12
10	8	9	11	14	7	14	12	9
13	10	9	12	12	13	10	10	9
12	11	13	10	10	9	8	15	17

   (a) Enter the data in a data sheet.
   (b) Make a stem-and-leaf display.

CHAPTER 3 AVERAGES AND VARIATION

AVERAGES AND STANDARD DEVIATION OF UNGROUPED DATA (SECTIONS 3.1 AND 3.2 OF *UNDERSTANDABLE STATISTICS*)

The menu options ➤**Analyze**➤**Descriptive Statistics** of SPSS gives many of the summary statistics described in *Understandable Statistics*. There are several sub options under ➤**Analyze**➤**Descriptive Statistics.** Here we describe the use of ➤**Analyze**➤**Descriptive Statistics** ➤ **Frequencies.**

➤**Analyze**➤**Descriptive Statistics** ➤ **Frequencies** gives the frequency table and also prints descriptive statistics for the variable (column).

Dialogue Box Response

Variables: enter the variable names that contain the data.

In case you do not want to see the frequency table, uncheck the button Display frequency tables.

Click the button Statistics. A new dialogue box will show up for the user to descriptive statistics to be displayed.

The choice for descriptive statistics to be displayed include

N	number of data in the variable (column)
	number of missing data in the variable (column)
Mean	arithmetic mean of the variable (column)
S. E. mean	standard error of the mean, **Std Deviation/SQRT(N)** (we will use it in Chapter 7)
Median	median or center of the data in the variable (column)
Std Deviation	the sample standard deviation of the variable, *s*
Minimum	minimum data value in the variable (column)
Maximum	maximum data value in the variable (column)
Quartiles	25 (Q1, that is, 1st quartile of distribution in the variable)
	50 (median)
	75 (Q3, that is, 3rd quartile of distribution in the variable)

(Q1 and Q3 are similar to Q_1 and Q_3 as discussed in Section 3.4 of *Understandable Statistics*. However, the computation process is slightly different and gives values slightly different from those in the text.)

Example

Let's again consider the data about the number and duration of ads during prime time TV. We will retrieve worksheet ads and use ➤**Analyze**➤**Descriptive Statistics** ➤ **Frequencies** on the variable Min_Per_Hr, the number of minutes per hour of ads during prime time TV.

First use ➤**File**➤**Open** ➤**Data** to open the data file ads.

Next use ➤**Analyze**➤**Descriptive Statistics** ➤ **Frequencies**. Select Min_Per_Hr and click on OK.

	Ad_Count	Min_Per_Hr
11	20	14.30
12	22	11.70
13	18	14.90
14	19	10.70
15	23	12.30
16	13	10.10
17	23	11.20
18	21	10.80
19	22	10.30
20	25	15.70

After that, uncheck the button Display frequency tables. Then click the button Statistics. A new dialogue box shows up:

	Ad_Count	Min_Per_Hr
11	20	14.30
12	22	11.70
13	18	14.90
14	19	10.70
15	23	12.30
16	13	10.10
17	23	11.20
18	21	10.80
19	22	10.30
20	25	15.70

Now, check the buttons Mean, Median, Std Deviation, Minimum, Maximum, S. E. mean, and Quartiles. Click on Continue. Then click on OK. The results follow.

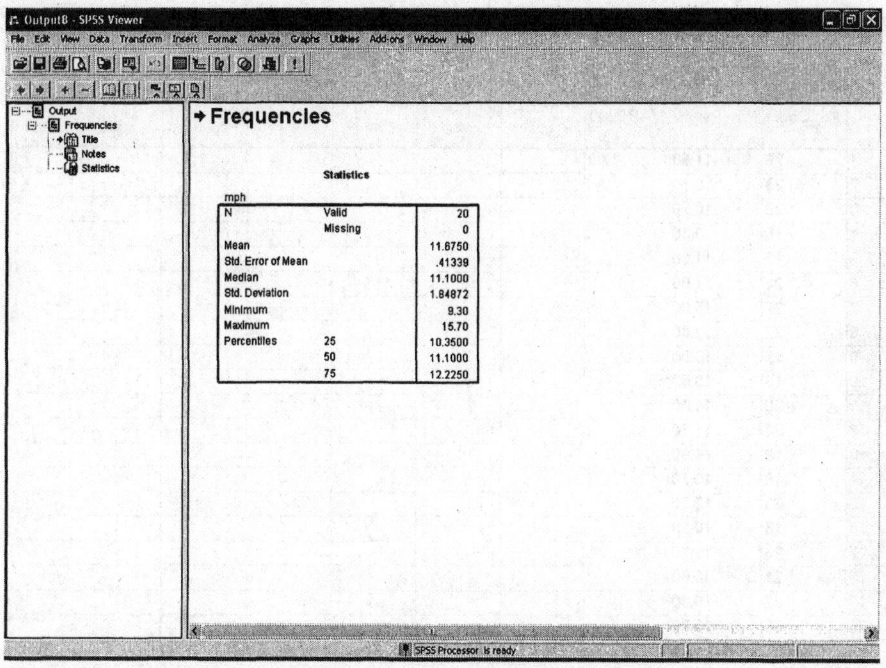

ARITHMETIC IN SPSS

The standard deviation given in Std. Deviation is the sample standard deviation

$$s = \sqrt{\frac{\sum(x - \bar{x})^2}{N - 1}}$$

We can compute the population standard deviation σ by multiplying s by the factor below:

$$\sigma = s\sqrt{\frac{N - 1}{N}}$$

MINITAB allows us to do such arithmetic. Use the built-in calculator under menu selection ➤**Transform**➤**Compute.** Note that * means multiply and ** means exponent. Also note that SPSS does computations by variables (columns). Therefore, when we want to compute a certain value and store it in a certain cell, we need to indicate that by using the "if" option under ➤**Transform**➤**Compute.**

Example

Let's use the arithmetic operations to evaluate the population standard deviation and population variance for the minutes per hour of TV ads. Notice that the sample standard deviation $s = 1.849$ and the sample size is 20.

First in the first cell of the third column, enter the number 1. We will use this in the "if" option to indicate that the result will be stored in the first cell of the fourth column. Note that SPSS automatically gives a default variable name VAR00001 to this column. Also note that since the sample size is 20 in this data file, the other 19 cells in the third column are filled with "." as missing values, as shown below.

Now use ➤**Transform**➤**Compute.** Enter a name for the Target Variable, say x. Then enter the expression for the population standard deviation as the Numeric Expression.

Now click on the button If. Another dialogue box shows up. Select "Include if case satisfies condition". Then enter VAR00001 = 1 in the condition box.

Click on continue. Then click on OK. The result follows.

LAB ACTIVITIES FOR AVERAGES AND STANDARD DEVIATION OF UNGROUPED DATA

1. A random sample of 20 people were asked to dial 30 telephone numbers each. The incidences of numbers misdialed by these people follow:

| 3 | 2 | 0 | 0 | 1 | 5 | 7 | 8 | 2 | 6 |
| 0 | 1 | 2 | 7 | 2 | 5 | 1 | 4 | 5 | 3 |

Enter the data and use the menu selections ➤**Analyze**➤**Descriptive Statistics** ➤ **Frequencies** to find the mean, median, minimum value, maximum value, and standard deviation.

2. Consider the test scores of 30 students in a political science class.

85	73	43	86	73	59	73	84	100	62
75	87	70	84	97	62	76	89	90	83
70	65	77	90	94	80	68	91	67	79

(a) Use the menu selections ➤**Analyze**➤**Descriptive Statistics** ➤ **Frequencies** to find the mean, median, minimum value, maximum value, and standard deviation.

(b) Greg was in political science class. Suppose he missed a number of classes because of illness, but took the exam anyway and made a score of 30 instead of 85 as listed in the data set. Change the 85 (first entry in the data set) to 30 and use the above menu selections again. Compare the new mean, median and standard deviation with the ones in part (a). Which average was most affected: median or mean? What about the standard deviation?

3. Consider the 10 data values

| 4 | 7 | 3 | 15 | 9 | 12 | 10 | 2 | 9 | 10 |

(a) Use the menu selections ➤**Analyze**➤**Descriptive Statistics** ➤ **Frequencies** to find the sample standard deviation of these data values. Then, following the second example in this chapter as a model, find the population standard deviation of these data. Compare the two values.

(b) Now consider these 50 data values in the same range.

7	9	10	6	11	15	17	9	8	2
2	8	11	15	14	12	13	7	6	9
3	9	8	17	8	12	14	4	3	9
2	15	7	8	7	13	15	2	5	6
2	14	9	7	3	15	12	10	9	10

Again use the menu selections ➤**Analyze**➤**Descriptive Statistics** ➤ **Frequencies** to find the sample standard deviation of these data values. Then, as above, find the population standard deviation of these data. Compare the two values.

(c) Compare the results of parts (a) and (b). As the sample size increases, does it appear that the difference between the population and sample standard deviations decreases? Why would you expect this result from the formulas?

4. In this problem we will explore the effects of changing data values by multiplying each data value by a constant, or by adding the same constant to each data value.

(a) Make sure you start a new data sheet. Enter the following data into the first column with variable name C1:

| 1 | 8 | 3 | 5 | 7 | 2 | 10 | 9 | 4 | 6 | 32 |

Use the menu selections ➤**Analyze**➤**Descriptive Statistics** ➤ **Frequencies** to find the mean, median, minimum and maximum values and sample standard deviation.

(b) Now use the calculator box to create a new column of data C2 = 10*C1. Use menu selections again to find the mean, median, minimum and maximum values, and sample standard deviation of C2. Compare these results to those of C1. How do the means compare? How do the medians compare? How do the standard deviations compare? Referring to the formulas for these measures (see Sections 3.1 and 3.2 of *Understandable Statistics*), can you explain why these statistics behaved the way they did? Will these results generalize to the situation of multiplying each data entry by 12 instead of 10? Confirm your answer by creating a new variable C3 that has each datum of C1 multiplied by 12. Predict the corresponding statistics that would occur if we multiplied each datum of C1 by 1000. Again, create a new variable C4 that does this, and use ➤**Analyze**➤**Descriptive Statistics** ➤ **Frequencies** to confirm your prediction.

(c) Now suppose we add 30 to each data value in C1. We can do this by using ➤**Transform**➤**Compute** to create a new column of data C6 = C1 + 30. Use menu selection ➤**Analyze**➤**Descriptive Statistics** ➤ **Frequencies** on C6 and compare the mean, median, and standard deviation to those shown for C1. Which are the same? Which are different? Of those that are different, did each change by being 30 more than the corresponding value of part (a)? Again look at the formula for the standard deviation. Can you predict the observed behavior from the formulas? Can you generalize these results? What if we added 50 to each datum of C1? Predict the values for the mean, median, and sample standard deviation. Confirm your predictions by creating a variable C7 in which each datum is 50 more than that in the respective position of C1. Use menu selections ➤**Analyze**➤**Descriptive Statistics** ➤ **Frequencies** on C7.

(d) Rename C1 (renaming a variable can be done in the variable view of the data editor screen) as 'orig', C2 as 'T10', C3 as 'T12', C4 as 'T1000', C6 as 'P30', C7 as 'P50'. Now use the menu selections ➤**Analyze**➤**Descriptive Statistics** ➤ **Frequencies** on all these variables simultaneously and look at the display. Is it easier to compare the results this way?

BOX-AND-WHISKER PLOTS (SECTION 3.4 OF *UNDERSTANDABLE STATISTICS*)

The box-and-whisker plot is another of the explanatory data analysis techniques supported by SPSS. With SPSS Version 13, unusually large or small values are displayed beyond the whisker and labeled as outliers by a *. The upper whisker extends to the highest data value within the upper limit, here the upper limit = Q3 + 1.5 (Q3 − Q1). Similarly, the lower whisker extends to the lowest value within the lower limit, and the lower limit = Q1− 1.5 (Q3 − Q1). By default, the top of the box is the third quartile (Q3) and the bottom of the box is the first quartile (Q1). The line in the box indicates the value of median.

The menu selections are

➤**Graphs**➤**Interactive**➤**Boxplot**

Dialogue Box Responses:

Choose display of boxplot: vertical or horizontal, etc.

Drag the variable into the corresponding (vertical or horizontal) variable box.

Titles: get into the Titles tab and you can title the graph.

Boxes: get into the Boxes tab and check Outliers, Extremes, Median line, and select Whisker Caps.

There are other options available with ➤**Graphs**➤**Interactive**➤**Boxplot**. See Help to learn more about these options.

Example

Now let's make a box-and-whisker plot of the data stored in worksheet ads. Ad_Count contains the number of ads per hour of prime time TV, while Min_Per_Hr contains the duration per hour of the ads.

Use the menu selection ➤**Graphs**➤**Interactive**➤**Boxplot.** Choose the vertical display. Drag the variable Ad_Count into the corresponding variable box.

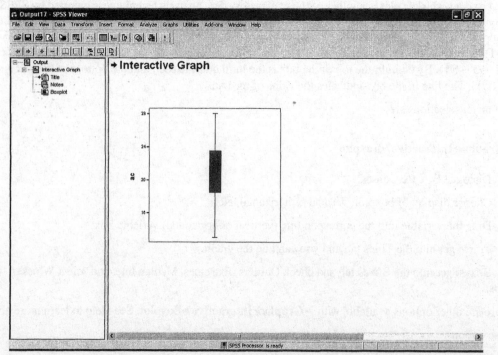

Get into the Boxes tab and check Outliers, Extremes, Median line, and select a Whisker Cap display. Click OK. The result follows.

LAB ACTIVITIES FOR BOX-AND-WHISKER PLOTS

1. State-regulated nursing homes have a requirement that there be a minimum of 132 minutes of nursing care per resident per 8-hr shift. During an audit of Easy Life Nursing home, a random sample of 30 shifts showed the number of minutes of nursing care per resident per shift to be

200	150	190	150	175	90	195	115	170	100
140	270	150	195	110	145	80	130	125	115
90	135	140	125	120	130	170	125	135	110

    (a) Enter the data.

    (b) Make a box-and-whisker plot. Are there any unusual observations?

    (c) Make a stem-and-leaf plot. Compare the two ways of presenting the data.

    (d) Make a histogram. Compare the information in the histogram with that in the other two displays.

    (e) Use the ➤**Analyze**➤**Descriptive Statistics** ➤ **Frequencies** menu selections to find descriptive statistics for this data.

    (f) Now remove any data beyond the outer fences. Do this by using ➤**Data**➤**Select Cases**. Then use the menu selections ➤**Analyze**➤**Descriptive Statistics** ➤ **Frequencies** on this data. How do the means compare?

    (h) Pretend you are writing a brief article for a newspaper. Describe the information about the time nurses spend with residents of a nursing home. Use non-technical terms. Be sure to make some comments about the "average" of the data measurements and some comments about the spread of the data.

2. Select one of these data files from the CD-ROM and repeat parts (b) and (h).

    Disney Stock Volume: **Svls01.sav**

    Weights of Pro Football Players: **Svls02.sav**

    Heights of Pro Basketball Players: **Svls03.sav**

    Miles per Gallon Gasoline Consumption: **Svls04.sav**

    Fasting Glucose Blood Tests: **Svls05.sav**

    Number of Children in Rural Canadian Families: **Svls06.sav**

CHAPTER 4 ELEMENTARY PROBABILITY THEORY

RANDOM VARIABLES AND PROBABILITY

SPSS supports random samples from a column of numbers or from many probability distributions. By using some of the same techniques shown in Chapter 1 of this guide for random samples, you can simulate a number of probability experiments.

Example

Simulate the experiment of tossing a fair coin 200 times. Look at the percent of heads and the percent of tails. How do these compare with the expected 50% of each?

Assign the outcome heads to digit 1 and tails to digit 2. We will follow the procedure described in Chapter 1 to generate a random sample of size 200 from the integers {1,2} under the assumption that each of them is equally likely to occur. We can do this through these steps:

1) In the data editor, choose a variable name, say Coin, for the first column. Move the curser to the 200th cell of this column and just enter a number (any number). This simply defines the data size of this column to be 200. SPSS will sign "." (missing value) to the 199 cells above it.

2) Use the menu options **➤Transform➤Compute.** In the dialog box, first type in Coin as the Target Variable, then use **TRUNC(RV.UNIFORM(1, 3)).**

The result follows.

	Coin	var	var	var	var	var	var	var	var	var
1	1.00									
2	1.00									
3	2.00									
4	1.00									
5	1.00									
6	2.00									
7	1.00									
8	1.00									
9	1.00									
10	1.00									
11	1.00									
12	1.00									
13	2.00									
14	1.00									
15	2.00									
16	2.00									
17	2.00									
18	1.00									
19	2.00									
20	1.00									
21	1.00									

To tally the results use **➤Analyze➤Descriptive Statistics ➤ Frequencies** as described in Chapter 2. The results follow. (Your results will vary.)

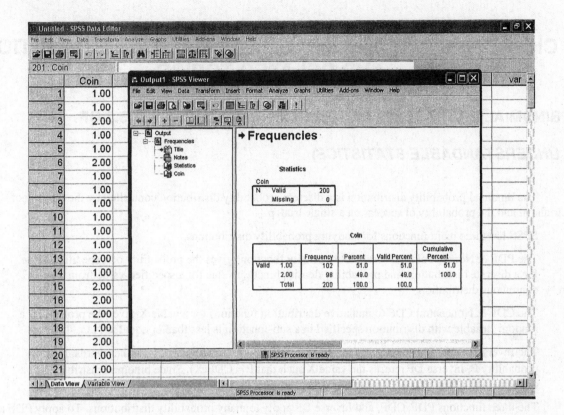

LAB ACTIVITIES FOR RANDOM VARIABLES AND PROBABILITY

1. Use the steps described in this chapter to simulate 50 tosses of a fair coin. Use ➤**Analyze**➤**Descriptive Statistics** ➤ **Frequencies** to find the percent of each outcome. Compare the result with the theoretical expected percents (50% heads, 50% tails). Repeat the process for 1000 trials. Are these outcomes closer the results predicted by the theory?

2. We can use the steps given in this chapter with integer values from 1 to 6 to simulate the experiment of rolling two dice 50 times and recording each sum. Use variable Die1 (in the first column) to store outcomes of die 1, and Die2 to store outcomes of die 2. Use ➤**Transform**➤**Compute** to compute the sum of the dice and store the sum into the third column under variable name Sum. Then use ➤**Analyze**➤**Descriptive Statistics** ➤ **Frequencies** to record the percent of each outcome. Repeat the process for 1000 rolls of the dice.

CHAPTER 5 THE BINOMIAL PROBABILITY DISTRIBUTION AND RELATED TOPICS

BINOMIAL PROBABILITY DISTRIBUTIONS (SECTIONS 5.2, 5.3 OF

UNDERSTANDABLE STATISTICS)

The binomial probability distribution is a discrete probability distribution controlled by the number of trials, n, and the probability of success on a single trial, p.

SPSS has three main functions for studying probability distributions.

The PDF & Noncentral PDF (probability density function) gives the probability of a specified value for a discrete distribution, and probability density function value for a specified value from a continuous distribution.

The CDF & Noncentral CDF (cumulative distribution function) for a value X gives the probability a random variable with distribution specified in a sub-function is less than or equal to X.

The Inverse DF gives the inverse of the CDF for continuous distributions. In other words, for a probability P, Inverse DF returns the value X such that $P \approx CDF(X)$. Since binomial distribution is a discrete distribution, it is not covered by this function.

The three functions PDF, CDF, and Inverse DF apply to many probability distributions. To apply PDF and CDF to a binomial distribution, we need to use the menu selections **Transform➤Compute** followed by function selections.

Transform➤Compute

Dialog Box Responses

Enter name of the Target Variable

Enter the function formula into the Numeric Expression box:

Under the Function group, select PDF & Noncentral PDF for probability; CDF & Noncentral CDF for cumulative probability for CDF; Inverse DF for inverse cumulative probability.

Then under Functions and Special Variables, select the sub-function for the specified distribution.

For example, the PDF function for a binomial distribution is **PDF.BINOM(quant, n, prob)** and the CDF function for a binomial distribution is **CDF.BINOM(quant, n, prob)**. Here **n** is the number of trials, that is, the value of n in a binomial experiment. **prob** is the probability of success, that is, the value of p, the probability of success on a single trial. The input **quant** is the values of r, the number of successes in a binomial experiment. You may enter a value for **quant**, or you may store the values for **quant** in a variable (column) and enter the variable name for **quant**.

Example

A surgeon performs a difficult spinal column operation. The probability of success of the operation is $p = 0.73$. Ten such operations are scheduled. Find the probability of success for 0 through 10 successes out of these ten operations.

First enter the possible values of r, 0 through 10, in the first column and name this variable r (Choose 0 under Decimals in the Variable View to have integers for r). We will put the probabilities in the second column, so name the column *prob* (Choose 6 under Decimals in the Variable View to give enough precision.) Fill in the dialog box as shown below.

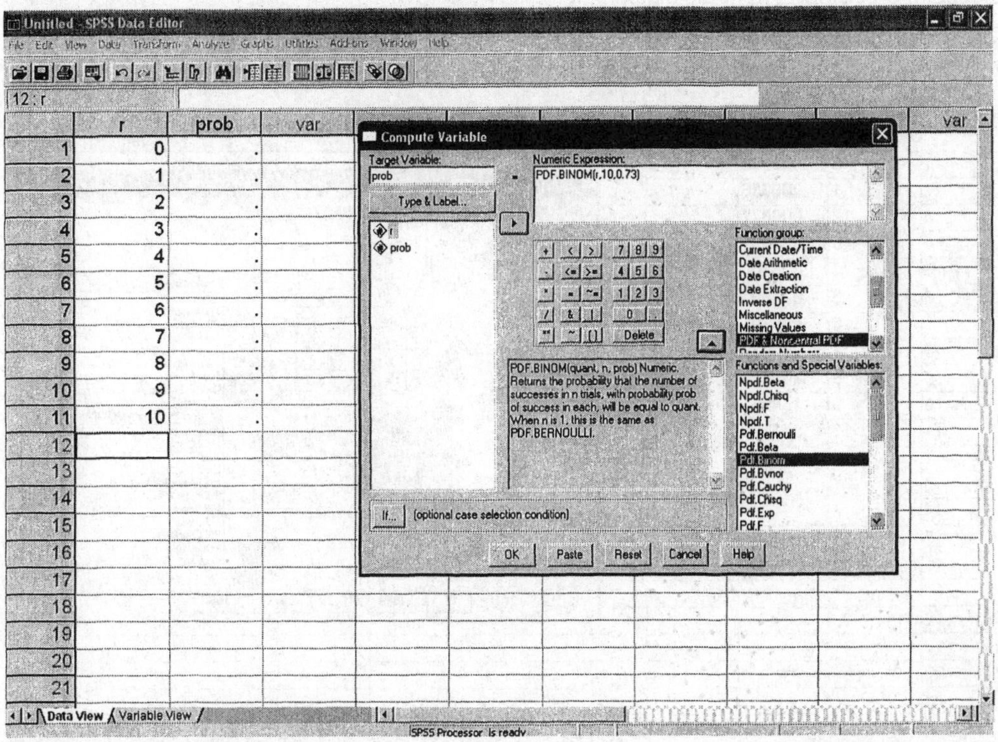

Click on OK. The results follow.

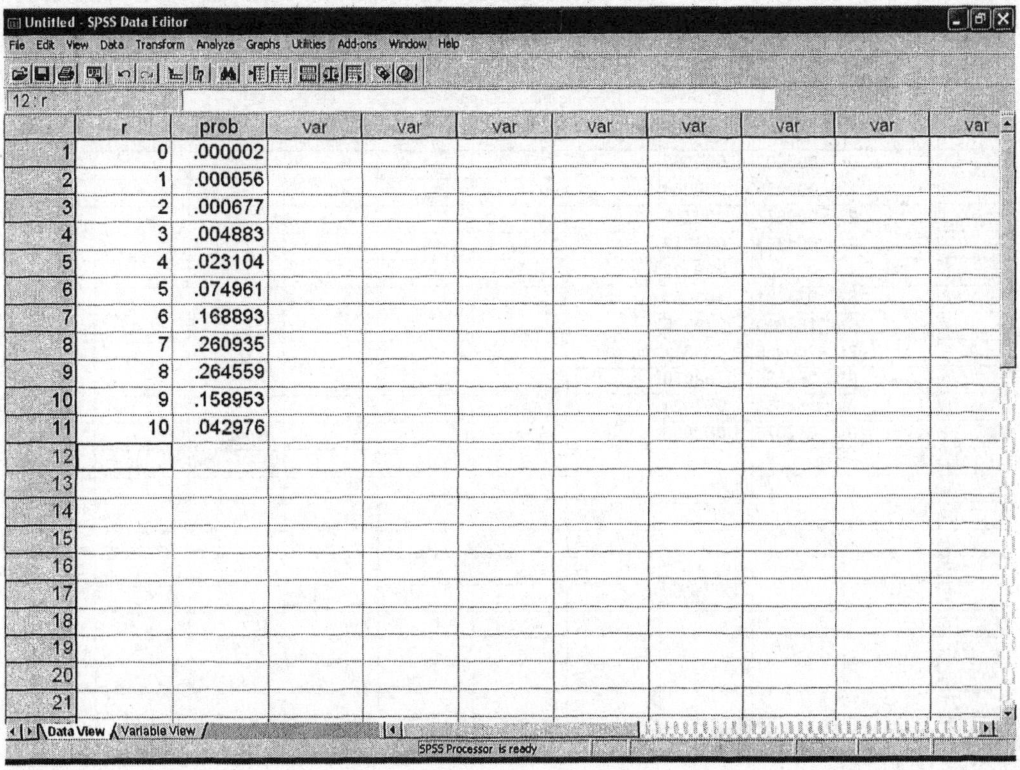

Next use the **CDF.BINOM** function to find the probability of *r* or fewer successes. Let us put the probabilities in the third column and name it *cprob* (Choose 6 under Decimals in the Variable View to give enough precision.) Fill in the dialog box as shown below.

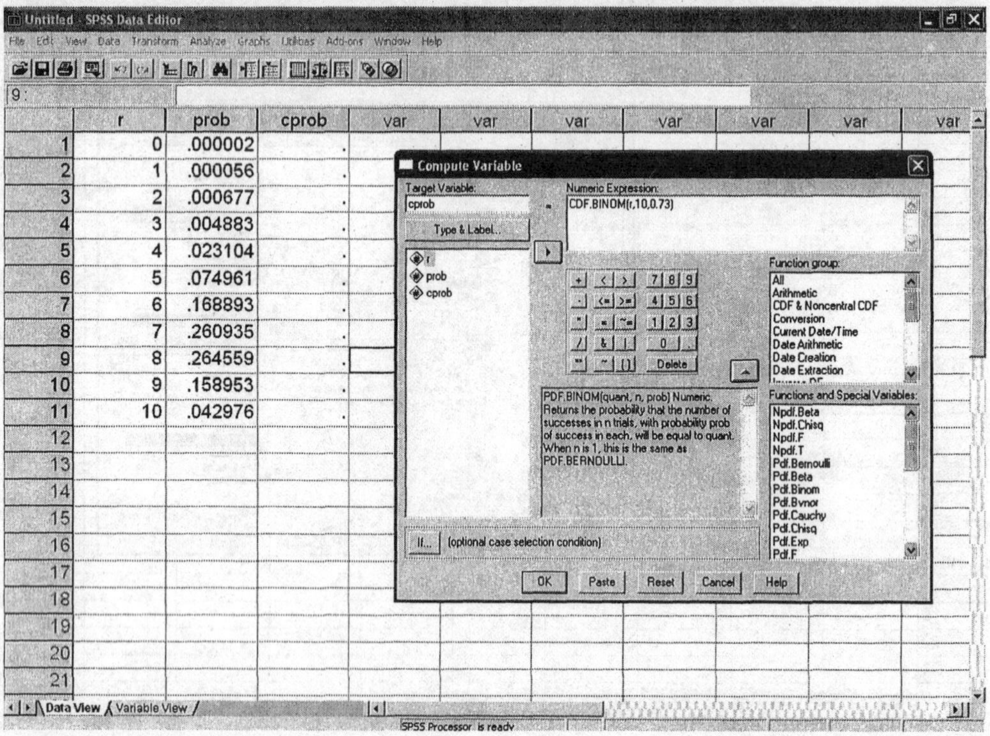

	r	prob	cprob
1	0	.000002	.
2	1	.000056	.
3	2	.000677	.
4	3	.004883	.
5	4	.023104	.
6	5	.074961	.
7	6	.168893	.
8	7	.260935	.
9	8	.264559	.
10	9	.158953	.
11	10	.042976	.

Click on OK. The results are shown below. From this screen we see, for example, $P(r \le 5) = 0.103683$.

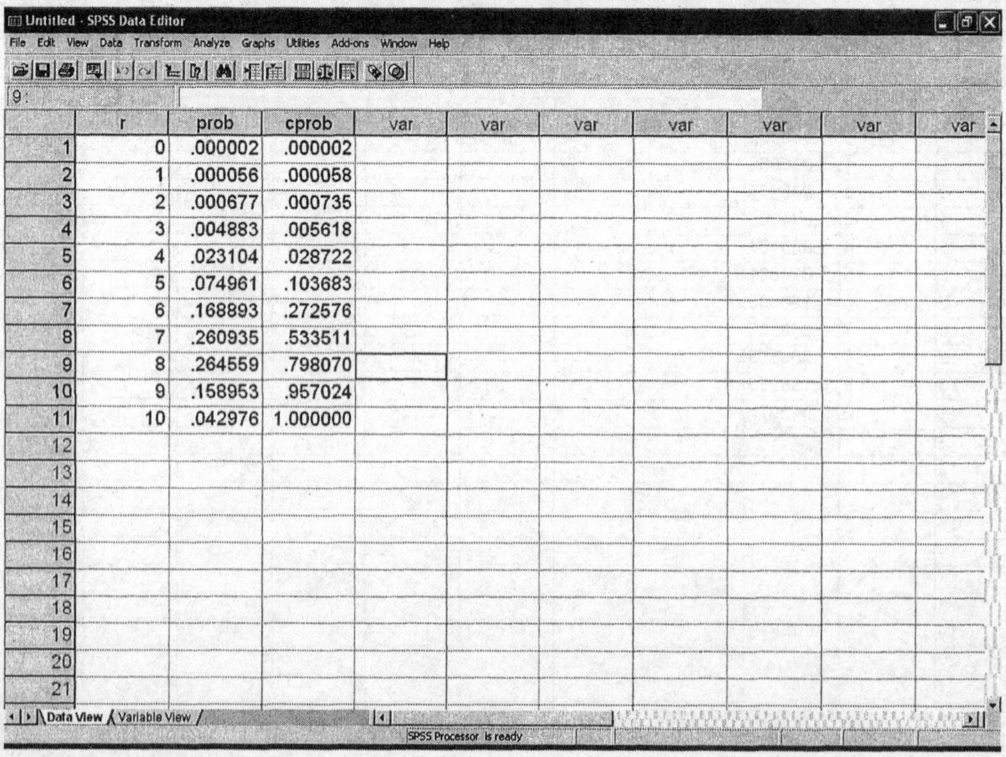

	r	prob	cprob
1	0	.000002	.000002
2	1	.000056	.000058
3	2	.000677	.000735
4	3	.004883	.005618
5	4	.023104	.028722
6	5	.074961	.103683
7	6	.168893	.272576
8	7	.260935	.533511
9	8	.264559	.798070
10	9	.158953	.957024
11	10	.042976	1.000000

LAB ACTIVITIES FOR BINOMIAL PROBABILITY DISTRIBUTIONS

1. You toss a coin 8 times. Call heads success. If the coin is fair, the probability of success P is 0.5. What is the probability of getting exactly 5 heads out of 8 tosses? of exactly 20 heads out of 100 tosses?

2. A bank examiner's record shows that the probability of an error in a statement for a checking account at Trust Us Bank is 0.03. The bank statements are sent monthly. What is the probability that exactly two of the next 12 monthly statements for our account will be in error? Now use the CDF function to find the probability that <u>at least</u> two of the next 12 statements contain errors. Use this result with subtraction to find the probability that <u>more than</u> two of the next 12 statements contain errors.

3. Some tables for the binomial distribution give values only up to 0.5 for the probability of success p. There is symmetry to the values for p greater than 0.5 with those values of p less than 0.5.

   (a) Consider the binomial distribution with $n = 10$ and $p = .75$. Since there are 0–10 successes possible, put $0 - 10$ in the first column. Use PDF function with this column and store the distribution probabilities in the second column. Name the second column as P.75. We will use the results in part (c).

   (b) Now consider the binomial distribution with $n = 10$ and $p = .25$. Use PDF function with the first column as **quant** and store the distribution probabilities in the third column. Name the third column as P.25.

   (c) Now compare the second and third column and see if you can discover the symmetries of P.75 with P.25. How does $P(K = 4$ successes with $p = .75)$ compare to $P(K = 6$ successes with $p = .25)$?

   (d) Now consider a binomial distribution with $n = 20$ and $p = .35$. Use PDF on the number 5 to get $P(K = 5$ successes out of 20 trials with $p = .35)$. Predict how this result will compare to the probability $P(K = 15$ successes out of 20 trials with $p = .65)$. Check your prediction by using the PDF on 15 with the binomial distribution $n = 20, p = .65$.

CHAPTER 6 NORMAL DISTIBUTIONS

GRAPHS OF NORMAL DISTRIBUTIONS (SECTION 6.1 OF *UNDERSTANDABLE STATISTICS*)

The normal distribution is a continuous probability distribution determined by the value of μ and σ. Similar to the binomial distribution, the normal distribution can be studied with three main functions of SPSS: The PDF & Noncentral PDF function which gives the probability density function value for a value X; The CDF & Noncentral CDF function which, for a value X, gives the probability less than or equal to X; and the Inverse DF which gives the inverse of the CDF. To apply PDF and CDF to a normal distribution, we need to use the menu selections **Transform➤Compute** followed by function selections.

> **Transform➤Compute**
>
> Dialog Box Responses
>
> > Enter name of the Target Variable
> >
> > Enter the function formula into the Numeric Expression box:
> >
> > > Under the Function group, select PDF & Noncentral PDF for probability; CDF & Noncentral CDF for cumulative probability for CDF; Inverse DF for inverse cumulative probability.
> > >
> > > Then under Functions and Special Variables, select the sub-function for the normal distribution.

For example, the PDF function for a normal distribution is **PDF.Normal(quant, mean, stddev)** and the CDF function for a binomial distribution is **CDF.Normal(quant, mean, stddev)**. Here **mean** is the mean value of the normal distribution, **stddev** is the standard deviation of the normal distribution. The input **quant** is the value of X, for which we want to find the PDF value or CDF value. You may enter a value for **quant**, or you may store the values for **quant** in a variable (column) and enter the variable name for **quant**. The Inverse DF function for a normal distribution is **Idf(prob, mean, stddev).** It returns the value of X such that CDF(X) = prob.

We can sketch the graph of a normal distribution by following these steps:

1. Name the first column X. Enter the values of X for which we want to compute PDF(X).

2. Name the second column pdf_X. Use **Transform➤Compute** to compute the values of PDF(X) and store them in the second column.

3. Sketch the graph by using the following menu options for graphs.

Menu Options for Graphs

To graph functions in SPSS, we use the menu **➤Graphs➤Interactive➤Line**

Dialog Box Responses

Choose "2D Coordinate"

Drag the variable for horizontal axis into the variable box for the horizontal axis.

Drag the variable for vertical axis into the variable box for the vertical axis.

Click on "Dots and Lines" to get into the Dots and Lines tab.

Under "Interpolation" click on "Spline"

Click on OK. The graph will show up.

There are other options available. See the Help menu for more information.

Example

Graph the normal distribution with mean $\mu = 10$ and standard deviation $\sigma = 2$.

Since most of the normal curve occurs over the values $\mu - 3\sigma$ to $\mu + 3\sigma$, we will start the graph at $10 - 3(2) = 4$ and end it at $10 + 3(2) = 16$. We will let SPSS set the scale on the vertical axis.

To graph a normal distribution, use X as the variable name for the first column and enter X values in this column. Let the first value of X be 4, last value be 16, and increment be 0.25. Altogether there are 49 X values. Use pdf_X as the variable name for the second column. Allow more decimal digits for pdf_X, say 6. (This can be set in the variable view of data editor.) Use **Transform▶Compute**. Let pdf_X be the target variable. Under Numeric Expression enter **PDF.Normal(X, 10, 2)**.

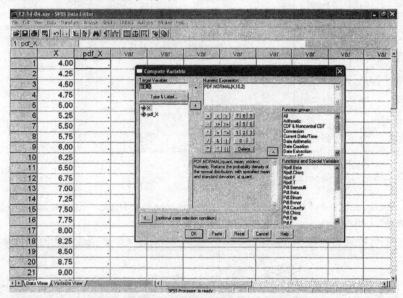

Click on OK. The PDF(X) are computed and stored in the second column.

	X	pdf_X	var	var	var	var	var	var	var	var
1	4.00	.002216								
2	4.25	.003199								
3	4.50	.004547								
4	4.75	.006362								
5	5.00	.008764								
6	5.25	.011886								
7	5.50	.015870								
8	5.75	.020860								
9	6.00	.026995								
10	6.25	.034393								
11	6.50	.043139								
12	6.75	.053269								
13	7.00	.064759								
14	7.25	.077506								
15	7.50	.091325								
16	7.75	.105938								
17	8.00	.120985								
18	8.25	.136027								
19	8.50	.150569								
20	8.75	.164080								
21	9.00	.176033								

Now use the menu options ➤**Graphs**➤**Interactive**➤**Line.** In the dialog box, choose "2D Coordinate". Then drag X into the variable box for the horizontal axis, pdf_X into the variable box for the vertical axis.

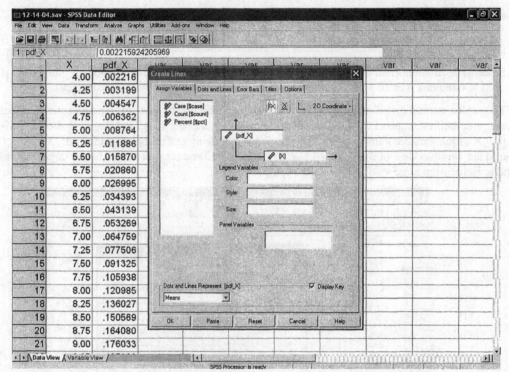

Now click on "Dots and Lines" to get into the Dots and Lines tab. Under "Interpolation" click on "Spline" as shown below.

Click on OK. The graph follows.

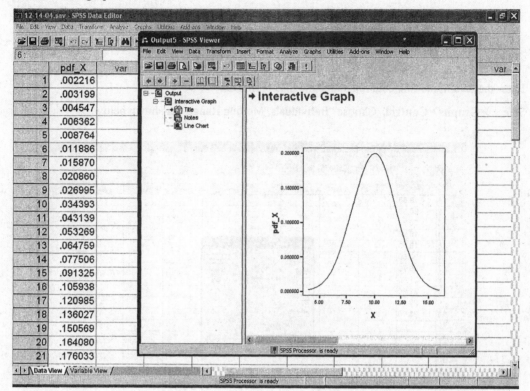

CONTROL CHARTS (SECTION 6.1 OF *UNDERSTANDABLE STATISTICS*)

SPSS supports a variety of control charts. The type discussed in Section 6.1 of *Understandable Statistics* is called an individual chart. The menu selection is ➤**Graphs**➤**Control** followed by the sequence of dialogue box responses described below.

Dialogue Box Responses

Choose "Individuals, Moving Range" and click on Define. This brings a new dialogue box.

Enter the variable into the Process Measurement box. Click on "Individuals" under Charts.

Click on "Options". In the Options dialog box, enter the Number of Sigmas. In the chart, the lines indicating sample mean, and sample mean plus (and minus) Number of Sigmas times sample standard deviation will be displayed. Click on Continue.

Click on "Statistics". In this dialog box, specify upper limit and lower limit. The lines indicating these two limits will also be displayed. This will help you to see the data lying out of these limits. You may also check the button "Actual % outside specification limits" to show the percentage of data outside these limits.

For information about the other options, see the Help menu.

Example

In a packaging process, the weight of popcorn that is to go in a bag has a normal distribution with $\mu = 20.7$ oz and $\sigma = 0.7$ oz. During one session of packaging, eleven samples were taken. Use an individual control chart to show these observations. The weights were (in oz).

19.5   20.3   20.7   18.9   19.5   20.5

20.7   21.4   21.9   22.7   23.8

Enter the data in the first column, and name the column oz.

Select **≻Graphs≻Control.** Choose "Individuals, Moving Range" as shown below.

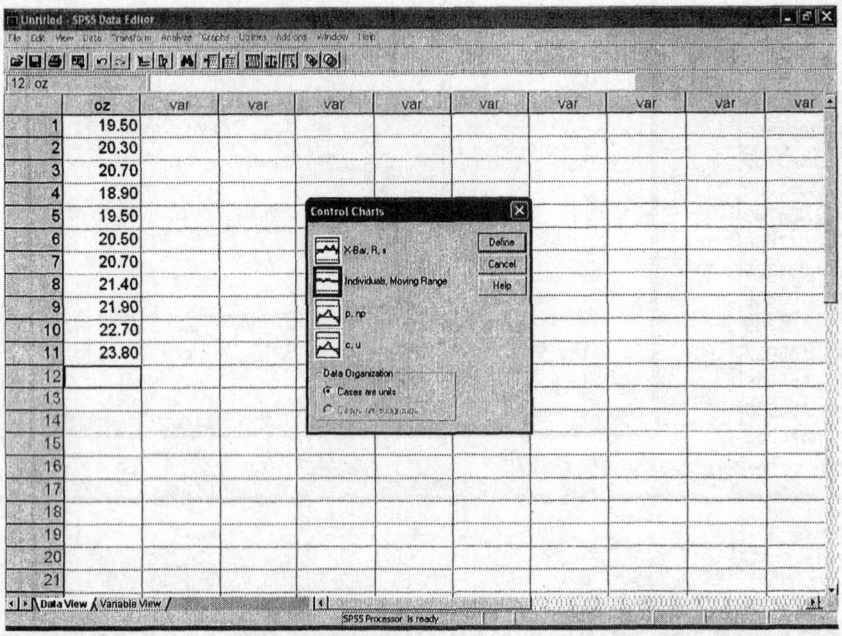

Click on Define. Enter oz into the Process Measurement box. Click on "Individuals" under Charts.

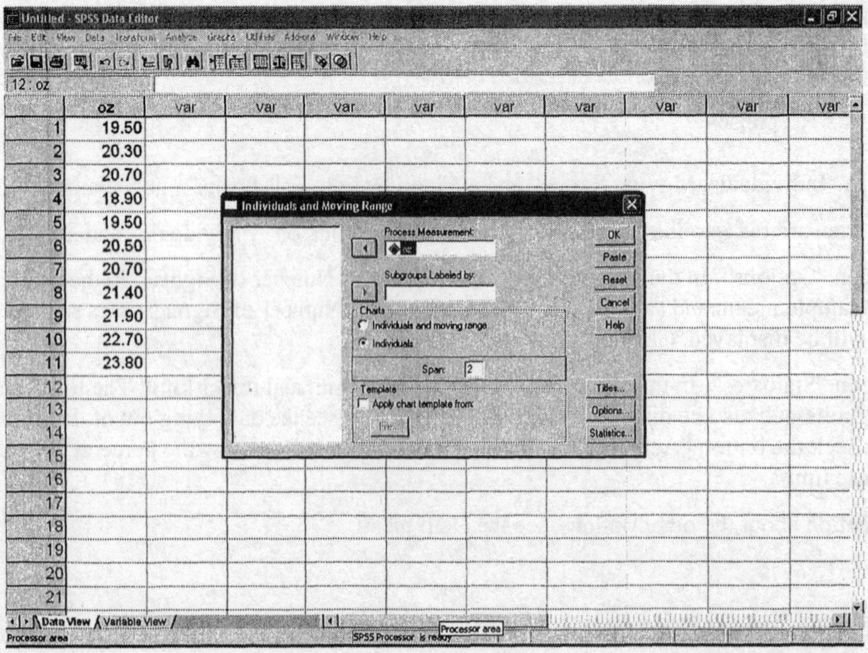

Click on "Options". In the Options dialog box, enter 3 as the Number of Sigmas. In the chart, the lines indicating sample mean, sample mean plus 3(sample standard deviation), and sample mean minus 3(sample standard deviation) will be displayed. After that click on Continue.

Click on "Statistics". In this dialog box, specify upper limit and lower limit. Let us use the population statistics $\mu + 3\sigma = 20.7 + 3(0.7) = 22.8$ as the upper limit and $\mu - 3\sigma = 20.7 - 3(0.7) = 18.6$ as the lower limit. The lines indicating these two limits will also be displayed in the chart. Also check the button "Actual % outside specification limits".

Click on Continue. Then click on OK. The graph follows. We see that one observation (or 9.1% of the data) is outside the set limits.

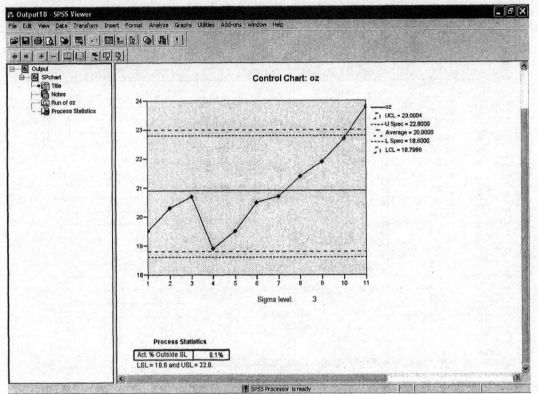

LAB ACTIVITIES FOR GRAPHS OF NORMAL DISTRIBUTIONS AND CONTROL CHARTS

1. (a) Sketch a graph of the standard normal distribution with $\mu = 0$ and $\sigma = 1$. Enter into the first column the data from -3 to 3 in increments of 0.5. Then generate the graph.

(b) Sketch a graph of a normal distribution with $\mu = 10$ and $\sigma = 1$. Enter into the first column the data from -7 to 7 in increments of 0.5. Compare this graph to that of part (a). Do the height and spread of the graphs appear to be the same? What is different? Why would you expect this difference?

(c) Sketch a graph of a normal distribution with $\mu = 0$ and $\sigma = 2$. Enter into the first column the data from -6 to 6 in increments of 0.5. Compare this graph to that of part (a). Do the height and spread of the graphs appear to be the same? What is different? Why would you expect this difference? Note, to really compare the graphs, it is best to graph them using the same scales. Redo the graph of part (a) using X from -6 to 6. Then redo the graph in this part using the same X values as in part (a) and Y values ranging from 0 to the high value of part (a).

2. Use one of the following SPSS data files found on the CD-ROM to draw a control chart. In each of the files the target value for the mean μ is stored in the Column2-Row1 position and the target value for the standard deviation is stored in the Column3-Row1 position. Use the targeted MU and SIGMA values to set upper and lower limits.

Yield of Wheat: Tscc01.sav

PepsiCo Stock Closing Prices: Tscc02.sav

PepsiCo Stock Volume of Sales: Tscc03.sav

Futures Quotes for the Price of Coffee Beans: Tscc04.sav

Incidence of Melanoma Tumors: Tscc05.sav

Percent Change in Consumer Price Index: Tscc06.sav

CHAPTER 7

INTRODUCTION TO SAMPLING DISTRIBUTIONS

CENTRAL LIMIT THEOREM (SECTION 7.2 OF *UNDERSTANDABLE STATISTICS*)

The Central Limit Theorem says that if x is a random variable with <u>any</u> distribution having mean μ and standard deviation σ, then the distribution of sample means $\bar{x}$ based on random samples of size n is such that for sufficiently large n:

(a) The mean of the $\bar{x}$ distribution is approximately the same as the mean of the x distribution.

(b) The standard deviation of the x distribution is approximately $\sigma/\sqrt{n}$.

(c) The $\bar{x}$ distribution is approximately a normal distribution.

Furthermore, as the sample size n becomes larger and larger, the approximations mentions in (a), (b) and (c) become better.

We can use SPSS to demonstrate the Central Limit Theorem. The computer does not prove the theorem. A proof of the Central Limit Theorem requires advanced mathematics and is beyond the scope of an introductory course. However, we can use the computer to gain a better understanding of the theorem.

To demonstrate the Central Limit Theorem, we need a specific x distribution. One of the simplest is the <u>uniform probability distribution</u>.

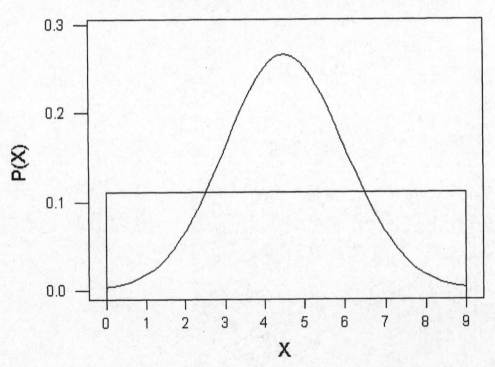

Uniform Distribution and a Normal Distribution

The normal distribution is the usual bell-shaped curve, but the uniform distribution is the rectangular or box-shaped graph. The two distributions are very different.

The uniform distribution has the property that all subintervals of the same length inside the interval 0 to 9 have the same probability of occurrence no matter where they are located. This means that the uniform distribution on the interval from 0 to 9 could be represented on the computer by selecting random numbers from 0 to 9. Since all numbers from 0 to 9 would be equally likely to be chosen, we say we are dealing with a uniform (equally likely) probability distribution. Note that when we say we are selecting random numbers from 0 to 9, we do not just mean whole numbers or integers; we mean real numbers in decimal form such as 2.413912, and so forth.

Because the interval from 0 to 9 is 9 units long and because the total area under the probability graph must by 1, the height of the uniform probability graph must be 1/9. The mean of the uniform distribution on the interval from 0 to 9 is the balance point. Looking at the Figure, it is fairly clear that the mean is 4.5. Using advanced methods of statistics, it can be shown that for the uniform probability distribution x between 0 and 9,

$$\mu = 4.5 \text{ and } \sigma = 3\sqrt{3}/2 \approx 2.598$$

The figure shows us that the uniform x distribution and the normal distribution are quite different. However, using the computer we will construct one hundred sample means $\bar{x}$ from the x distribution using a sample size of $n = 40$. We will use 100 rows (for the 100 samples) and 40 columns (sample size is 40). We can vary the number of samples as well as the sample size n according to how many rows and columns we use.

We will see that even though the uniform distribution is very different from the normal distribution, the histogram of the sample means is somewhat bell shaped. We will also see that the mean or the $\bar{x}$ distribution is close to the predicted mean of 4.5 and that the standard deviation is close to $\sigma/\sqrt{n}$ or $2.598/\sqrt{40}$ or 0.411.

Example

In order for us to get familiar with the procedure, let us first work with 100 samples using a sample size of $n = 5$. Follow these steps. Also note that your results will rary.

First, name the first column (variable) x1. Enter a number (any number) in the 100th cell of the first column to define the variable size (that is, the number of samples). Then use **Transform➤Compute** for five times (since our sample size $n = 5$). Note that **Transform➤Compute** works with one target variable at a time. Since our sample size is 5, we need to generate random numbers from the uniform distribution in 5 columns (that is, 5 variables). That is why we need to use **Transform➤Compute** for five times. Each time we use the formula

   xi = RV.UNIFORM(0, 9),   here i = 1, 2, 3, 4, 5.

Note that the **Transform➤Compute** dialog box preserves the numeric expression used most recently. Therefore the expression **RV.UNIFORM(0, 9)** only needs to be entered once. After that, all you have to do in the **Transform➤Compute** dialog box is to change the target variable name, that is, to change the value of i. Displayed below is our fifth use of **Transform➤Compute** with this formula. Here i = 5. Therefore the formula reads **x5 = RV.UNIFORM(0, 9).**

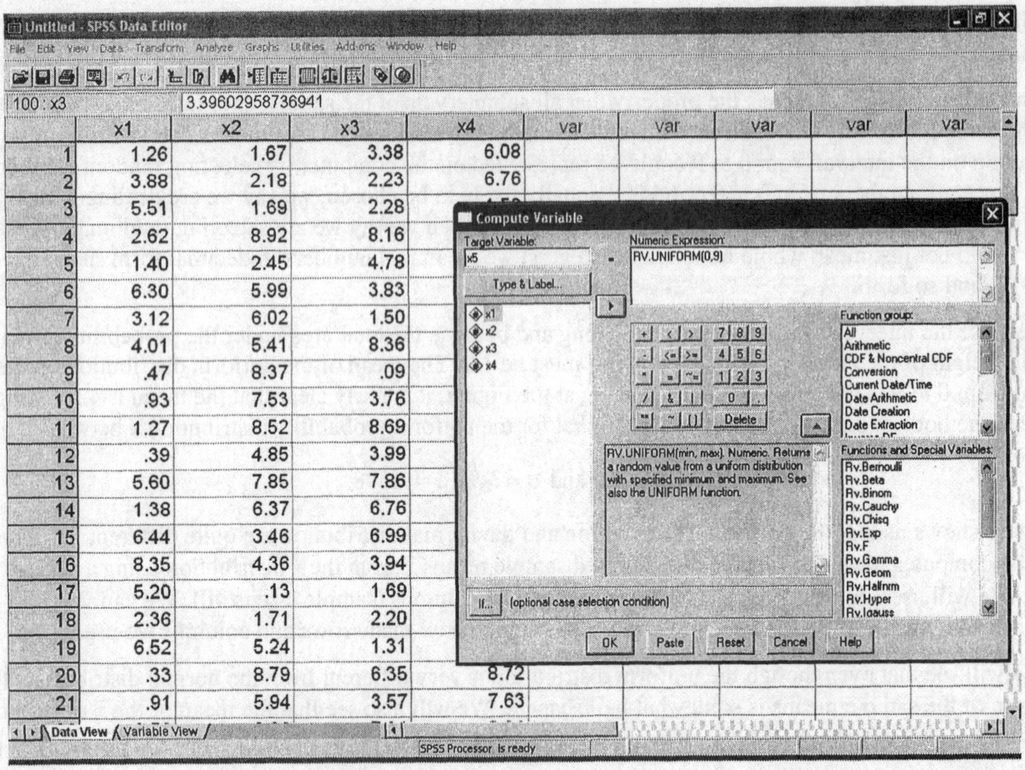

Click on OK. Another hundred of random numbers will be generated in the fifth column under variable name x5. So 100 random samples of size 5 from the uniform distribution on (0, 9) are generated.

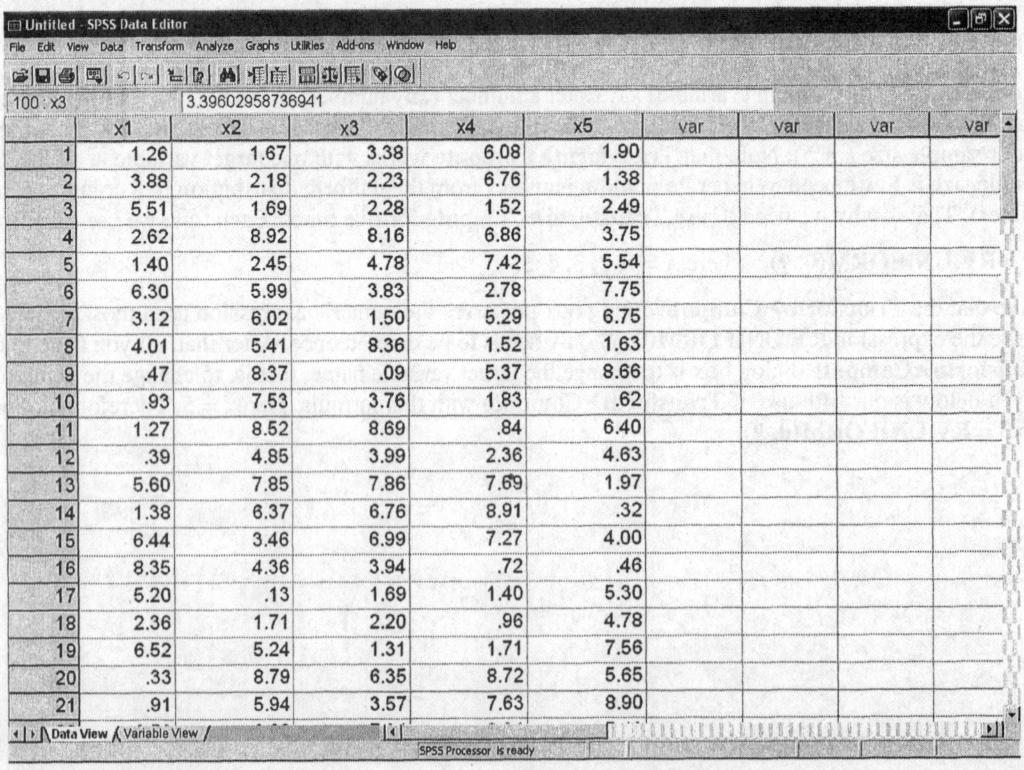

Next, let us take the mean of each of the 100 rows (5 columns across) and store the values under the variable name xbar. Use **Transform➤Compute** with the formula **xbar = MEAN(x1, x2, x3, x4, x5)** as shown below.

Click on OK. The results follow.

Let us now look at the mean and standard deviation of xbar (the sample means) as well as its histogram, using the menu options ➤**Analyze**➤**Descriptive Statistics** ➤ **Frequencies**. Uncheck "Display frequency table", click on "Charts" and select "Histogram", then click on "Statistics" and select "Mean" and "Std deviation". Click on OK. The results follow.

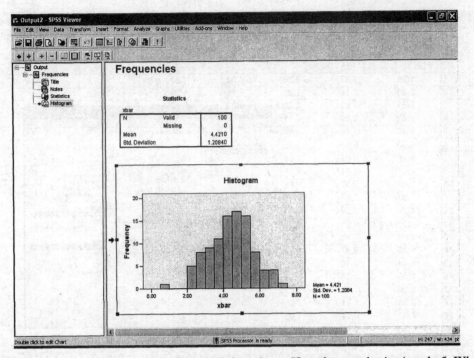

Note that the histogram is already quite close to a bell shaped one. Here the sample size is only 5. When the sample size is sufficiently large, the histogram will look more like a normal distribution.

Now let us draw 100 random samples of size 40 from the uniform distribution on the interval from 0 to 9. The steps will be the same as above, only that now we need to repeat **Transform➤Compute** for 40 times with the formula

xi = RV.UNIFORM(0, 9),   here **i = 1, 2 . . . 40.**

After that we compute the sample mean by **xbar = MEAN(x1, x2, x3, x4, x5, x6, x7, x8, x9, x10, x11, x12, x13, x14, x15, x16, x17, x18, x19, x20, x21, x22, x23, x24, x25, x26, x27, x28, x29, x30, x31, x32, x33, x34, x35, x36, x37, x38, x39, x40).** Do these, and the results follow. (Your results will vary.)

	x36	x37	x38	x39	x40	xbar	var	var
1	5.56	6.03	8.42	8.44	4.78	5.21		
2	8.54	5.24	1.55	1.90	1.00	4.47		
3	3.95	7.28	8.45	4.36	.44	4.63		
4	2.36	.06	1.33	8.67	1.74	4.07		
5	4.43	4.26	6.04	3.76	6.36	4.46		
6	6.15	3.73	4.78	7.26	2.02	4.21		
7	8.49	3.05	8.65	7.60	1.53	5.15		
8	.26	2.57	3.92	1.28	4.81	4.31		
9	6.11	.72	3.27	.19	5.35	4.71		
10	6.11	8.05	1.85	7.91	8.26	3.85		
11	5.48	4.54	6.18	6.04	4.58	4.51		
12	.25	5.06	6.18	3.61	3.10	4.42		
13	.71	.76	5.49	3.57	3.18	4.28		
14	.69	1.11	.17	3.70	2.01	3.92		
15	5.52	4.84	6.55	7.58	7.09	4.24		
16	.72	1.44	2.88	1.75	4.18	3.79		
17	4.48	6.34	7.86	2.88	3.35	4.80		
18	4.67	3.46	8.34	2.93	1.81	4.10		
19	.29	6.65	3.22	5.81	2.09	4.15		
20	4.17	.87	.75	1.19	3.03	4.71		
21	2.88	1.46	2.49	1.49	6.56	3.90		

1 : x6   8.47064486145973

Now look at the mean and standard deviation of xbar (the sample means) as well as its histogram, using the menu options ➤**Analyze**➤**Descriptive Statistics** ➤ **Frequencies**. Uncheck "Display frequency table", click on "Charts" and select "Histogram", then click on "Statistics" and select "Mean" and "Std deviation". Click on OK. The results follow.

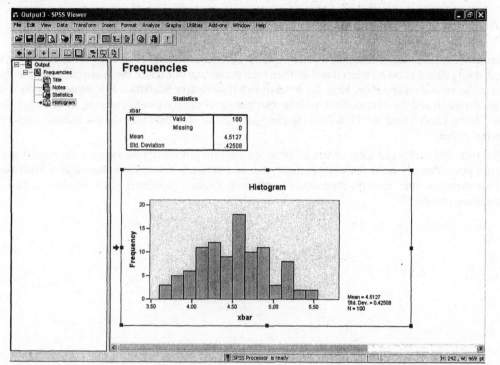

Note the Mean and Std Dev are very close to the values predicted by the Central Limit Theorem. The histogram for this sample does not appear very similar to a normal distribution. Let's try another sample. The following are the results.

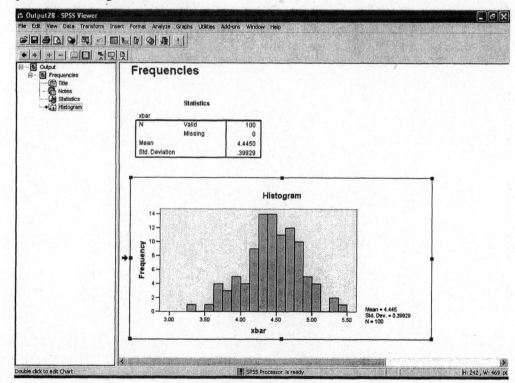

This histogram looks more like a normal distribution. You will get slightly different results each time you draw 100 samples.

LAB ACTIVITIES FOR CENTRAL LIMIT THEOREM

1. Repeat the experiment of Example 1. That is, draw 100 random samples of size 40 each from the uniform probability distribution between 0 and 9. Then take the means of each of these samples and put the results under the variable name xbar. Next use ➤**Analyze**➤**Descriptive Statistics** ➤ **Frequencies** on xbar. How does the mean and standard deviation of the distribution of sample means compare to those predicted by the Central Limit Theorem? How does the histogram of the distribution of sample means compare to a normal curve?

2. Next take 100 random samples of size 20 from the uniform probability distribution between 0 and 9. Again put the means under the variable name xbar and then use ➤**Analyze**➤**Descriptive Statistics** ➤ **Frequencies** on xbar. How do these results compare to those in problem 1? How do the standard deviations compare?

CHAPTER 8 ESTIMATION

CONFIDENCE INTERVALS FOR A MEAN (SECTIONS 8.1–8.2 OF *UNDERSTANDABLE STATISTICS*)

Student's *t* Distribution

SPSS uses a Student's *t* distribution to generate confidence intervals for μ, regardless of the sample size or the knowledge about the standard deviation σ. In Section 8.2, Student's *t* distribution is introduced. SPSS construct confidence intervals using the Student's *t* distribution with degrees of freedom $(n - 1)$.

There is a different Student's *t* distribution for every degree of freedom. SPSS includes Student's *t* distribution in its library of probability distributions. The three main functions are: the PDF & Noncentral PDF function which gives the probability density function value for a value X; The CDF & Noncentral CDF function which, for a value X, gives the probability less than or equal to X; and the Inverse DF which gives the inverse of the CDF. To apply these functions to a Student's *t* distribution, we need to use the menu selections **Transform➤Compute** followed by function selections.

Menu selection: **Transform➤Compute**

Dialog Box Responses

Enter name of the Target Variable

Enter the function formula into the Numeric Expression box:

Under the Function group, select PDF & Noncentral PDF for probability; CDF & Noncentral CDF for cumulative probability for CDF; Inverse DF for inverse cumulative probability.

Then under Functions and Special Variables, select the sub-function for the normal distribution.

For example, the PDF function for a Student's *t* distribution is **PDF.T(quant, df)** and the CDF function for a Student's *t* distribution is **CDF.T(quant, df)**. Here **df** is the degree of freedom of this particular *t* distribution, and **quant** is the value of X, for which we want to find the PDF value or CDF value. You may enter a value for **quant**, or you may store the values for **quant** in a variable (column) and enter the variable name for **quant**. The Inverse DF function for a Student's *t* distribution is **Idf.T(prob, df).** It returns the value of X such that CDF(X) = prob.

We can sketch the graph of a Student's *t* distribution by following these steps:

1. Name the first column X. Enter the values of X for which we want to compute PDF(X).

2. Name the second column pdf_X. Use **Transform➤Compute** to compute the values of PDF(X) and store them in the second column.

3. Sketch the graph by using the following menu options for graphs.

Menu Options for Graphs

To graph functions in SPSS, we use the menu **➤Graphs➤Interactive➤Line**

Dialog Box Responses

Choose "2D Coordinate"

Drag the variable for horizontal axis into the variable box for the horizontal axis.

Drag the variable for vertical axis into the variable box for the vertical axis.

Click on "Dots and Lines" to get into the Dots and Lines tab.

Under "Interpolation" click on "Spline"

Click on OK. The graph will show up.

There are other options available. See the Help menu for more information.

Example

Graph the *t*-distributions with degree of freedom 10 for X values from -4 to 4. Follow steps similar to those given in Chapter 6 for graphing a normal distribution. Use X as the variable name for the first column and enter X values in this column. Let the first value of X be -4, last value be 4, and increment be 0.1. Use pdf_X as the variable name for the second column. Allow more decimal digits for pdf_X, for example, six. (This can be set in the variable view of data editor.) Use **Transform**▶**Compute**. Let pdf_X be the target variable. Under Numeric Expression enter **PDF.T(X, 10)**. By following above steps, we get the Student's *t* Distribution with 10 Degrees of Freedom:

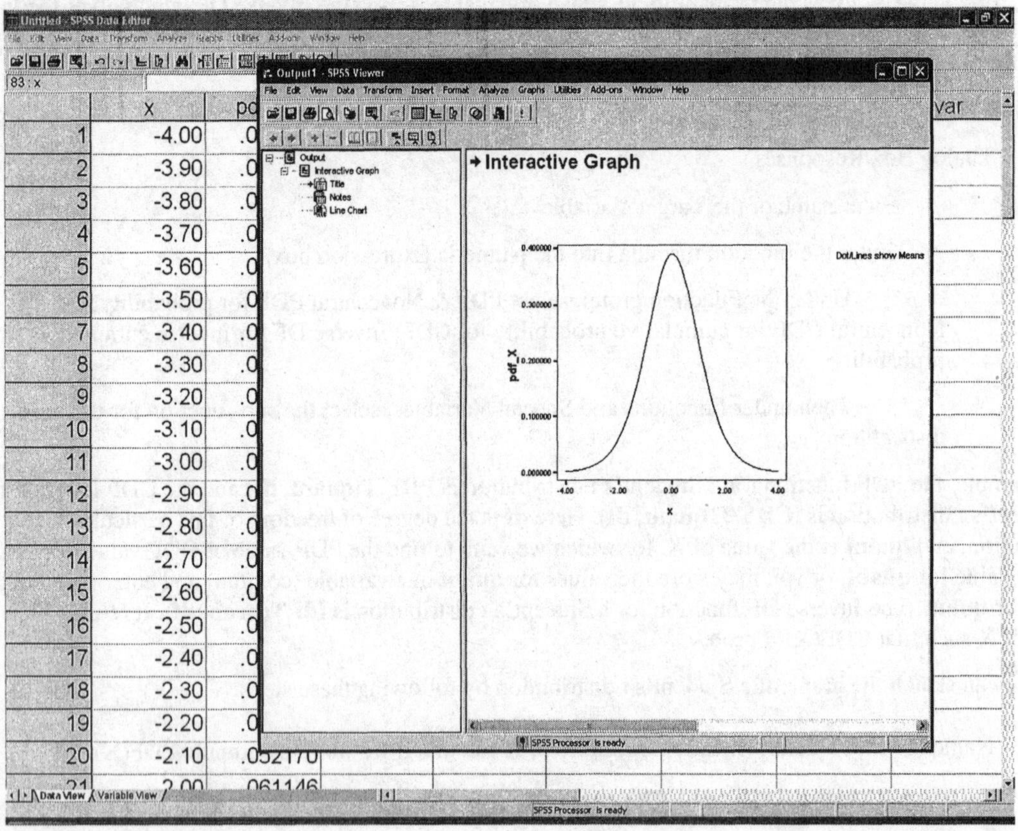

Confidence Intervals for Means

SPSS uses a Student's *t* distribution to generate confidence intervals for μ, regardless of the sample size or the knowledge about the standard deviation σ. The following menu selections are used to generate confidence intervals.

▶**Analyze**▶**Compare Means**▶**One-Sample T Test**

Dialog Box Responses

Test Variables: Enter the variable (column) name that contains the data.

For confidence Interval: Click on [Options], then enter the confidence level, such as 95%.

Test Value: Leave blank at this time. We will use the option in Chapter 9.

Example

The manager of First National Bank wishes to know the average waiting times for student loan application action. A random sample of 20 applications showed the waiting times from application submission (in days) to be

3	7	8	24	6	9	12	25	18	17
4	32	15	16	21	14	12	5	18	16

Find a 90% confidence interval for the population mean of waiting times.

We use the menu selection ➤**Analyze**➤**Compare Means**➤**One-Sample T Test.**

The results are

	Days
1	3.00
2	7.00
3	8.00
4	24.00
5	6.00
6	9.00
7	12.00
8	25.00
9	18.00
10	17.00
11	4.00
12	32.00
13	15.00
14	16.00
15	21.00
16	14.00
17	12.00
18	5.00
19	18.00
20	16.00

One-Sample Statistics

	N	Mean	Std. Deviation	Std. Error Mean
Days	20	14.1000	7.70441	1.72276

One-Sample Test

	Test Value = 0					
					90% Confidence Interval of the Difference	
	t	df	Sig. (2-tailed)	Mean Difference	Lower	Upper
Days	8.185	19	.000	14.10000	11.1211	17.0789

Confidence Intervals for Difference of Means

In SPSS, confidence intervals for difference of means are included in the menu selection for tests of hypothesis for difference of means. Student's *t* distributions are used. Menu selections are ➤**Analyze** ➤**Compare Means**➤**Paired-Samples T Test** or ➤**Analyze** ➤**Compare Means**➤**Independent-Samples T Test.** These menu selections with their dialog boxes will be discussed in Chapter 9.

LAB ACTIVITIES FOR CONFIDENCE INTERVALS FOR A MEAN

1. Snow King Ski resorts is considering opening a downhill ski slope in Montana. To determine if there would be an adequate snow base in November in the particular region under consideration, they studied snowfall records for the area over the last 100 years. They took a random sample of 15 years. The snowfall during November for the sample years was (in inches)

   26   35   42   18   29   42   28   35
   47   29   38   27   21   35   30

   (a) Find a 90% confidence interval for the mean snowfall.
   (b) Find a 95% confidence interval for the mean snowfall.
   (c) Compare the intervals of parts (a) and (b). Which one is narrower? Why would you expect this?

2. Retrieve the worksheet Svls01.sav from the CD-ROM. This worksheet contains the number of shares of Disney Stock (in hundreds of shares) sold for a random sample of 60 trading days in 1993 and 1994. The data is in column C1.

   (a) Find a 99% confidence interval for the population mean volume.

   (b) Find a 95% confidence interval for the population mean volume.

   (c) Find a 90% confidence interval for the population mean volume.

   (d) Find an 85% confidence interval for the population mean volume.

   (d) What do you notice about the lengths of the intervals as the confidence level decreases?

CHAPTER 9 HYPOTHESIS TESTING

TESTING A SINGLE POPULATION MEAN (SECTIONS 9.1–9.2 OF *UNDERSTANDABLE STATISTICS*)

Tests involving a single mean are found in Sections 9.2. In SPSS, the user concludes the test by comparing the P value of the test statistic to the level of significance α. The method of using P values to conclude tests of hypotheses is explained in Section 9.2. SPSS uses a Student's t distribution to conduct the test regardless of the sample size or the knowledge of population standard deviation σ. The P value produced by SPSS is the one for the two-tailed test, that is, $H_0: \mu = k$ versus $H_1: \mu \neq k$. For a one-tailed test, you may convert the P value produced by SPSS into the P value for the corresponding one-tailed test, based on the definition of P values given in Section 9.2.

Use the menu selections ➤**Analyze**➤**Compare Means**➤**One-Sample T Test**

Dialog Box Responses

Test Variables: Enter the variable (column) name that contains the data.

Test Value: Enter the value of k.

The tests gives the two-tailed test P value of the sample statistic $\bar{x}$. The user can then compare the P value to α, the level of significance of the test. If

P value $\leq \alpha$ we reject the null hypothesis;

P value $> \alpha$ we do not reject the null hypothesis.

Example

Many times patients visit a health clinic because they are ill. A random sample of 12 patients visiting a health clinic had temperatures (in °F) as follows:

97.4	99.3	99.0	100.0	98.6
97.1	100.2	98.9	100.2	98.5
98.8	97.3			

Dr. Tafoya believes that patients visiting a health clinic have a higher temperature than normal. The normal temperature is 98.6 degrees. Test the claim at the $\alpha = 0.01$ level of significance.

Enter the data in the first column and name the column Temp. Then select ➤**Analyze**➤**Compare Means**➤**One-Sample T Test**. Use 98.6 as the test value.

The results follow.

The P value produced by SPSS in the output screen is "Sig. (2-tailed)", which equals 0.587 in this case. To convert it into the P value for this up-tailed test (H_0: $\mu = 98.6$ versus H_1: $\mu > 98.6$), we notice that with this sample, the sample mean 98.775 is greater than the test value 98.6. Hence the P value for this up-tailed test equals $0.5(0.587) = 0.2935$. Since $0.2935 > 0.01$ we do not reject the null hypothesis.

LAB ACTIVITIES FOR TESTING A SINGLE POPULATION MEAN

1. A new catch-and-release policy was established for a river in Pennsylvania. Prior to the new policy, the average number of fish caught per fisherman hour was 2.8. Two years after the policy went into effect, a random sample of 12 fisherman hours showed the following catches per hour.

3.2	1.1	4.6	3.2	2.3	2.5
1.6	2.2	3.7	2.6	3.1	3.4

   Test the claim that the per hour catch has increased, at the 0.05 level of significance.

2. Open or retrieve the worksheet **Svls04.sav** from the CD-ROM. The data in the first column represent the miles per gallon gasoline consumption (highway) for a random sample of 55 makes and models of passenger cars (source: Environmental Protection Agency).

30	27	22	25	24	25	24	15
35	35	33	52	49	10	27	18
20	23	24	25	30	24	24	24
18	20	25	27	24	32	29	27
24	27	26	25	24	28	33	30
13	13	21	28	37	35	32	33
29	31	28	28	25	29	31	

   (a) Test the hypothesis that the population mean miles per gallon gasoline consumption for such cars is not equal to than 25 mpg, at the 0.05 level of significance.

   (b) Using the same data, test the claim that the average mpg for these cars is greater than 25. How should you find the new P value? Compare the new P value to α. Do we reject the null hypothesis or not?

3. Open or retrieve the worksheet **Svss01.sav** from the CD-ROM. The data in the first column represent the number of wolf pups per den from a sample of 16 wolf dens (source: *The Wolf in the Southwest: The Making of an Endangered Species* by D.E. Brown, University of Arizona Press).

5	8	7	5	3	4	3	9
5	8	5	6	5	6	4	7

   Test the claim that the population mean number of wolf pups in a den is greater than 5.4, at the 0.01 level of significance.

TESTS INVOLVING PAIRED DIFFERENCES (DEPENDENT SAMPLES)
(SECTION 9.4 OF *UNDERSTANDABLE STATISTICS*)

To perform a paired difference test, we put our paired data into two columns. Select

>**Analyze**>**Compare Means**>**Paired-Samples T Test**

Dialog Box Responses

Paired Variables: Highlight both variables and enter them as the paired variables.

In case a confidence interval is desired, click [Options] and enter a confidence level such as 95%.

SPSS produces the P value for the two-tailed test $H_0: \mu_1 = \mu_2$. versus $H_1: \mu_1 \neq \mu_2$, and the user needs to convert the P value in case a one-tailed test is being conducted.

Example

Promoters of a state lottery decided to advertise the lottery heavily on television for one week during the middle of one of the lottery games. To see if the advertising improved ticket sales, the promoters surveyed a random sample of 8 ticket outlets and recorded weekly sales for one week before the television campaign and for one week after the campaign. The results follow (in ticket sales) where B stands for "before" and A for "after" the advertising campaign.

B:	3201	4529	1425	1272	1784	1733	2563	3129
A:	3762	4851	1202	1131	2172	1802	2492	3151

We want to test to see if $D = B - A$ is less than zero, since we are testing the claim that the lottery ticket sales are greater after the television campaign. We will put the before data in the first column, the after data in the second column. Select >**Analyze**>**Compare Means**>**Paired-Samples T Test.** Use a level of significance 0.05.

The results follow.

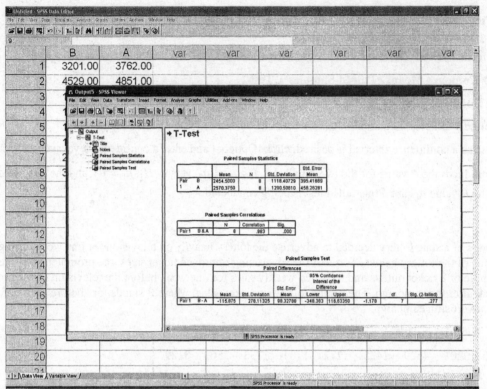

Note that the sample mean of B − A is less than 0. Hence the *P* value for this lower-tailed test equals half of the two-tailed test *P* value 0.277 provided by SPSS under "Sig. (2-tailed). That is, the *P* value for this lower-tailed test equals 0.1385, which is larger than the level of significance of 0.05. Thus we do not reject the null hypothesis.

LAB ACTIVITIES FOR TESTS INVOLVING PAIRED DIFFERENCES (DEPENDENT SAMPLES)

1. Open or retrieve the worksheet **Tvds01.sav** from the CD-ROM. The data are pairs of values, where the entry in the first column represents the average salary ($1000/yr) for male faculty members at an institution and the second column represents the average salary for female faculty members ($1000/yr) at the same institution. A random sample of 22 U.S. colleges and universities was used (source: *Academe, Bulletin of the American Association of University Professors*).

(34.5, 33.9)	(30.5, 31.2)	(35.1, 35.0)	(35.7, 34.2)	(31.5, 32.4)
(34.4, 34.1)	(32.1, 32.7)	(30.7, 29.9)	(33.7, 31.2)	(35.3, 35.5)
(30.7, 30.2)	(34.2, 34.8)	(39.6, 38.7)	(30.5, 30.0)	(33.8, 33.8)
(31.7, 32.4)	(32.8, 31.7)	(38.5, 38.9)	(40.5, 41.2)	(25.3, 25.5)
(28.6, 28.0)	(35.8, 35.1)			

(a) Test the hypothesis that there is a difference in salaries. What is the *P* value of the sample test statistic? Do we reject or fail to reject the null hypothesis at the 5% level of significance? What about at the 1% level of significance?

(b) Test the hypothesis that female faculty members have a lower average salary than male faculty members. What is the test conclusion at the 5% level of significance? At the 1% level of significance?

2. An audiologist is conducting a study on noise and stress. Twelve subjects selected at random were given a stress test in a room that was quiet. Then the same subjects were given another stress test, this time in a room with high-pitched background noise. The results of the stress tests were scores 1 through 20, with 20 indicating the greatest stress. The results follow, where B represents the score of the test administered in the quiet room and A represents the scores of the test administered in the room with the high-pitched background noise.

Subject		1	2	4	5	6	7	8	9	10	11	12
	B	13	12	16	19	7	13	9	15	17	6	14
	A	18	15	14	18	10	12	11	14	17	8	16

Test the hypothesis that the stress level was greater during exposure to noise. Look at the *P* value. Should you reject the null hypothesis at the 1% level of significance? At the 5% level?

TESTS OF DIFFERENCE OF MEANS (INDEPENDENT SAMPLES) (SECTION 9.5 OF *UNDERSTANDABLE STATISTICS*)

We consider the $\bar{x}_1 - \bar{x}_2$ distribution. The null hypothesis is that there is no difference between means, so $H_0: \mu_1 = \mu_2$, or $H_0: \mu_1 - \mu_2 = 0$. SPSS uses a Student's *t* distribution to conduct the test. The menu selections are **➤Analyze➤Compare Means➤Independent-Samples T Test.** For each variable the output screen displays sample size, mean, standard deviation, and standard error of the mean. For the difference in means, it provides mean, standard error, and confidence interval (you can specify the confidence level). It also produces the *P* values for the two-tailed tests, including the pooled-variances t tests as well as the separate-variances t tests.

To use the menu selections **➤Analyze➤Compare Means➤Independent-Samples T Test,** data needs to be entered in a special way. Two columns are to be used. In one column you enter all data, from both samples. In the other column you enter the sample number for each data in the first column. You may name the first column Data, in which you enter all the data. The second column may be named Sample where you enter the sample number for each value in the column Data. Here is an example:

Example

Sellers of microwave French fry cookers claim that their process saves cooking time. McDougle Fast Food Chain is considering the purchase of these new cookers, but wants to test the claim. Six batches of French fries were cooked in the traditional way. Cooking times (in minutes) are

> 15   17   14   15   16   13

Six batches of French fries of the same weight were cooked using the new microwave cooker. These cooking times (in minutes) are

> 11   14   12   10   11   15

Test the claim that the microwave process takes less time. Use $\alpha = 0.05$. Note that this is an upper-tailed test since the alternative hypothesis here is $H_1: \mu_1 > \mu_2$.

First, let's enter the data into two columns as shown below.

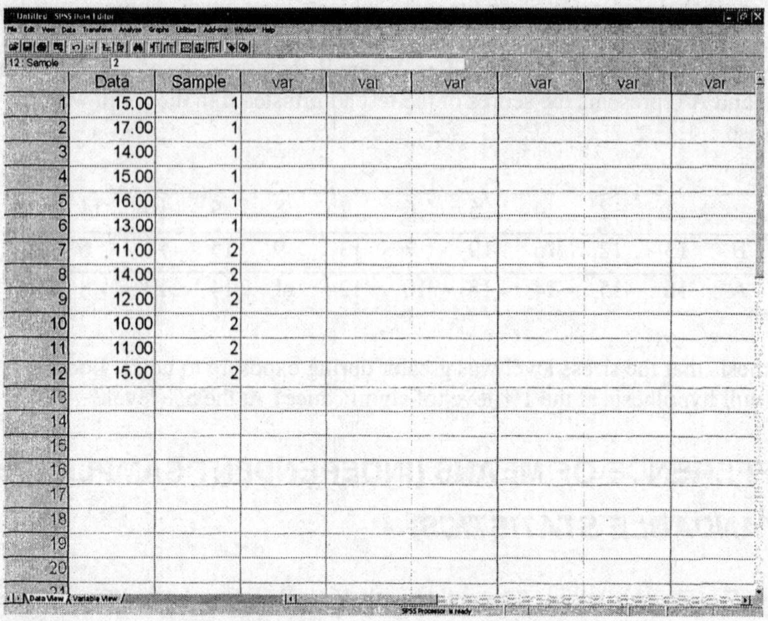

	Data	Sample	var	var	var	var	var	var
1	15.00	1						
2	17.00	1						
3	14.00	1						
4	15.00	1						
5	16.00	1						
6	13.00	1						
7	11.00	2						
8	14.00	2						
9	12.00	2						
10	10.00	2						
11	11.00	2						
12	15.00	2						
13								
14								
15								
16								
17								
18								
19								
20								

Values in the column "Sample" show that the first 6 numbers in the column "Data" are from the first sample while the rest of that column forms the second data. Use the menu sections

> **Analyze➤Compare Means➤Independent-Samples T Test**

Dialog Box Responses

Test Variable: Enter the variable Data.

Grouping Variable: Enter Sample as the grouping variable.

Click on [Define Groups] then choose Use specified values. Enter 1 for Group 1, and 2 for Group 2.

If you also want to produce a confidence interval for the difference of mean, click on [Options] and enter a confidence level such as 95%.

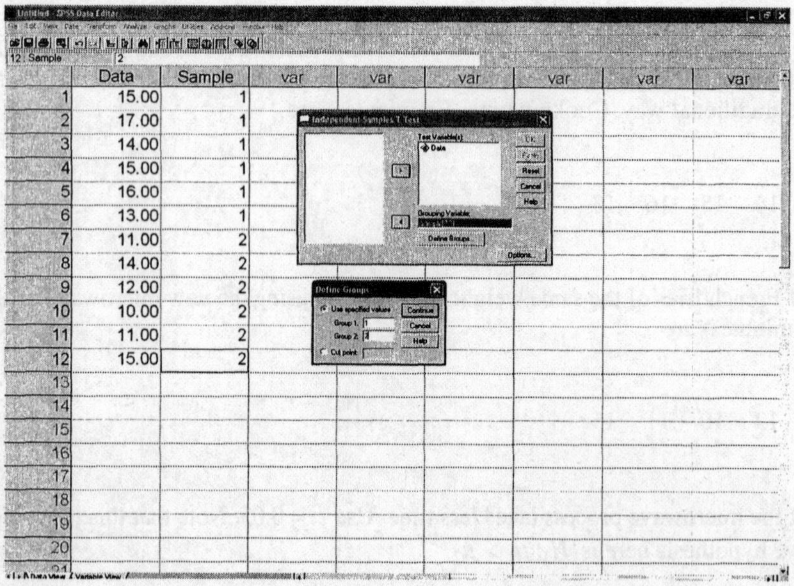

The results follow.

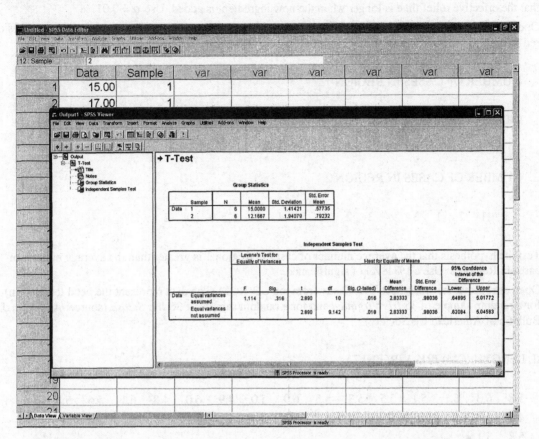

Since the sample mean of the first data (15) is greater than that of the second data (12.1667), the P value of this upper-tailed test equals half of the two-tailed test P values provided by SPSS. From the output screen we see that the P value of the test assuming equal variance is 0.008 (half of 0.016), and the P value assuming unequal variance is 0.009 (half of 0.018). Since P value is less than $\alpha = 0.05$, we reject the null hypothesis and conclude that the microwave method takes less time to cook French fries.

LAB ACTIVITIES USING DIFFERENCE OF MEANS (INDEPENDENT SAMPLES)

1. Calm Cough Medicine is testing a new ingredient to see if its addition will lengthen the effective cough relief time of a single dose. A random sample of 15 doses of the standard medicine were tested, and the effective relief times were (in minutes):

   42   35   40   32   30   26   51   39   33   28
   37   22   36   33   41

   A random sample of 20 doses was tested when the new ingredient was added. The effective relief times were (in minutes):

   43   51   35   49   32   29   42   38   45   74
   31   31   46   36   33   45   30   32   41   25

Assume that the standard deviations of the relief times are equal for the two populations. Test the claim that the effective relief time is longer when the new ingredient is added. Use $\alpha = 0.01$.

2. Open or retrieve the worksheet **Tvis06.sav** from the CD-ROM. The data represent the number of cases of red fox rabies for a random sample of 16 areas in each of two different regions of southern Germany.

NUMBER OF CASES IN REGION 1

10   2   2   5   3   4   3   3   4   0   2   6   4   8   7   4

NUMBER OF CASES IN REGION 2

1   1   2   1   3   9   2   2   4   5   4   2   2   0   0   2

Test the hypothesis that the average number of cases in Region 1 is greater than the average number of cases in Region 2. Use a 1% level of significance.

3. Open or retrieve the worksheet **Tvis02.sav** from the CD-ROM. The data represent the petal length (cm) for a random sample of 35 *Iris Virginica* and for a random sample of 38 *Iris Setosa* (source: Anderson, E., Bulletin of American Iris Society).

PETAL LENGTH (CM) IRIS VIRGINICA

5.1	5.8	6.3	6.1	5.1	5.5	5.3	5.5	6.9	5.0	4.9	6.0	4.8	6.1	5.6	5.1
5.6	4.8	5.4	5.1	5.1	5.9	5.2	5.7	5.4	4.5	6.1	5.3	5.5	6.7	5.7	4.9
4.8	5.8	5.1

PETAL LENGTH (CM) IRIS SETOSA

1.5	1.7	1.4	1.5	1.5	1.6	1.4	1.1	1.2	1.4	1.7	1.0	1.7	1.9	1.6	1.4
1.5	1.4	1.2	1.3	1.5	1.3	1.6	1.9	1.4	1.6	1.5	1.4	1.6	1.2	1.9	1.5
1.6	1.4	1.3	1.7	1.5	1.7

Test the hypothesis that the average petal length for the *Iris Setosa* is shorter than the average petal length for the *Iris Virginica*. Assume that the two populations have unequal variances.

CHAPTER 10 REGRESSION AND CORRELATION

SIMPLE LINEAR REGRESSION: TWO VARIABLES (SECTIONS 10.1–10.3 OF *UNDERSTANDABLE STATISTICS*)

Chapter 10 of *Understandable Statistics* introduces linear regression. The formula for the correlation coefficient r is given in Section 10.1. Formulas to find the equation of the least squares line,

$$y = a + bx$$

are given in Section 10.2. This section also contains the formula for the coefficient of determination r^2. The equation for the standard error of estimate and the procedure to find a confidence interval for the predicted value of y are given in Section 10.3.

The menu selection ➤**Analyze**➤**Regression**➤**Linear** gives the equation of the least-squares line, the value of the standard error of estimate (Std. Error of the Estimate), the value of the correlation coefficient r, the value of the coefficient of determination r^2 (R Square), as well as several other values such as Adjusted R Square (an unbiased estimate of the population r^2).

The standard deviation (Std Error), t-ratio and P values (Sig.) of the coefficients are also given. The P value is useful for testing the coefficients to see that the population coefficient is not zero (see Section 10.4 of *Understandable Statistics* for a discussion about testing the coefficients). For the time being we do not use these values. An analysis of variance chart is also given. We do not use the analysis of variance chart in the introduction to regression. However, in more advanced treatments of regression, it will be useful.

To find the equation of the least-squares line and the value of the correlation coefficient, use the menu options

➤**Analyze**➤**Regression**➤**Linear**

Dialog Box Responses

Dependent: Enter the variable name of the column containing the responses (that is Y values).

Independent: Enter the variable name of the column containing the explanatory variables (that

is, X values).

[Options]: Check the option [Include Constant in Equation].

[Save]: This is used to find predicted values and their confidence intervals.

To graph the scatter plot and show the least-squares line on the graph, use the menu options

➤**Analyze**➤**Regression**➤**Curve Estimate**

Dialog Box Responses

Dependent: Enter the variable name of the column containing the Y values.

Independent: Enter the variable name of the column containing the X values.

Models: Select Linear.

Check on the options [Include constant in equation] and [Plot models].

We may also use ➤**Graph**➤**Interactive**➤**Scatter Plot** to show the least-square line, least-square equation, as well as optional prediction bands.

Dialog Box Responses

Move the dependent variable to the vertical axis box, independent variable to the horizontal axis box.

Click the [fit] tab, select [Regression], and check the option [Include constant in the equation]. For optional prediction band, check the options [individual], enter the confidence level, and check [total].

Example

Merchandise loss due to shoplifting, damage, and other causes is called shrinkage. Shrinkage is a major concern to retailers. The managers of H.R. Merchandise think there is a relationship between shrinkage and number of clerks on duty. To explore this relationship, a random sample of 7 weeks was selected. During each week the staffing level of sales clerks was kept constant and the dollar value (in hundreds of dollars) of the shrinkage was recorded.

X	10	12	11	15	9	13	8	
Y	19	15	20	9	25	12	31	(in hundreds)

Store the value of X in the first column and name it X. Store the values of Y in the second column with variable name Y.

Use menu choices to give descriptive statistics regarding the values of X and Y. Use commands to draw an (X, Y) scatter plot together with the least-squares line and then to find the equation of the regression line. Find the value of the correlation coefficient, and test to see if it is significant.

(a) First we will use ➤**Analyze**➤**Descriptive Statistics**➤**Explore** and each of the columns X, and Y. The results follow.

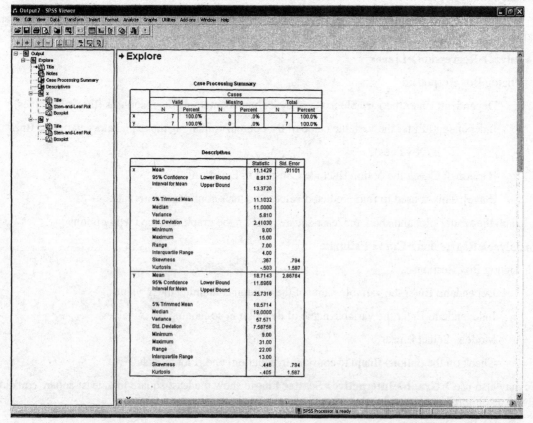

(b) Next we will use **>Analyze>Regression>Curve Estimate** to graph the scatter plot and show the least-squares line on the graph.

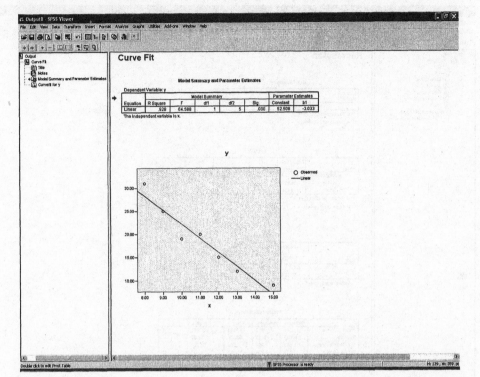

Notice that the coefficients of the equation of the regression line are given in the table above the figure, as well as the value of r^2.

(c) To find out more information about the linear regression model, we use the menu selection **>Analyze>Regression>Linear.** Use Y as Dependent and X as Independent.

The results follow.

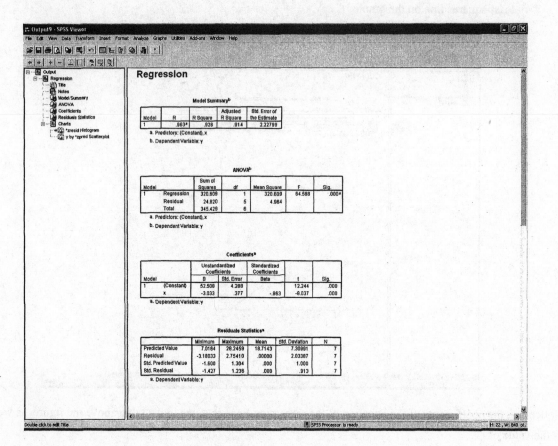

Notice that of the coefficients of the regression equation are given in the table Coefficients. Thus the regression equation is

$$y = 52.5 - 3.03x$$

The value of the standard error of estimate S_e is given as Std. Error of the Estimate = 2.228. The value of r^2 (R Square in the screen) is 92.8%, and the value of r is 96.3%.

(d) Next, let's use the prediction option to find the shrinkage when 14 clerks are available. To do that, we first add the value 14 to the column for variable X, but leave the corresponding cell in the Y column blank, as shown below.

Next, in the **>Analyze>Regression>Linear** dialog box, click on [Save]. In the [Save] tab, check [Unstandardized] under "Predicted values", check [Individual] under "Prediction intervals" with a confidence level 95%.

Click [Continue] and then [OK]. The results follow.

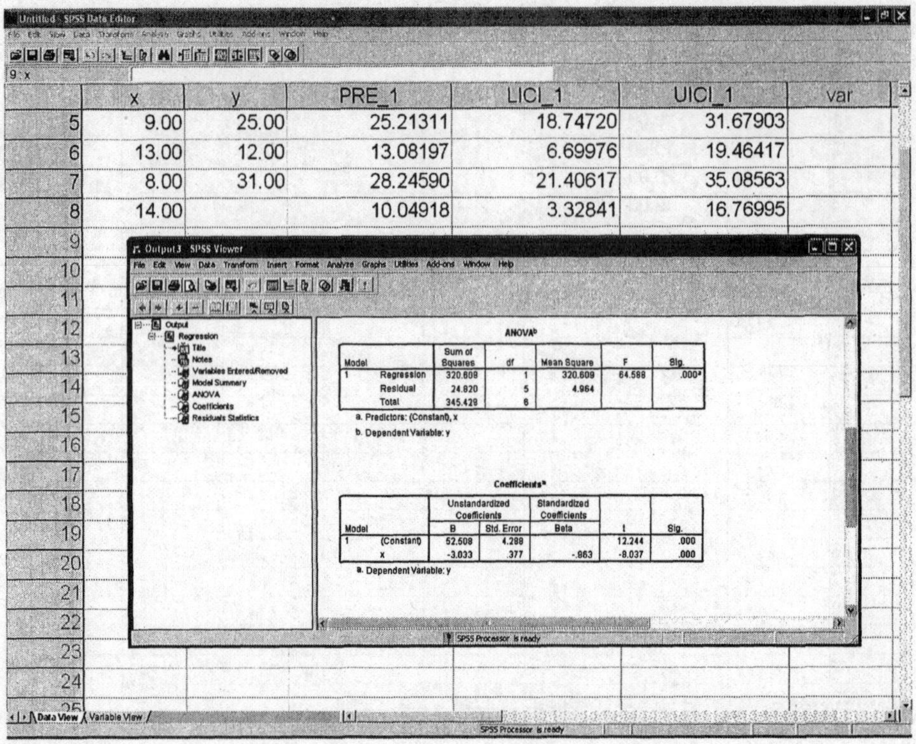

	x	y	PRE_1	LICI_1	UICI_1	var
5	9.00	25.00	25.21311	18.74720	31.67903	
6	13.00	12.00	13.08197	6.69976	19.46417	
7	8.00	31.00	28.24590	21.40617	35.08563	
8	14.00		10.04918	3.32841	16.76995	

The predicted value (PRE_1) of the shrinkage when 14 clerks are on duty is 10.05 hundred dollars, or $1005. A 95% prediction interval goes from 3.33 hundred dollars (LICI_1) to 16.77 hundred dollars (UICI_1) — that is, from $333 to $1677.

(e) Graph a prediction band at a confidence level 95% for predicted values. For this we use **➤Graph➤Interactive➤Scatter Plot.** First we move the dependent variable to the vertical axis box, and independent variable to the horizontal axis box, as shown below.

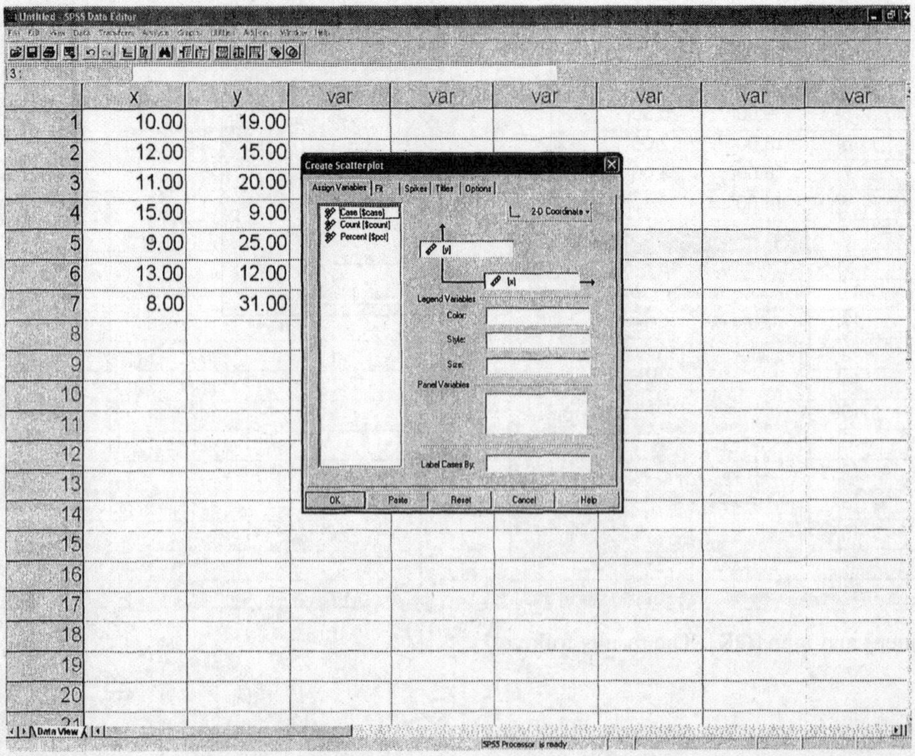

Next, we click the [fit] tab, select [Regression], and check the option [Include constant in the equation]. Then we check the options [individual], enter the confidence level 95%, and check [total].

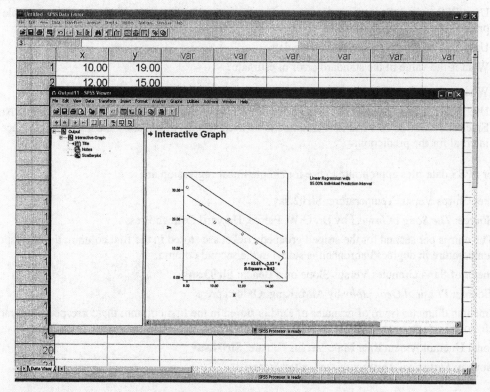

The results follow.

LAB ACTIVITIES FOR SIMPLE LINEAR REGRESSION: TWO VARIABLES

1. Open or retrieve the worksheet **Slr01.sav** from the CD-ROM. This worksheet contains the following data, with the list price in the first column and the best price in the second column. The best price is the best price negotiated by a team from the magazine.

   List Price versus Best Price for a New GMC Pickup Truck

   In the following data pairs (X, Y)

   X = List Price (in $1000) for a GMC Pickup Truck

   Y = Best Price (in $1000) for a GMC Pickup Truck

   SOURCE: CONSUMERS DIGEST, FEBRUARY 1994

(12.4, 11.2)	(14.3, 12.5)	(14.5, 12.7)
(14.9, 13.1)	(16.1, 14.1)	(16.9, 14.8)
(16.5, 14.4)	(15.4, 13.4)	(17.0, 14.9)
(17.9, 15.6)	(18.8, 16.4)	(20.3, 17.7)
(22.4, 19.6)	(19.4, 16.9)	(15.5, 14.0)
(16.7, 14.6)	(17.3, 15.1)	(18.4, 16.1)
(19.2, 16.8)	(17.4, 15.2)	(19.5, 17.0)
(19.7, 17.2)	(21.2, (18.6)	

   (a) Use SPSS to find the least-squares regression line using the best price as the response variable and list price as the explanatory variable.

   (b) Use SPSS to draw a scatter plot of the data.

   (c) What is the value of the standard error of estimate?

   (d) What is the value of the coefficient of determination r^2? of the correlation coefficient r?

   (e) Use the least-squares model to predict the best price for a truck with a list price of $20,000. Note: Enter this value as 20 since X is assumed to be in thousands of dollars. Find a 95% confidence interval for the prediction.

2. Other SPSS data files appropriate to use for simple linear regression are

   Cricket Chirps Versus Temperature: **Slr02.sav**

   Source: *The Song of Insects* by Dr. G.W. Pierce, Harvard College Press

   The chirps per second for the striped grouped cricket are stored in the first column; the corresponding temperature in degrees Fahrenheit is stored in the second column.

   Diameter of Sand Granules Versus Slope on a Beach: **Slr03.sav**

   Source: *Physical Geography* by A.M. King, Oxford press

   The median diameter (mm) of granules of sand is stored in the first column; the corresponding gradient of beach slope in degrees is stored in the second column.

   National Unemployment Rate Male Versus Female: **Slr04.sav**

   Source: *Statistical Abstract of the United States*

The national unemployment rate for adult males is stored in the first column; the corresponding unemployment rate for adult females for the same period of time is stored in the second column.

The data in these worksheets are described in the Appendix of this *Guide*. Select these worksheets and repeat parts (a)–(e) of problem 1, using the first column as the explanatory variable and the second column as the response variable.

3. A psychologist interested in job stress is studying the possible correlation between interruptions and job stress. A clerical worker who is expected to type, answer the phone and do reception work has many interruptions. A store manager who has to help out in various departments as customers make demands also has interruptions. An accountant who is given tasks to accomplish each day and who is not expected to interact with other colleagues or customers except during specified meeting times has few interruptions. The psychologist rated a group of jobs for interruption level. The results follow, with X being interruption level of the job on a scale of 1 to 20, with 20 having the most interruptions, and Y the stress level on a scale of 1 to 50, with 50 the most stressed.

Person	1	2	3	4	5	6	7	8	9	10	11	12
X	9	15	12	18	20	9	5	3	17	12	17	6
Y	20	37	45	42	35	40	20	10	15	39	32	25

(a) Enter the X values into the first column and the Y values into the second column. Use the menu selections ➤**Analyze**➤**Descriptive Statistics**➤**Explore** on the two columns. What is the mean of the Y-values? Of the X-values? What are the respective standard deviations?

(b) Make a scatter plot of the data using the ➤**Analyze**➤**Regression**➤**Curve Estimate** menu selection. From the diagram, do you expect a positive or negative correlation?

(c) Use the ➤**Analyze**➤**Regression**➤**Linear** menu choices to get the value of r. Is this value consistent with your response in part (b)? Also, find the equation of the regression line.

(d) Use the ➤**Analyze**➤**Regression**➤**Linear** menu choices with Y as the response variable and X as the explanatory variable. Get the predicted stress level of jobs with interruption levels of 5, 10, 15, 20. Look at the 95% P.I. intervals. Which are the longest? Why would you expect these results? Find the standard error of estimate. Is value of r the same as that you found in part (c)? What is the equation of the least-squares line?

(e) Use the ➤**Analyze**➤**Regression**➤**Linear** menu option again, this time using X as the response variable and Y as the explanatory variable. Is the equation different than that of part (d)? What about the value of the standard error of estimate (s on your output)? Did it change? Did R Square change?

4. The researcher of problem 3 was able to add to her data. Another random sample of 11 people had their jobs rated for interruption level and were then evaluated for stress level.

Person	13	14	15	16	17	18	19	20	21	22	23
X	4	15	19	13	10	9	3	11	12	15	4
Y	20	35	42	37	40	23	15	32	28	38	12

Add this data to the data in problem 3, and repeat parts (a) through (e). Compare the values of s, the standard error of estimate in parts (c). Did more data tend to reduce the value of s? Look at the 95% P.I. intervals. How do they compare to the corresponding ones of problem 3? Are they shorter or longer? Why would you expect this result?

MULTIPLE REGRESSION (SECTION 10.4 OF *UNDERSTANDABLE STATISTICS*)

The **➤Analyze➤Regression➤Linear** menu choices also do multiple regression.

➤Analyze➤Regression➤Linear

Dialog Box Responses

Dependent: Enter the variable name of the column containing the responses (that is Y values).

Independent: Enter the variable names of the columns containing the explanatory variables.

[Options]: Check the option [Include Constant in Equation].

[Save]: This is used to find predicted values and their confidence intervals.

Example

Bowman Brothers is a large sporting goods store in Denver that has a giant ski sale every year during the month of October. The chief executive officer at Bowman Brothers is studying the following variables regarding the ski sale:

X_1 = Total dollar receipts from October ski sale

X_2 = Total dollar amount spent advertising ski sale on local TV

X_3 = Total dollar amount spent advertising ski sale on local radio

X_4 = Total dollar amount spent advertising ski sale in Denver newspapers

Data for the past eight years is shown below (in thousands of dollars):

Year	1	2	3	4	5	6	7	8
X1	751	768	801	832	775	718	739	780
X2	19	23	27	32	25	18	20	24
X3	14	17	20	24	19	9	10	19
X4	11	15	16	18	12	5	7	14

(a) Enter the data in the first four columns and name them as X1, X2, X3, and X4. Use **➤Analyze➤Descriptive Statistics➤Explore** to study the data.

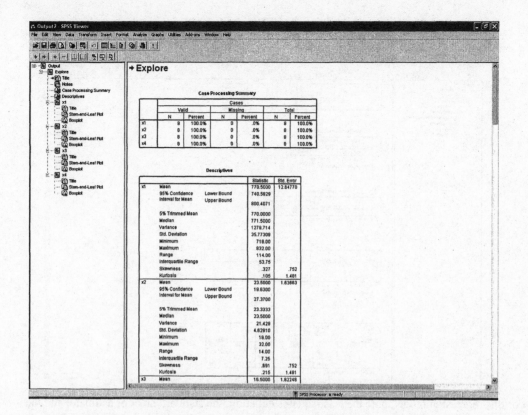

(b) Next use **➤Analyze➤Regression➤Linear** menu option to conduct a linear regression, using X1 as dependent variable, X2, X3 and X4 as independent variables. We may also obtain the correlation between each pair of variables by clicking on [Statistics] and selecting [Descriptives]. The results are shown below (in two screen images).

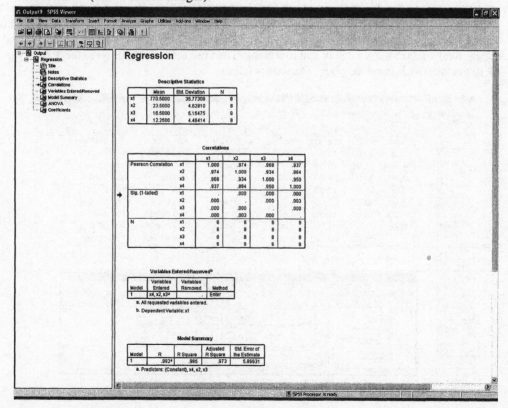

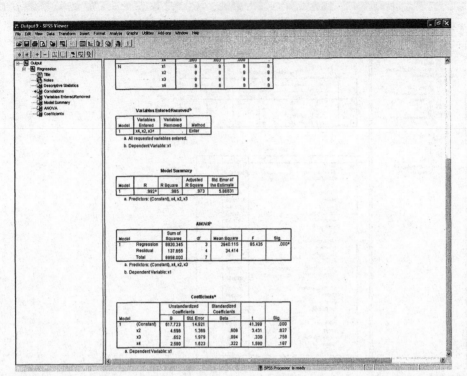

For this regression model, note the least-squares equation, the standard error of estimate, and the coefficient of multiple determination R Square. Look at the P values of the coefficients. Remember we are testing the null hypothesis H_0: $\beta_1 = 0$. against the alternate hypothesis H_1: $\beta_1 \neq 0$. A P value less than α is evidence to reject H_0.

(c) Finally, we use ➤**Analyze**➤**Regression**➤**Linear.** Use X1 as the response variable with 3 predictors X2, X3, X4. Find the predicted value of X1 for X2 = 21, X3 = 11, and X4 = 8. The procedure is similar to that for simple linear regression. Add values 21, 11 and 8 to the columns X2, X3 and X4, respectively. Leave the X4 cell in that row blank. Then use the [Save] option in the ➤**Analyze** ➤**Regression**➤**Linear** dialog box. The results follow.

	x1	x2	x3	x4	PRE_1	LICI_1	UICI_1	var
1	751.00	19.00	14.00	11.00	744.50221	724.74366	764.26076	
2	768.00	23.00	17.00	15.00	775.57248	755.40058	795.74439	
3	801.00	27.00	20.00	16.00	798.90219	780.47990	817.32447	
4	832.00	32.00	24.00	18.00	830.16250	809.03679	851.28821	
5	775.00	25.00	19.00	12.00	778.53276	757.79204	799.27348	
6	718.00	18.00	9.00	5.00	721.06196	700.92764	741.19628	
7	739.00	20.00	10.00	7.00	736.27101	715.89263	756.64938	
8	780.00	24.00	19.00	14.00	778.99489	760.29826	797.69152	
9		21.00	11.00	8.00	744.20162	723.91063	764.49260	
10								
11								

LAB ACTIVITIES FOR MULTIPLE REGRESSION

Solve problems 3–6 of Section 10.4. Each of these problems have SPSS data files stored on the CD-ROM.

Section 10.4 problem #3 (Systolic Blood Pressure Data)

SPSS data file: **Mlr02.sav**

Section 10.4 problem #4 (Test Scores for General Psychology)

SPSS data file: **Mlr03.sav**

Section 10.4 problem #5 (Hollywood Movies data)

SPSS data file: **Mlr04.sav**

Section 10.4 problem #6 (All Greens Franchise Data)

SPSS data file: **Mlr05.sav**

Two additional case studies are available on the CD-ROM. The data are listed in the Appendix. For each of these studies, explore the relationships among the variables.

SPSS data file **Mlr07.sav**

This is file contains data of a case study of public health, income, and population density for small cities in eight midwestern states: Ohio, Indiana, Illinois, Iowa, Missouri, Nebraska, Kansas, and Oklahoma. The data is for a sample of 53 small cities in these states.

X1 = Death Rate per 1000 Residents

X2 = Doctor Availability per 100,000 Residents

X3 = Hospital Availability per 100,000 Residents

X4 = Annual per Capita Income in Thousands of Dollars

X5 = Population Density People per Square Mile

SPSS data file **Mlr06.sav**

This file contains data of a case study of education, crime, and police funding for small cities in ten eastern and southeastern states. The states are New Hampshire, Connecticut, Rhode Island, Maine, New York, Virginia, North Carolina, South Carolina, Georgia, and Florida. The data is for a sample of 50 small cities in these states.

X1 = Total Overall Reported Crime Rate per 1 Million Residents

X2 = Reported Violent Crime Rate per 100,000 Residents

X3 = Annual Police Funding in Dollars per Resident

X4 = Percent of People 25 Years and Older that have had 4 years

of High School

X5 = Percent of 16 to 19 Year-Olds Not in High School and Not High School

Graduates

X6 = Percent of 18 to 24 Year-Olds Enrolled in College

X7 = Percent of People 25 Years and Older with at Least 4 Years of College

CHAPTER 11 CHI-SQUARE AND *F* DISTRIBUTIONS

CHI-SQUARE TESTS OF INDEPENDENCE (SECTION 11.1 OF *UNDERSTANDABLE STATISTICS*)

In chi-square tests of independence we use the hypotheses.

H_0: The two variables are independent

H_1: The two variables are not independent

To use SPSS for tests of independence of two variables, we need to enter the original occurrence records into the data editing screen (or retrieve it from a data file). The command Chi-square then prints a contingency table showing both the observed and expected counts. It computes the sample chi-square value using the following formula, in which E stands for the expected count in a cell and O stands for the observed count in that same cell. The sum takes over all cells.

$$\chi^2 = \sum \frac{(O - E)^2}{E}$$

Then SPSS gives the number of degrees of the chi-square distribution. To conclude the test, use the P value of the sample chi-square statistic if your version of SPSS provides it. Otherwise, compare the calculated chi-square value to a table of the chi-square distribution with the indicated degrees of freedom. Use Table 8 of Appendix II of *Understandable Statistics*. If the calculated sample chi-square value is larger than the value in Table 8 for a specified level of significance, reject H_0.

Use the menu selection

➤**Analyze➤Descriptive Statistics➤Crosstabs**

Dialog Box Responses

Enter one variable as the row variable. Enter the other variable as the column variable.

Click on [Cells] and check [Observed] as well as [Expected] for Counts. Then click on [Continue].

Click on [Statistics] and check [Chi-square]. Then click on [Continue].

Click on [OK].

Example

Let us first use a small sample to illustrate the procedure. Suppose among ten students four are male and six are female. When they vote on a certain issue, one male gives "yes", other three male students vote "no", two female students vote "yes", and other four vote "no". Use the Chi-square test at the 5% level of significance to determine whether the two variables gender and votes are independent of each other.

First, enter the data under two variables Gender and Vote (both are of the type "string") as shown below.

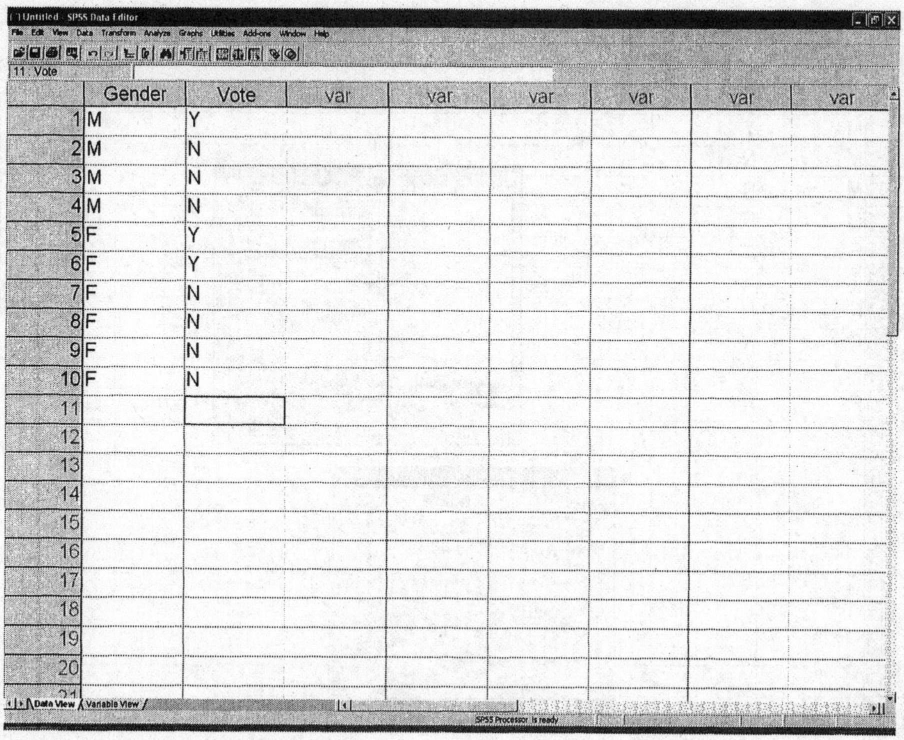

Now, use the menu options **>Analyze>Descriptive Statistics>Crosstabs.** Use Gender as the row variable, and Vote as the column variable. Click on [Cells] and check [Observed] as well as [Expected] for Counts, as shown below.

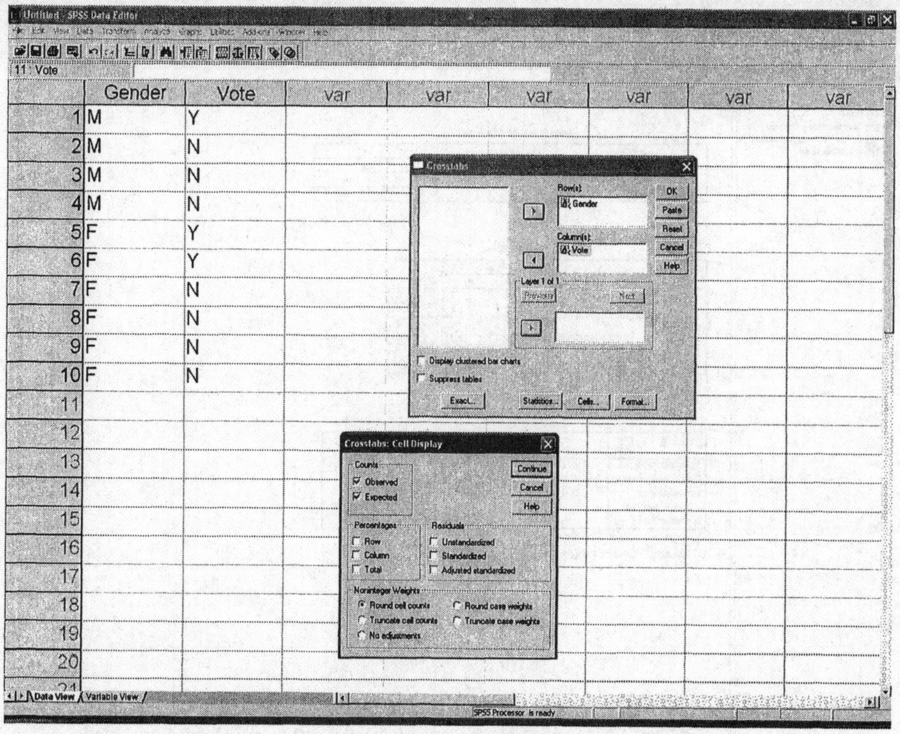

Click on [Continue]. Then click on [Statistics] and check [Chi-square], as shown below.

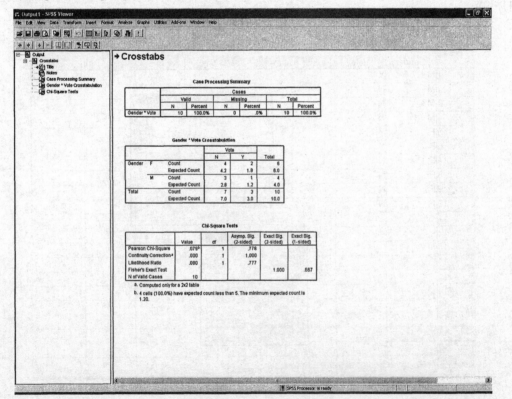

Click [Continue] and then [OK]. The results follow.

Since the *P* value (Asymp. Sig.) equals 0.778 which is greater than 0.05, we do not reject the null hypothesis.

Example

Consider an example that involves a relatively large data. A computer programming aptitude test has been developed for high school seniors. The test designers claim that scores on the test are independent of the type of school the student attends: rural, suburban, urban. A study involving a random sample of students from these types of institutions yielded the following contingency table. Use the Chi-Square command to compute the sample chi-square value, and to determine the degrees of freedom of the chi-square distribution. Then determine if type or school and test score are independent at the $\alpha = 0.05$ level of significance.

School Type

Score	Rural	Suburban	Urban
200–299	33	65	83
300–399	45	79	95
400–500	21	47	63

SPSS conducts the Chi-square test on the original occurrence records data, as illustrated in the previous example. Therefore, first create a data file containing original records under two variables Score and Region. As shown below, using the above contingency table we can use ▶**Transform**▶**Compute** to see that there are 181 scores between 200 and 299, 219 scores between 300 and 399, and 131 scores between 400 and 500. Altogether there are 531 scores.

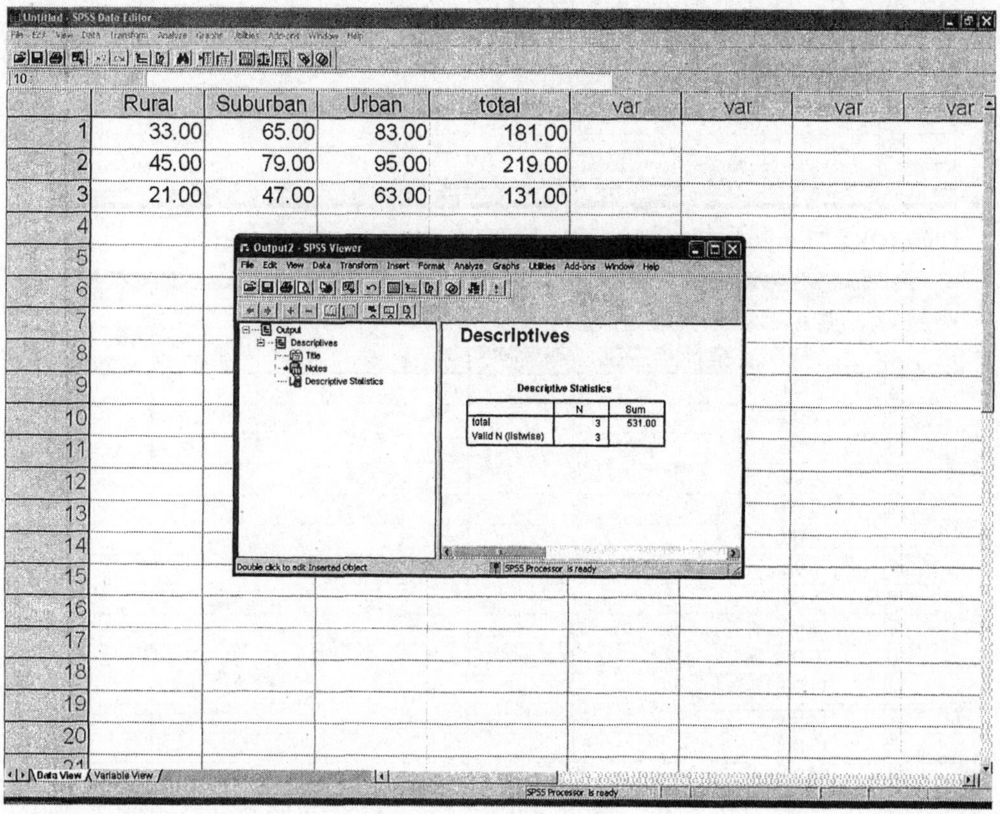

Next, in a new data editing screen, define three variables: id (type: numeric), Score (type: string), and Region (type: string). The variable id contains the record number, which equals the row number, and is used to make the entering of data more convenient as described below. We create the data following these steps:

1. Roll the screen down to row 531 and enter a value (any value) for the variable id. This defines the length of the data.

2. Use **➤Transform➤Compute.** Select id as the target variable. Under the Function group select All. In that subgroup select the function $Casenum, which will assign the case number (row number) to the variable id.

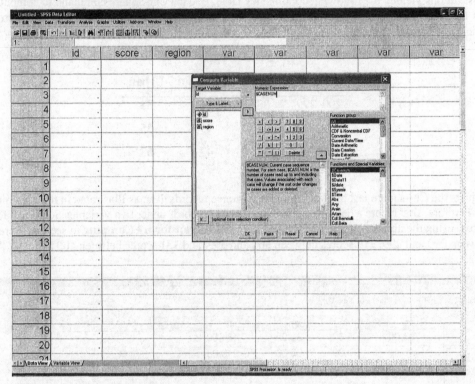

Click [OK]. The results follow.

3. Now use ▶**Transform**▶**Compute** to enter data for the variables Score and Region. For the variable Score, we enter "200-299" when 1 <= id <= 181, enter "300-399" when 182 <= id <= 400, enter "400-500" when 401 <= id <= 531. For the variable Region, we enter "rural" when 1 <= id <= 33 or 182 <= id <= 226 or 401 <= id <=421, and similarly, we enter "suburban" when 34 <= id <= 98 or 227 <= id <= 305 or 422 <= id <=468, enter "urban" when 99 <= id <= 181 or 306 <= id <= 400 or 469 <= id <=531. Here is how to enter "200-299" for Score when 1 <= id <= 181. Use ▶**Transform**▶**Compute**, enter Score as the target variable, enter "200-299" as string expression, click on [if], then choose [include if case satisfies condition] and enter the condition id >= 1 & id <= 181.

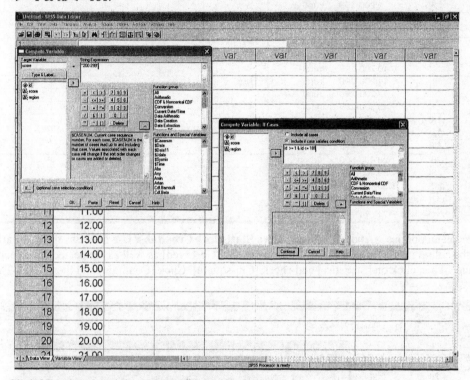

Click [Continue] and then [OK]. The results follow.

	id	score	region	var	var	var	var	var
1	1.00	200-299						
2	2.00	200-299						
3	3.00	200-299						
4	4.00	200-299						
5	5.00	200-299						
6	6.00	200-299						
7	7.00	200-299						
8	8.00	200-299						
9	9.00	200-299						
10	10.00	200-299						
11	11.00	200-299						
12	12.00	200-299						
13	13.00	200-299						
14	14.00	200-299						
15	15.00	200-299						
16	16.00	200-299						
17	17.00	200-299						
18	18.00	200-299						
19	19.00	200-299						
20	20.00	200-299						
21	21.00	200-299						

Similarly enter the rest of the data. The finished data will appear as below (window split feature is used to display the beginning and the end of data.)

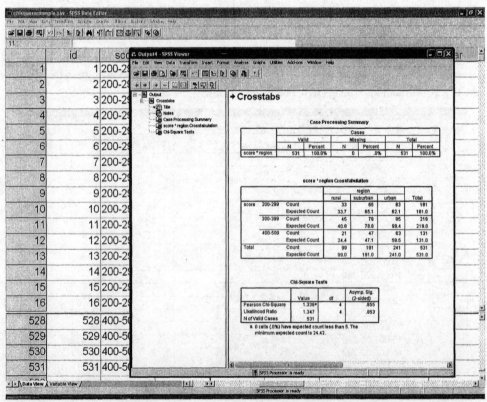

Now use the menu options **➤Analyze➤Descriptive Statistics➤Crosstabs.** Use Score as the row variable, and Region as the column variable. Click on [Cells] and check [Observed] as well as [Expected] for Counts. Click on [Continue]. Then click on [Statistics] and check [Chi-square]. Click [Continue] and then [OK]. The results follow.

Since the *P* value, 0.855, is greater than $\alpha = 0.05$, we do not reject the null hypothesis.

LAB ACTIVITIES FOR CHI-SQUARE TESTS OF INDEPENDENCE

Use SPSS to compute the sample chi square value. If your version of SPSS produces the P value of the sample chi-square statistic, conclude the test using P values. Otherwise, use Table 8 of *Understandable Statistics* to find the chi-square value for the given α and degrees of freedom. Compare the sample chi-square value to the value found in Table 8 to conclude the test.

1. *We Care Auto Insurance* had its staff of actuaries conduct a study to see if vehicle type and loss claim are independent. A random sample of auto claims over six months gives the information in the contingency table.

Total Loss Claims per Year per Vehicle

Type of vehicle	$0–999	$1000–2999	$3000–5999	$6000+
Sports car	20	10	16	8
Truck	16	25	33	9
Family Sedan	40	68	17	7
Compact	52	73	48	12

Test the claim that car type and loss claim are independent. Use $\alpha = 0.05$.

2. An educational specialist is interested in comparing three methods of instruction.

SL–standard lecture with discussion

TV–video taped lectures with no discussion

IM–individualized method with reading assignments

and tutoring, but no lectures.

The specialist conducted a study of these methods to see if they are independent. A course was taught using each of the three methods and a standard final exam was given at the end. Students were put into the different method sections at random. The course type and test results are shown in the next contingency table.

Final Exam Score

Course Type	< 60	60–69	70–79	80–89	90–100
SL	10	4	70	31	25
TV	8	3	62	27	23
IM	7	2	58	25	22

Test the claim that the instruction method and final exam test scores are independent, using $\alpha = 0.01$.

ANALYSIS OF VARIANCE (ANOVA) (SECTION 11.5 OF *UNDERSTANDABLE STATISTICS*)

Section 11.5 of *Understandable Statistics* introduces single factor analysis of variance (also called one-way ANOVA). We consider several populations which are each assumed to follow a normal distribution. The standard deviations of the populations are assumed to be approximately equal. ANOVA provides a method to compare several different populations to see if the means are the same. Let population 1 have mean μ_1, population 2 have mean μ_2, and so forth. The hypotheses of ANOVA are

$H_0: u_1 = u_2 = \ldots = u_n$

H_1: not all the means are equal.

In SPSS use the menu selection **≻Analyze≻Compare Means≻One-Way ANOVA** to perform one-way ANOVA. Use two variables. One variable (column) contains data from all populations. The other variable contains population numbers (called levels) to indicate which population the corresponding data is from. An analysis of variance table is printed, as well a confidence interval for the mean of each level.

≻Analyze≻Compare Means≻One-Way ANOVA

Dialog Box Responses

Dependent list: Enter the columns containing the data.

Factor: Enter the column containing levels.

Click on [Options] and choose [Descriptive]

Example

A psychologists has developed a series of tests to measure a person's level of depression. The composite scores range from 50 to 100 with 100 representing the most severe depression level. A random sample of 12 patients with approximately the same depression level, as measured by the tests, was divided into 3 different treatment groups. Then, one month after treatment was completed, the depression level of each patient was again evaluated. The after-treatment depression levels are given below.

Treatment 1	70	65	82	83	71
Treatment 2	75	62	81		
Treatment 3	77	60	80	75	

First we enter the data as shown below.

Now use ➤Analyze➤Compare Means➤One-Way ANOVA. Enter Depression as the dependent variable. Enter Treatment as the factor. Click on [Options] and choose [Descriptive].

Click [Continue] and then [OK]. The results follow.

Since the level of significance $\alpha = 0.05$ is less than the P value of 0.965, we do not reject H_0.

LAB ACTIVITIES FOR ANALYSIS OF VARIANCE

1. A random sample of 20 overweight adults were randomly divided into 4 groups. Each group was given a different diet plan, and the weight loss for each individual after 3 months follows:

Plan 1	18	10	20	25	17
Plan 2	28	12	22	17	16
Plan 3	16	20	24	8	17
Plan 4	14	17	18	5	16

Test the claim that the population mean weight loss is the same for the four diet plans, at the 5% level of significance.

2. A psychologist is studying the time it takes rats to respond to stimuli after being given doses of different tranquilizing drugs. A random sample of 18 rats were divided into 3 groups. Each group was given a different drug. The response time to stimuli was measured (in seconds). The results follow.

Drug A	3.1	2.5	2.2	1.5	0.7	2.4
Drug B	4.2	2.5	1.7	3.5	1.2	3.1
Drug C	3.3	2.6	1.7	3.9	2.8	3.5

Test the claim that the population mean response times for the three drugs is the same, at the 5% level of significance.

3. A research group is testing various chemical combinations designed to neutralize and buffer the effects of acid rain on lakes. A random sample of 18 lakes of similar size in the same region have all been affected in the same way by acid rain. The lakes are divided into four groups and each group of lakes is sprayed with a different chemical combination. An acidity index is then take after treatment. The index ranges from 60 to 100, with 100 indicating the greatest acid rain pollution. The results follow.

Combination I	63	55	72	81	75
Combination II	78	56	75	73	82
Combination III	59	72	77	60	
	72	81	66	71	

Test the claim that the population mean acidity index after each of the four treatments is the same at the 0.01 level of significance.

APPENDIX

Descriptions of Data Sets

on the

HM StatPass CD-ROM

FOR

UNDERSTANDABLE STATISTICS

EIGHTH EDITION

I.   General

There are over 100 data sets saved in Excel, Minitab Portable, SPSS, TI-83 Plus and TI-84 Plus/ASCII formats to accompany *Understandable Statistics,* 8th edition. These files can be found on the HM StatPass CD-ROM packaged with the student textbook. You may also visit the Brase/Brase statistics site at http://math.college.hmco.com/students to view these data sets online. The data sets are organized by category.

C. The formats are
    1. Excel files in subdirectory Excel_8e. These files have suffix .xls
    2. Minitab portable files in subdirectory Minitab_8e. These files have suffix .mtp
    3. TI-83 Plus and TI-84 Plus/ASCII files in subdirectory TI8384_8e. These files have suffix .txt
    4. SPSS files in subdirectory SPSS_8e. These files have suffix .sav

II. Suggestions for using the data sets

1. Single variable large sample (file name prefix Svls)
   These data sets are appropriate for
       Graphs: Histograms, box plots
       Descriptive statistics: Mean, median, mode, variance, standard deviation, coefficient of variation, 5 number summary
       Inferential statistics: Confidence intervals for the population mean, hypothesis tests of a single mean

2. Single variable small sample (file name prefix Svss)
       Graphs: Histograms, box plots,
       Descriptive statistics: Mean, median, mode, variance, standard deviation, coefficient of variation, 5 number summary
       Inferential statistics: Confidence intervals for the population mean, hypothesis tests of a single mean

3. Time series data (file name prefix Tscc)
       Graphs: Time plots, control charts about the mean utilizing individual data for the data sets so designated, P charts for the data sets so designated

4. Two independent data sets (file name prefix Tvis)
       Graphs: Histograms, box plots for each data set
       Descriptive statistics: Mean, median, mode, variance, standard deviation, coefficient of variation, 5-number summary for each data set
       Inferential statistics: Confidence intervals for the difference of means, hypothesis tests for the difference of means

5. Paired data, dependent samples (file name prefix Tvds)
       Descriptive statistics: Mean, median, mode, variance, standard deviation, coefficient of variation, 5 number summary for the difference of the paired data values.
       Inferential statistics: Hypothesis tests for the difference of means (paired data)

6. Data pairs for simple linear regression (file name prefix Slr)
       Graphs: Scatter plots, for individual variables histograms and box plots
       Descriptive statistics:
         (a) Mean, median, mode, variance, standard deviation, coefficient of variation, 5 number summary for individual variables.
         (b) Least squares line, sample correlation coefficient, sample coefficient of determination
       Inferential statistics: Testing ρ, confidence intervals for β, testing β

7. Data for multiple linear regression (file name prefix Mlr)
       Graphs:
       Descriptive statistics: Histograms, box plots for individual variables
         (a) Mean, median, mode, variance, standard deviation, coefficient of variation, 5 number summary for individual variables.
         (b) Least squares line, sample coefficient of determination
       Inferential statistics: confidence intervals for coefficients, testing coefficients

8. Data for one-way ANOVA (file name prefix Owan)
  Graphs: Histograms, box plots for individual samples
  Descriptive statistics: Mean, median, mode, variance, standard deviation, coefficient of variation, 5 number summary for individual samples.
  Inferential statistics: One-way ANOVA

9. Data for two-way ANOVA (file name prefix Twan)
  Graphs: Histograms, box plots for individual samples
  Descriptive statistics: Mean, median, mode, variance, standard deviation, coefficient of variation, 5 number summary for data in individual cells.
  Inferential statistics: Two-way ANOVA

III. Hints for transferring ASCII text files (suffix .txt) to the calculators

For TI-83 Plus, the TI-GRAPH LINK™ (sold separately) is an accessory that links a TI-83Plus with a personal computer. This link cable and software is necessary for transferring files from a computer to the calculator.

From the TI-Graph Link 83Plus main menu, use the following selections.
  Tools ➤ Import ➤ ASCII data
In the resulting dialog box, select a list (L1, L2, L3, L4, L5,L6) and click **OK**.

Then the **Import ASCII List** dialog box appears. Select the drive and folders that contain the data file you wish to import. Note that the folder containing ASCII text files is called **TI8384_8e.** The file type is **Text files(*.txt)**. Select the desired file and click **OK.**

The next dialog box is titled **Save As**. This is the dialog box that converts the text file (*.txt) to a TI-83Plus list file. Select the computer drive and folder where the new TI-83Plus file is to be stored. Click **OK**. If necessary, overwrite existing file.

Now the data in your file will be displayed in the TI-83Plus list box on the right side of the computer screen. Click on **Send to RAM**. Replace contents of list if necessary. Now the original data is stored in the designated list on your calculator. Use the appropriate statistical functions on the TI-83Plus calculator to analyze your data.

For TI-84 Plus, the TI Connect™ software and a USB computer cable are included with the TI-84 Plus calculator. Follow the instructions in the TI-84 Plus guidebook to install the TI Connect™ software and to transfer files from a computer to the calculator.

Single Variable Large Sample ($n \geq 30$)
File name prefix: Svls followed by the number of the data file

01. Disney Stock Volume (Single Variable Large Sample $n \geq 30$)
The following data represents the number of shares of Disney stock (in hundreds of shares) sold for a random sample of 60 trading days
Reference: *The Denver Post*, Business section

12584	9441	18960	21480	10766	13059	8589	4965
4803	7240	10906	8561	6389	14372	18149	6309
13051	12754	10860	9574	19110	29585	21122	14522
17330	18119	10902	29158	16065	10376	10999	17950
15418	12618	16561	8022	9567	9045	8172	13708
11259	10518	9301	5197	11259	10518	9301	5197
6758	7304	7628	14265	13054	15336	14682	27804
16022	24009	32613	19111				

File names   Excel: Svls01.xls
                             Minitab: Svls01.mtp
                             SPSS: Svls01.sav
                             TI-83 Plus and TI-84 Plus/ASCII: Svls01.txt

02. Weights of Pro Football Players (Single Variable Large Sample $n \geq 30$)
The following data represents weights in pounds of 50 randomly selected pro football linebackers.
Reference: The Sports Encyclopedia Pro Football

225	230	235	238	232	227	244	222
250	226	242	253	251	225	229	247
239	223	233	222	243	237	230	240
255	230	245	240	235	252	245	231
235	234	248	242	238	240	240	240
235	244	247	250	236	246	243	255
241	245						

File names   Excel: Svls02.xls
                             Minitab: Svls02.mtp
                             SPSS: Svls02.sav
                             TI-83 Plus and TI-84 Plus/ASCII: Svls02.txt

03. Heights of Pro Basketball Players (Single Variable Large Sample $n \geq 30$)
The following data represents heights in feet of 65 randomly selected pro basketball players.
Reference: All-Time Player Directory, The Official NBA Encyclopedia

6.50	6.25	6.33	6.50	6.42	6.67	6.83	6.82
6.17	7.00	5.67	6.50	6.75	6.54	6.42	6.58
6.00	6.75	7.00	6.58	6.29	7.00	6.92	6.42
5.92	6.08	7.00	6.17	6.92	7.00	5.92	6.42
6.00	6.25	6.75	6.17	6.75	6.58	6.58	6.46
5.92	6.58	6.13	6.50	6.58	6.63	6.75	6.25
6.67	6.17	6.17	6.25	6.00	6.75	6.17	6.83
6.00	6.42	6.92	6.50	6.33	6.92	6.67	6.33
6.08							

File names
Excel: Svls03.xls
Minitab: Svls03.mtp
SPSS: Svls03.sav
TI-83 Plus and TI-84 Plus/ASCII: Svls03.txt

04. Miles per Gallon Gasoline Consumption (Single Variable Large Sample $n \geq 30$)

The following data represents miles per gallon gasoline consumption (highway) for a random sample of 55 makes and models of passenger cars.
Reference: Environmental Protection Agency

30	27	22	25	24	25	24	15
35	35	33	52	49	10	27	18
20	23	24	25	30	24	24	24
18	20	25	27	24	32	29	27
24	27	26	25	24	28	33	30
13	13	21	28	37	35	32	33
29	31	28	28	25	29	31	

File names
Excel: Svls04.xls
Minitab: Svls04.mtp
SPSS: Svls04.sav
TI-83 Plus and TI-84 Plus/ASCII: Svls04.txt

05. Fasting Glucose Blood Tests (Single Variable Large Sample $n \geq 30$)

The following data represents glucose blood level (mg/100ml) after a 12-hour fast for a random sample of 70 women.
Reference: American J. Clin. Nutr. Vol. 19, 345-351

45	66	83	71	76	64	59	59
76	82	80	81	85	77	82	90
87	72	79	69	83	71	87	69
81	76	96	83	67	94	101	94
89	94	73	99	93	85	83	80
78	80	85	83	84	74	81	70
65	89	70	80	84	77	65	46
80	70	75	45	101	71	109	73
73	80	72	81	63	74		

File names
Excel: Svls05.xls
Minitab: Svls05.mtp
SPSS: Svls05.sav
TI-83 Plus and TI-84 Plus/ASCII: Svls05.txt

06. Number of Children in Rural Canadian Families (Single Variable Large Sample $n \geq 30$)

The following data represents the number of children in a random sample of 50 rural Canadian families.
Reference: American Journal Of Sociology Vol. 53, 470-480

11	13	4	14	10	2	5	0
0	3	9	2	5	2	3	3
3	4	7	1	9	4	3	3
2	6	0	2	6	5	9	5
4	3	2	5	2	2	3	5
14	7	6	6	2	5	3	4
6	1						

File names
Excel: Svls06.xls

Minitab: Svls06.mtp
SPSS: Svls06.sav
TI-83 Plus and TI-84 Plus/ASCII: Svls06.txt

07.　Children as a % of Population (Single Variable Large Sample $n \geq 30$)
The following data represent percentage of children in the population for a random sample of 72 Denver neighborhoods.
Reference: The Piton Foundation, Denver, Colorado

```
30.2  18.6  13.6  36.9  32.8  19.4  12.3  39.7  22.2  31.2
36.4  37.7  38.8  28.1  18.3  22.4  26.5  20.4  37.6  23.8
22.1  53.2   6.8  20.7  31.7  10.4  21.3  19.6  41.5  29.8
14.7  12.3  17.0  16.7  20.7  34.8   7.5  19.0  27.2  16.3
24.3  39.8  31.1  34.3  15.9  24.2  20.3  31.2  30.0  33.1
29.1  39.0  36.0  31.8  32.9  26.5   4.9  19.5  21.0  24.2
12.1  38.3  39.3  20.2  24.0  28.6  27.1  30.0  60.8  39.2
21.6  20.3
```

File names　　　　Excel: Svls07.xls
　　　　　　　　　Minitab: Svls07.mtp
　　　　　　　　　SPSS: Svls07.sav
　　　　　　　　　TI-83 Plus and TI-84 Plus/ASCII: Svls07.txt

08.　Percentage Change in Household Income (Single Variable Large Sample $n \geq 30$)
The following data represent the percentage change in household income over a five-year period for a random sample of $n = 78$ Denver neighborhoods.
Reference: The Piton Foundation, Denver, Colorado

```
27.2  25.2  25.7  80.9  26.9  20.2  25.4  26.9  26.4  26.3
27.5  38.2  20.9  31.3  23.5  26.0  35.8  30.9  15.5  24.8
29.4  11.7  32.6  32.2  27.6  27.5  28.7  28.0  15.6  20.0
21.8  18.4  27.3  13.4  14.7  21.6  26.8  20.9  32.7  29.3
21.4  29.0   7.2  25.7  25.5  39.8  26.6  24.2  33.5  16.0
29.4  26.8  32.0  24.7  24.2  29.8  25.8  18.2  26.0  26.2
21.7  27.0  23.7  28.0  11.2  26.2  21.6  23.7  28.3  34.1
40.8  16.0  50.5  54.1   3.3  23.5  10.1  14.8
```

File names　　　　Excel: Svls08.xls
　　　　　　　　　Minitab: Svls08.mtp
　　　　　　　　　SPSS: Svls08.sav
　　　　　　　　　TI-83 Plus and TI-84 Plus/ASCII: Svls08.txt

09.　Crime Rate per 1,000 Population (Single Variable Large Sample $n \geq 30$)
The following data represent the crime rate per 1,000 population for a random sample of 70 Denver neighborhoods.
Reference: The Piton Foundation, Denver, Colorado

```
 84.9   45.1  132.1  104.7  258.0   36.3   26.2  207.7
 58.5   65.3   42.5   53.2  172.6   69.2  179.9   65.1
 32.0   38.3  185.9   42.4   63.0   86.4  160.4   26.9
154.2  111.0  139.9   68.2  127.0   54.0   42.1  105.2
 77.1  278.0   73.0   32.1   92.7  704.1  781.8   52.2
 65.0   38.6   22.5  157.3   63.1  289.1   52.7  108.7
 66.3   69.9  108.7   96.9   27.1  105.1   56.2   80.1
 59.6   77.5   68.9   35.2   65.4  123.2  130.8   70.7
 25.1   62.6   68.6  334.5   44.6   87.1
```

File names   Excel: Svls09.xls
          Minitab: Svls09.mtp
          SPSS: Svls09.sav
          TI-83 Plus and TI-84 Plus/ASCII: Svls09.txt

10. **Percentage Change in Population (Single Variable Large Sample $n \geq 30$)**
The following data represent the percentage change in population over a nine-year period for a random sample of 64 Denver neighborhoods.
Reference: The Piton Foundation, Denver, Colorado

6.2	5.4	8.5	1.2	5.6	28.9	6.3	10.5	-1.5	17.3
21.6	-2.0	-1.0	3.3	2.8	3.3	28.5	-0.7	8.1	32.6
68.6	56.0	19.8	7.0	38.3	41.2	4.9	7.8	7.8	97.8
5.5	21.6	32.5	-0.5	2.8	4.9	8.7	-1.3	4.0	32.2
2.0	6.4	7.1	8.8	3.0	5.1	-1.9	-2.6	1.6	7.4
10.8	4.8	1.4	19.2	2.7	71.4	2.5	6.2	2.3	10.2
1.9	2.3	-3.3	2.6						

File names   Excel: Svls10.xls
          Minitab: Svls10.mtp
          SPSS: Svls10.sav
          TI-83 Plus and TI-84 Plus/ASCII: Svls10.txt

11. **Thickness of the Ozone Column (Single Variable Large Sample $n \geq 30$)**
The following data represent the January mean thickness of the ozone column above Arosa, Switzerland (Dobson units: one milli-centimeter ozone at standard temperature and pressure). The data is from a random sample of years from 1926 on.
Reference: Laboratorium fuer Atmosphaerensphysik, Switzerland

324	332	362	383	335	349	354	319	360	329
400	341	315	368	361	336	349	347	338	332
341	352	342	361	318	337	300	352	340	371
327	357	320	377	338	361	301	331	334	387
336	378	369	332	344					

File names   Excel: Svls11.xls
          Minitab: Svls11.mtp
          SPSS: Svls11.sav
          TI-83 Plus and TI-84 Plus/ASCII: Svls11.txt

12. **Sun Spots (Single Variable Large Sample $n \geq 30$)**
The following data represent the January mean number of sunspots. The data is taken from a random sample of Januarys from 1749 to 1983.
Reference: Waldmeir, M, *Sun Spot Activity*, International Astronomical Union Bulletin

12.5	14.1	37.6	48.3	67.3	70.0	43.8	56.5	59.7	24.0
12.0	27.4	53.5	73.9	104.0	54.6	4.4	177.3	70.1	54.0
28.0	13.0	6.5	134.7	114.0	72.7	81.2	24.1	20.4	13.3
9.4	25.7	47.8	50.0	45.3	61.0	39.0	12.0	7.2	11.3
22.2	26.3	34.9	21.5	12.8	17.7	34.6	43.0	52.2	47.5
30.9	11.3	4.9	88.6	188.0	35.6	50.5	12.4	3.7	18.5
115.5	108.5	119.1	101.6	59.9	40.7	26.5	23.1	73.6	165.0
202.5	217.4	57.9	38.7	15.3	8.1	16.4	84.3	51.9	58.0
74.7	96.0	48.1	51.1	31.5	11.8	4.5	78.1	81.6	68.9

File names   Excel: Svls12.xls
                              Minitab: Svls12.mtp
                              SPSS: Svls12.sav
                              TI-83 Plus and TI-84 Plus/ASCII: Svls12.txt

13. **Motion of Stars (Single Variable Large Sample $n \geq 30$)**
     The following data represent the angular motions of stars across the sky due to the stars own velocity. A random sample of stars from the M92 global cluster was used. Units are arc seconds per century.
     Reference: Cudworth, K.M., Astronomical Journal, Vol. 81, p 975-982

     0.042 0.048 0.019 0.025 0.028 0.041 0.030 0.051 0.026
     0.040 0.018 0.022 0.048 0.045 0.019 0.028 0.029 0.018
     0.033 0.035 0.019 0.046 0.021 0.026 0.026 0.033 0.046
     0.023 0.036 0.024 0.014 0.012 0.037 0.034 0.032 0.035
     0.015 0.027 0.017 0.035 0.021 0.016 0.036 0.029 0.031
     0.016 0.024 0.015 0.019 0.037 0.016 0.024 0.029 0.025
     0.022 0.028 0.023 0.021 0.020 0.020 0.016 0.016 0.016
     0.040 0.029 0.025 0.025 0.042 0.022 0.037 0.024 0.046
     0.016 0.024 0.028 0.027 0.060 0.045 0.037 0.027 0.028
     0.022 0.048 0.053

     File names   Excel: Svls13.xls
                              Minitab: Svls13.mtp
                              SPSS: Svls13.sav
                              TI-83 Plus and TI-84 Plus/ASCII: Svls13.txt

14. **Arsenic and Ground Water (Single Variable Large Sample $n \geq 30$)**
     The following data represent (naturally occurring) concentration of arsenic in ground water for a random sample of 102 Northwest Texas wells. Units are parts per billion.
     Reference: Nichols, C.E. and Kane, V.E., Union Carbide Technical Report K/UR-1

      7.6 10.4 13.5   4.0 19.9 16.0 12.0 12.2 11.4 12.7
      3.0 10.3 21.4 19.4   9.0   6.5 10.1   8.7   9.7   6.4
      9.7 63.0 15.5 10.7 18.2   7.5   6.1   6.7   6.9   0.8
     73.5 12.0 28.0 12.6   9.4   6.2 15.3   7.3 10.7 15.9
      5.8   1.0   8.6   1.3 13.7   2.8   2.4   1.4   2.9 13.1
     15.3   9.2 11.7   4.5   1.0   1.2   0.8   1.0   2.4   4.4
      2.2   2.9   3.6   2.5   1.8   5.9   2.8   1.7   4.6   5.4
      3.0   3.1   1.3   2.6   1.4   2.3   1.0   5.4   1.8   2.6
      3.4   1.4 10.7 18.2   7.7   6.5 12.2 10.1   6.4 10.7
      6.1   0.8 12.0 28.1   9.4   6.2   7.3   9.7 62.1 15.5
      6.4   9.5

     File names   Excel: Svls14.xls
                              Minitab: Svls14.mtp
                              SPSS: Svls14.sav
                              TI-83 Plus and TI-84 Plus/ASCII: Svls14.txt

15. **Uranium In Ground Water (Single Variable Large Sample $n \geq 30$)**
     The following data represent (naturally occurring) concentrations of uranium in ground water for a random sample of 100 Northwest Texas wells. Units are parts per billion.
     Reference: Nichols, C.E. and Kane, V.E., Union Carbide Technical Report K/UR-1

      8.0 13.7   4.9   3.1 78.0   9.7   6.9 21.7 26.8
     56.2 25.3   4.4 29.8 22.3   9.5 13.5 47.8 29.8
     13.4 21.0 26.7 52.5   6.5 15.8 21.2 13.2 12.3

5.7	11.1	16.1	11.4	18.0	15.5	35.3	9.5	2.1
10.4	5.3	11.2	0.9	7.8	6.7	21.9	20.3	16.7
2.9	124.2	58.3	83.4	8.9	18.1	11.9	6.7	9.8
15.1	70.4	21.3	58.2	25.0	5.5	14.0	6.0	11.9
15.3	7.0	13.6	16.4	35.9	19.4	19.8	6.3	2.3
1.9	6.0	1.5	4.1	34.0	17.6	18.6	8.0	7.9
56.9	53.7	8.3	33.5	38.2	2.8	4.2	18.7	12.7
3.8	8.8	2.3	7.2	9.8	7.7	27.4	7.9	11.1
24.7								

File names   Excel: Svls15.xls
                    Minitab: Svls15.mtp
                    SPSS: Svls15.sav
                    TI-83 Plus and TI-84 Plus/ASCII: Svls15.txt

16. **Ground Water pH (Single Variable Large Sample $n \geq 30$)**
A pH less than 7 is acidic, and a pH above 7 is alkaline. The following data represent pH levels in ground water for a random sample of 102 Northwest Texas wells.
Reference: Nichols, C.E. and Kane, V.E., Union Carbide Technical Report K/UR-1

7.6	7.7	7.4	7.7	7.1	8.2	7.4	7.5	7.2	7.4
7.2	7.6	7.4	7.8	8.1	7.5	7.1	8.1	7.3	8.2
7.6	7.0	7.3	7.4	7.8	8.1	7.3	8.0	7.2	8.5
7.1	8.2	8.1	7.9	7.2	7.1	7.0	7.5	7.2	7.3
8.6	7.7	7.5	7.8	7.6	7.1	7.8	7.3	8.4	7.5
7.1	7.4	7.2	7.4	7.3	7.7	7.0	7.3	7.6	7.2
8.1	8.2	7.4	7.6	7.3	7.1	7.0	7.0	7.4	7.2
8.2	8.1	7.9	8.1	8.2	7.7	7.5	7.3	7.9	8.8
7.1	7.5	7.9	7.5	7.6	7.7	8.2	8.7	7.9	7.0
8.8	7.1	7.2	7.3	7.6	7.1	7.0	7.0	7.3	7.2
7.8	7.6								

File names   Excel: Svls16.xls
                    Minitab: Svls16.mtp
                    SPSS: Svls16.sav
                    TI-83 Plus and TI-84 Plus/ASCII: Svls16.txt

17. **Static Fatigue 90% Stress Level (Single Variable Large Sample $n \geq 30$)**
Kevlar Epoxy is a material used on the NASA space shuttle. Strands of this epoxy were tested at 90% breaking strength. The following data represent time to failure in hours at the 90% stress level for a random sample of 50 epoxy strands.
Reference: R.E. Barlow University of California, Berkeley

0.54	1.80	1.52	2.05	1.03	1.18	0.80	1.33	1.29	1.11
3.34	1.54	0.08	0.12	0.60	0.72	0.92	1.05	1.43	3.03
1.81	2.17	0.63	0.56	0.03	0.09	0.18	0.34	1.51	1.45
1.52	0.19	1.55	0.02	0.07	0.65	0.40	0.24	1.51	1.45
1.60	1.80	4.69	0.08	7.89	1.58	1.64	0.03	0.23	0.72

File names   Excel: Svls17.xls
                    Minitab: Svls17.mtp
                    SPSS: Svls17.sav
                    TI-83 Plus and TI-84 Plus/ASCII: Svls17.txt

18. **Static Fatigue 80% Stress Level (Single Variable Large Sample $n \geq 30$)**

Kevlar Epoxy is a material used on the NASA space shuttle. Strands of this epoxy were tested at 80% breaking strength. The following data represent time to failure in hours at the 80% stress level for a random sample of 54 epoxy strands.
Reference: R.E. Barlow University of California, Berkeley

152.2	166.9	183.8	8.5	1.8	118.0	125.4	132.8	10.6
29.6	50.1	202.6	177.7	160.0	87.1	112.6	122.3	124.4
131.6	140.9	7.5	41.9	59.7	80.5	83.5	149.2	137.0
301.1	329.8	461.5	739.7	304.3	894.7	220.2	251.0	269.2
130.4	77.8	64.4	381.3	329.8	451.3	346.2	663.0	49.1
31.7	116.8	140.2	334.1	285.9	59.7	44.1	351.2	93.2

File names   Excel: Svls18.xls
                        Minitab: Svls18.mtp
                        SPSS: Svls18.sav
                        TI-83 Plus and TI-84 Plus/ASCII: Svls18.txt

19. Tumor Recurrence (Single Variable Large Sample $n \geq 30$)
Certain kinds of tumors tend to recur. The following data represents the length of time in months for a tumor to recur after chemotherapy (sample size, 42).
Reference: Byar, D.P, *Urology* Vol. 10, p 556-561

19	18	17	1	21	22	54	46	25	49
50	1	59	39	43	39	5	9	38	18
14	45	54	59	46	50	29	12	19	36
38	40	43	41	10	50	41	25	19	39
27	20								

File names   Excel: Svls19.xls
                        Minitab: Svls19.mtp
                        SPSS: Svls19.sav
                        TI-83 Plus and TI-84 Plus/ASCII: Svls19.txt

20. Weight of Harvest (Single Variable Large Sample $n \geq 30$)
The following data represent the weights in kilograms of maize harvest from a random sample of 72 experimental plots on the island of St Vincent (Caribbean).
Reference: Springer, B.G.F. *Proceedings, Caribbean Food Corps. Soc.* Vol. 10 p 147-152

24.0	27.1	26.5	13.5	19.0	26.1	23.8	22.5	20.0
23.1	23.8	24.1	21.4	26.7	22.5	22.8	25.2	20.9
23.1	24.9	26.4	12.2	21.8	19.3	18.2	14.4	22.4
16.0	17.2	20.3	23.8	24.5	13.7	11.1	20.5	19.1
20.2	24.1	10.5	13.7	16.0	7.8	12.2	12.5	14.0
22.0	16.5	23.8	13.1	11.5	9.5	22.8	21.1	22.0
11.8	16.1	10.0	9.1	15.2	14.5	10.2	11.7	14.6
15.5	23.7	25.1	29.5	24.5	23.2	25.5	19.8	17.8

File names   Excel: Svls20.xls
                        Minitab: Svls20.mtp
                        SPSS: Svls20.sav
                        TI-83 Plus and TI-84 Plus/ASCII: Svls20.txt

21. Apple Trees (Single Variable Large Sample $n \geq 30$)
The following data represent trunk girth (mm) of a random sample of 60 four-year-old apple trees at East Malling Research Station (England)
Reference: S.C. Pearce, University of Kent at Canterbury

108	99	106	102	115	120	120	117	122	142
106	111	119	109	125	108	116	105	117	123
103	114	101	99	112	120	108	91	115	109
114	105	99	122	106	113	114	75	96	124
91	102	108	110	83	90	69	117	84	142
122	113	105	112	117	122	129	100	138	117

File names   Excel: Svls21.xls
                    Minitab: Svls21.mtp
                    SPSS: Svls21.sav
                    TI-83 Plus and TI-84 Plus/ASCII: Svls21.txt

22. **Black Mesa Archaeology (Single Variable Large Sample $n \geq 30$)**
The following data represent rim diameters (cm) of a random sample of 40 bowls found at Black Mesa archaeological site. The diameters are estimated from broken pot shards.
Reference: Michelle Hegmon, Crow Canyon Archaeological Center, Cortez, Colorado

17.2	15.1	13.8	18.3	17.5	11.1	7.3	23.1	21.5	19.7
17.6	15.9	16.3	25.7	27.2	33.0	10.9	23.8	24.7	18.6
16.9	18.8	19.2	14.6	8.2	9.7	11.8	13.3	14.7	15.8
17.4	17.1	21.3	15.2	16.8	17.0	17.9	18.3	14.9	17.7

File names   Excel: Svls22.xls
                    Minitab: Svls22.mtp
                    SPSS: Svls22.sav
                    TI-83 Plus and TI-84 Plus/ASCII: Svls22.txt

23. **Wind Mountain Archaeology (Single Variable Large Sample $n \geq 30$)**
The following data represent depth (cm) for a random sample of 73 significant archaeological artifacts at the Wind Mountain excavation site.
Reference: Woosley, A. and McIntyre, A. *Mimbres Mogolion Archaology*, University New Mexico press.

85	45	75	60	90	90	115	30	55	58
78	120	80	65	65	140	65	50	30	125
75	137	80	120	15	45	70	65	50	45
95	70	70	28	40	125	105	75	80	70
90	68	73	75	55	70	95	65	200	75
15	90	46	33	100	65	60	55	85	50
10	68	99	145	45	75	45	95	85	65
65	52	82							

File names   Excel: Svls23.xls
                    Minitab: Svls23.mtp
                    SPSS: Svls23.sav
                    TI-83 Plus and TI-84 Plus/ASCII: Svls23.txt

24. **Arrow Heads (Single Variable Large Sample $n \geq 30$)**
The following data represent length (cm) of a random sample of 61 projectile points found at the Wind Mountain Archaeological site.
Reference: Woosley, A. and McIntyre, A. *Mimbres Mogolion Archaology*, University New Mexico press.

3.1	4.1	1.8	2.1	2.2	1.3	1.7	3.0	3.7	2.3
2.6	2.2	2.8	3.0	3.2	3.3	2.4	2.8	2.8	2.9
2.9	2.2	2.4	2.1	3.4	3.1	1.6	3.1	3.5	2.3
3.1	2.7	2.1	2.0	4.8	1.9	3.9	2.0	5.2	2.2

2.6   1.9   4.0   3.0   3.4   4.2   2.4   3.5   3.1   3.7
3.7   2.9   2.6   3.6   3.9   3.5   1.9   4.0   4.0   4.6
1.9

File names   Excel: Svls24.xls
                          Minitab: Svls24.mtp
                          SPSS: Svls24.sav
                          TI-83 Plus and TI-84 Plus/ASCII: Svls24.txt

25. **Anasazi Indian Bracelets (Single Variable Large Sample $n \geq 30$)**
The following data represent the diameter (cm) of shell bracelets and rings found at the Wind Mountain archaeological site.
Reference: Woosley, A. and McIntyre, A. *Mimbres Mogolion Archaology*, University New Mexico press.

5.0   5.0   8.0   6.1   6.0   5.1   5.9   6.8   4.3   5.5
7.2   7.0   5.0   5.6   5.3   7.0   3.4   8.2   4.3   5.2
1.5   6.1   4.0   6.0   5.5   5.2   5.2   5.2   5.5   7.2
6.0   6.2   5.2   5.0   4.0   5.7   5.1   6.1   5.7   7.3
7.3   6.7   4.2   4.0   6.0   7.1   7.3   5.5   5.8   8.9
7.5   8.3   6.8   4.9   4.0   6.2   7.7   5.0   5.2   6.8
6.1   7.2   4.4   4.0   5.0   6.0   6.2   7.2   5.8   6.8
7.7   4.7   5.3

File names   Excel: Svls25.xls
                          Minitab: Svls25.mtp
                          SPSS: Svls25.sav
                          TI-83 Plus and TI-84 Plus/ASCII: Svls25.txt

26. **Pizza Franchise Fees (Single Variable Large Sample $n \geq 30$)**
The following data represent annual franchise fees (in thousands of dollars) for a random sample of 36 pizza franchises.
Reference: *Business Opportunities Handbook*

25.0   15.5   7.5   19.9   18.5   25.5   15.0   5.5   15.2   15.0
14.9   18.5   14.5   29.0   22.5   10.0   25.0   35.5   22.1   89.0
17.5   33.3   17.5   12.0   15.5   25.5   12.5   17.5   12.5   35.0
30.0   21.0   35.5   10.5   5.5   20.0

File names   Excel: Svls26.xls
                          Minitab: Svls26.mtp
                          SPSS: Svls26.sav
                          TI-83 Plus and TI-84 Plus/ASCII: Svls26.txt

27. **Pizza Franchise Start-up Requirement (Single Variable Large Sample $n \geq 30$)**
The following data represent annual start-up cost (in thousands of dollars) for a random sample of 36 pizza franchises.
Reference: *Business Opportunities Handbook*

40   25   50   129   250   128   110   142   25   90
75   100   500   214   275   50   128   250   50   75
30   40   185   50   175   125   200   150   150   120
95   30   400   149   235   100

File names   Excel: Svls27.xls
                          Minitab: Svls27.mtp
                          SPSS: Svls27.sav

TI-83 Plus and TI-84 Plus/ASCII: Svls27.txt

28. College Degrees (Single Variable Large Sample $n \geq 30$)
The following data represent percentages of the adult population with college degrees. The sample is from a random sample of 68 Midwest counties.
Reference: *County and City Data Book* 12th edition, U.S. Department of Commerce

9.9	9.8	6.8	8.9	11.2	15.5	9.8	16.8	9.9	11.6
9.2	8.4	11.3	11.5	15.2	10.8	16.3	17.0	12.8	11.0
6.0	16.0	12.1	9.8	9.4	9.9	10.5	11.8	10.3	11.1
12.5	7.8	10.7	9.6	11.6	8.8	12.3	12.2	12.4	10.0
10.0	18.1	8.8	17.3	11.3	14.5	11.0	12.3	9.1	12.7
5.6	11.7	16.9	13.7	12.5	9.0	12.7	11.3	19.5	30.7
9.4	9.8	15.1	12.8	12.9	17.5	12.3	8.2		

File names   Excel: Svls28.xls
                          Minitab: Svls28.mtp
                          SPSS: Svls28.sav
                          TI-83 Plus and TI-84 Plus/ASCII: Svls28.txt

29. Poverty Level (Single Variable Large Sample $n \geq 30$)
The following data represent percentages of all persons below the poverty level. The sample is from a random collection of 80 cities in the Western U.S.
Reference: *County and City Data Book* 12th edition, U.S. Department of Commerce

12.1	27.3	20.9	14.9	4.4	21.8	7.1	16.4	13.1
9.4	9.8	15.7	29.9	8.8	32.7	5.1	9.0	16.8
21.6	4.2	11.1	14.1	30.6	15.4	20.7	37.3	7.7
19.4	18.5	19.5	8.0	7.0	20.2	6.3	12.9	13.3
30.0	4.9	14.4	14.1	22.6	18.9	16.8	11.5	19.2
21.0	11.4	7.8	6.0	37.3	44.5	37.1	28.7	9.0
17.9	16.0	20.2	11.5	10.5	17.0	3.4	3.3	15.6
16.6	29.6	14.9	23.9	13.6	7.8	14.5	19.6	31.5
28.1	19.2	4.9	12.7	15.1	9.6	23.8	10.1	

File names   Excel: Svls29.xls
                          Minitab: Svls29.mtp
                          SPSS: Svls29.sav
                          TI-83 Plus and TI-84 Plus/ASCII: Svls29.txt

30. Working at Home (Single Variable Large Sample $n \geq 30$)
The following data represent percentages of adults whose primary employment involves working at home. The data is from a random sample of 50 California cities.
Reference: *County and City Data Book* 12th edition, U.S. Department of Commerce

4.3	5.1		3.1	8.7	4.0	5.2	11.8	3.4	8.5	3.0
4.3	6.0		3.7	3.7	4.0	3.3	2.8	2.8	2.6	4.4
7.0	8.0		3.7	3.3	3.7	4.9	3.0	4.2	5.4	6.6
2.4	2.5		3.5	3.3	5.5	9.6	2.7	5.0	4.8	4.1
3.8	4.8	14.3	9.2	3.8	3.6	6.5	2.6	3.5	8.6	

File names   Excel: Svls30.xls
                          Minitab: Svls30.mtp
                          SPSS: Svls30.sav
                          TI-83 Plus and TI-84 Plus/ASCII: Svls30.txt

Single Variable Small Sample (*n* < 30)
File name prefix: SVSS followed by the number of the data file

01.   Number of Pups in Wolf Den (Single Variable Small Sample *n* < 30)
The following data represent the number of wolf pups per den from a random sample of 16 wolf dens.
Reference: *The Wolf in the Southwest: The Making of an Endangered Species*, Brown, D.E., University of Arizona Press

```
5   8   7   5   3   4   3   9
5   8   5   6   5   6   4   7
```

File names   Excel: Svss01.xls
                    Minitab: Svss01.mtp
                    SPSS: Svss01.sav
                    TI-83 Plus and TI-84 Plus/ASCII: Svss01.txt

02.   Glucose Blood Level (Single Variable Small Sample *n* < 30)
The following data represent glucose blood level (mg/100ml) after a 12-hour fast for a random sample of 6 tests given to an individual adult female.
Reference: *American J. Clin. Nutr. Vol. 19*, p345-351

```
83   83   86   86   78   88
```

File names   Excel: Svss02.xls
                    Minitab: Svss02.mtp
                    SPSS: Svss02.sav
                    TI-83 Plus and TI-84 Plus/ASCII: Svss02.txt

03.   Length of Remission (Single Variable Small Sample *n* < 30)
The drug 6-mP (6-mercaptopurine) is used to treat leukemia. The following data represent the length of remission in weeks for a random sample of 21 patients using 6-mP.
Reference: E.A. Gehan, University of Texas Cancer Center

```
10    7   32   23   22    6   16   34   32   25
11   20   19    6   17   35    6   13    9    6
10
```

File names   Excel: Svss03.xls
                    Minitab: Svss03.mtp
                    SPSS: Svss03.sav
                    TI-83 Plus and TI-84 Plus/ASCII: Svss03.txt

04.   Entry Level Jobs (Single Variable Small Sample *n* < 30)
The following data represent percentage of entry-level jobs in a random sample of 16 Denver neighborhoods.
Reference: The Piton Foundation, Denver, Colorado

```
8.9   22.6   18.5    9.2    8.2   24.3   15.3    3.7
9.2   14.9    4.7   11.6   16.5   11.6    9.7    8.0
```

File names   Excel: Svss04.xls
                    Minitab: Svss04.mtp
                    SPSS: Svss04.sav

TI-83 Plus and TI-84 Plus/ASCII: Svss04.txt

05. Licensed Child Care Slots (Single Variable Small Sample *n* < 30)
The following data represents the number of licensed childcare slots in a random sample of 15 Denver neighborhoods.
Reference: The Piton Foundation, Denver, Colorado

```
523  106  184  121  357  319  656  170
241  226  741  172  266  423  212
```

File names   Excel: Svss05.xls
                       Minitab: Svss05.mtp
                       SPSS: Svss05.sav
                       TI-83 Plus and TI-84 Plus/ASCII: Svss05.txt

06. Subsidized Housing (Single Variable Small Sample *n* < 30)
The following data represent the percentage of subsidized housing in a random sample of 14 Denver neighborhoods.
Reference: The Piton Foundation, Denver, Colorado

```
10.2  11.8   9.7  22.3   6.8  10.4  11.0
 5.4   6.6  13.7  13.6   6.5  16.0  24.8
```

File names   Excel: Svss06.xls
                       Minitab: Svss06.mtp
                       SPSS: Svss06.sav
                       TI-83 Plus and TI-84 Plus/ASCII: Svss06.txt

07. Sulfate in Ground Water (Single Variable Small Sample *n* < 30)
The following data represent naturally occurring amounts of sulfate SO_4 in well water. Units: parts per million. The data is from a random sample of 24 water wells in Northwest Texas.
Reference: Union Carbide Corporation Technical Report K/UR-1

```
1850  1150  1340  1325  2500  1060  1220  2325  460
2000  1500  1775   620  1950   780   840  2650  975
 860   495  1900  1220  2125   990
```

File names   Excel: Svss07.xls
                       Minitab: Svss07.mtp
                       SPSS: Svss07.sav
                       TI-83 Plus and TI-84 Plus/ASCII: Svss07.txt

08. Earth's Rotation Rate (Single Variable Small Sample *n* < 30)
The following data represent changes in the earth's rotation (i.e. day length). Units 0.00001 second. The data is for a random sample of 23 years.
Reference: *Acta Astron. Sinica* Vol. 15, p79-85

```
-12  110   78  126  -35  104  111   22  -31   92
 51   36  231  -13   65  119   21  104  112  -15
137  139  101
```

File names   Excel: Svss08.xls
                       Minitab: Svss08.mtp
                       SPSS: Svss08.sav
                       TI-83 Plus and TI-84 Plus/ASCII: Svss08.txt

09. Blood Glucose (Single Variable Small Sample *n* < 30)
The following data represent glucose levels (mg/100ml) in the blood for a random sample of 27 non-obese adult subjects.
Reference: *Diabetologia*, Vol. 16, p 17-24

80	85	75	90	70	97	91	85	90	85
105	86	78	92	93	90	80	102	90	90
99	93	91	86	98	86	92			

File names   Excel: Svss09.xls
                    Minitab: Svss09.mtp
                    SPSS: Svss09.sav
                    TI-83 Plus and TI-84 Plus/ASCII: Svss09.txt

10. Plant Species (Single Variable Small Sample *n* < 30)
The following data represent the observed number of native plant species from random samples of study plots on different islands in the Galapagos Island chain.
Reference: *Science*, Vol. 179, p 893-895

23	26	33	73	21	35	30	16	3	17
9	8	9	19	65	12	11	89	81	7
23	95	4	37	28					

File names   Excel: Svss10.xls
                    Minitab: Svss10.mtp
                    SPSS: Svss10.sav
                    TI-83 Plus and TI-84 Plus/ASCII: Svss10.txt

11. Apples (Single Variable Small Sample *n* < 30)
The following data represent mean fruit weight (grams) of apples per tree for a random sample of 28 trees in an agricultural experiment.
Reference: *Aust. J. Agric res.*, Vol. 25, p783-790

85.3	86.9	96.8	108.5	113.8	87.7	94.5	99.9	92.9
67.3	90.6	129.8	48.9	117.5	100.8	94.5	94.4	98.9
96.0	99.4	79.1	108.5	84.6	117.5	70.0	104.4	127.1
135.0								

File names   Excel: Svss11.xls
                    Minitab: Svss11.mtp
                    SPSS: Svss11.sav
                    TI-83 Plus and TI-84 Plus/ASCII: Svss11.txt

Time Series Data for Control Charts or P Charts
File name prefix: Tscc followed by the number of the data file

01. Yield of Wheat (Time Series for Control Chart)
The following data represent annual yield of wheat in tonnes (one ton = 1.016 tonne) for an experimental plot of land at Rothamsted experiment station U.K. over a period of thirty consecutive years.
Reference: Rothamsted Experiment Station U.K.

We will use the following target production values:
target mu = 2.6 tonnes
target sigma = 0.40 tonnes

1.73	1.66	1.36	1.19	2.66	2.14	2.25	2.25	2.36	2.82
2.61	2.51	2.61	2.75	3.49	3.22	2.37	2.52	3.43	3.47
3.20	2.72	3.02	3.03	2.36	2.83	2.76	2.07	1.63	3.02

File names	Excel: Tscc01.xls
	Minitab: Tscc01.mtp
	SPSS: Tscc01.sav
	TI-83 Plus and TI-84 Plus/ASCII: Tscc01.txt

02. Pepsico Stock Closing Prices (Time Series for Control Chart)
The following data represent a random sample of 25 weekly closing prices in dollars per share of Pepsico stock for 25 consecutive days.
Reference: *The Denver Post*
The long term estimates for weekly closings are
target mu = 37 dollars per share
target sigma = 1.75 dollars per share

37.000	36.500	36.250	35.250	35.625	36.500	37.000	36.125
35.125	37.250	37.125	36.750	38.000	38.875	38.750	39.500
39.875	41.500	40.750	39.250	39.000	40.500	39.500	40.500
37.875							

File names	Excel: Tscc02.xls
	Minitab: Tscc02.mtp
	SPSS: Tscc02.sav
	TI-83 Plus and TI-84 Plus/ASCII: Tscc02.txt

03. Pepsico Stock Volume Of Sales (Time Series for Control Chart)
The following data represent volume of sales (in hundreds of thousands of shares) of Pepsico stock for 25 consecutive days.
Reference: The Denver Post, business section
For the long term mu and sigma use
target mu = 15
target sigma = 4.5

19.00	29.63	21.60	14.87	16.62	12.86	12.25	20.87
23.09	21.71	11.14	5.52	9.48	21.10	15.64	10.79
13.37	11.64	7.69	9.82	8.24	12.11	7.47	12.67
12.33							

File names	Excel: Tscc03.xls
	Minitab: Tscc03.mtp

SPSS: Tscc03.sav
TI-83 Plus and TI-84 Plus/ASCII: Tscc03.txt

04. Futures Quotes For The Price Of Coffee Beans (Time Series for Control Chart)
The following data represent futures options quotes for the price of coffee beans (dollars per pound) for 20 consecutive business days.
Use the following estimated target values for pricing
target mu = $2.15
target sigma = $0.12

2.300 2.360 2.270 2.180 2.150 2.180 2.120 2.090 2.150 2.200
2.170 2.160 2.100 2.040 1.950 1.860 1.910 1.880 1.940 1.990

File names   Excel: Tscc04.xls
Minitab: Tscc04.mtp
SPSS: Tscc04.sav
TI-83 Plus and TI-84 Plus/ASCII: Tscc04.txt

05. Incidence Of Melanoma Tumors (Time Series for Control Chart)
The following data represent number of cases of melanoma skin cancer (per 100,000 population) in Connecticut for each of the years 1953 to 1972.
Reference: *Inst. J. Cancer*, Vol. 25, p95-104
Use the following long term values (mu and sigma)
target mu = 3
target sigma = 0.9

2.4 2.2 2.9 2.5 2.6 3.2 3.8 4.2 3.9 3.7
3.3 3.7 3.9 4.1 3.8 4.7 4.4 4.8 4.8 4.8

File names   Excel: Tscc05.xls
Minitab: Tscc05.mtp
SPSS: Tscc05.sav
TI-83 Plus and TI-84 Plus/ASCII: Tscc05.txt

06. Percent Change In Consumer Price Index (Time Series for Control Chart)
The following data represent annual percent change in consumer price index for a sequence of recent years.
Reference: *Statistical Abstract Of The United States*
Suppose an economist recommends the following long term target values for mu and sigma.
target mu = 4.0%
target sigma = 1.0%

1.3   1.3   1.6   2.9   3.1   4.2   5.5   5.7   4.4   3.2
6.2   11.0 9.1   5.8   6.5   7.6   11.3 13.5 10.3 6.2
3.2   4.3   3.6   1.9   3.6   4.1   4.8   5.4   4.2   3.0

File names   Excel: Tscc06.xls
Minitab: Tscc06.mtp
SPSS: Tscc06.sav
TI-83 Plus and TI-84 Plus/ASCII: Tscc06.txt

07. Broken Eggs (Time Series for *P* Chart)
The following data represent the number of broken eggs in a case of 10 dozen eggs (120 eggs). The data represent 21 days or 3 weeks of deliveries to a small grocery store.

14   23   18   9   17   14   12   11   10   17

```
12   25   18   15   19   22   14   22   15   10
13
```

File names   Excel: Tscc07.xls
                           Minitab: Tscc07.mtp
                           SPSS: Tscc07.sav
                           TI-83 Plus and TI-84 Plus/ASCII: Tscc07.txt

08. Theater Seats (Time Series for *P* Chart)

The following data represent the number of empty seats at each show of a Community Theater production. The theater has 325 seats. The show ran 18 times.

```
28   19   41   38   32   47   53   17   29
32   31   27   25   33   26   62   15   12
```

File names   Excel: Tscc08.xls
                           Minitab: Tscc08.mtp
                           SPSS: Tscc08.sav
                           TI-83 Plus and TI-84 Plus/ASCII: Tscc08.txt

09. Rain (Time Series for *P* Chart)

The following data represents the number of rainy days at Waikiki beach, Hawaii during the prime tourist season of December and January (62 days). The data was taken over a 20-year period.

```
21   27   19   17    6    9   25   36   23   26
12   16   27   41   18    8   10   22   15   24
```

File names   Excel: Tscc09.xls
                           Minitab: Tscc09.mtp
                           SPSS: Tscc09.sav
                           TI-83 Plus and TI-84 Plus/ASCII: Tscc09.txt

10. Quality Control (Time Series for *P* Chart)

The following data represent the number of defective toys in a case of 500 toys coming off a production line. Every day for 35 consecutive days, a case was selected at random.

```
26   23   33   49   28   42   29   41   27   25
35   21   48   12    5   15   36   55   13   16
93    8   38   11   39   18    7   33   29   42
26   19   47   53   61
```

File names   Excel: Tscc10.xls
                           Minitab: Tscc10.mtp
                           SPSS: Tscc10.sav
                           TI-83 Plus and TI-84 Plus/ASCII: Tscc10.txt

Two Variable Independent Samples
File name prefix: Tvis followed by the number of the data file

01. Heights of Football Players Versus Heights of Basketball Players
   (Two variable independent large samples)
   The following data represent heights in feet of 45 randomly selected pro football players, and 40 randomly selected pro basketball players.
   Reference: Sports Encyclopedia Pro Football, and Official NBA Basketball Encyclopedia

   $X1$ = heights (ft.) of pro football players

6.33	6.50	6.50	6.25	6.50	6.33	6.25	6.17	6.42	6.33
6.42	6.58	6.08	6.58	6.50	6.42	6.25	6.67	5.91	6.00
5.83	6.00	5.83	5.08	6.75	5.83	6.17	5.75	6.00	5.75
6.50	5.83	5.91	5.67	6.00	6.08	6.17	6.58	6.50	6.25
6.33	5.25	6.67	6.50	5.83					

   $X2$ = heights (ft.) of pro basketball players

6.08	6.58	6.25	6.58	6.25	5.92	7.00	6.41	6.75	6.25
6.00	6.92	6.83	6.58	6.41	6.67	6.67	5.75	6.25	6.25
6.50	6.00	6.92	6.25	6.42	6.58	6.58	6.08	6.75	6.50
6.83	6.08	6.92	6.00	6.33	6.50	6.58	6.83	6.50	6.58

   File names   Excel: Tvis01.xls
                       Minitab: Tvis01.mtp
                       SPSS: Tvis01.sav
                       TI-83 Plus and TI-84 Plus/ASCII:
                                   X1 data is stored in Tvis01L1.txt
                                   X2 data is stored in Tvis01L2.txt

02. Petal Length for Iris Virginica Versus Petal Length for Iris Setosa
   (Two variable independent large samples)
   The following data represent petal length (cm.) for a random sample of 35 iris virginica and a random sample of 38 iris setosa
   Reference: Anderson, E., *Bull. Amer. Iris Soc.*

   $X1$ = petal length (c.m.) iris virginica

5.1	5.8	6.3	6.1	5.1	5.5	5.3	5.5	6.9	5.0	4.9	6.0	4.8	6.1	5.6	5.1
5.6	4.8	5.4	5.1	5.1	5.9	5.2	5.7	5.4	4.5	6.1	5.3	5.5	6.7	5.7	4.9
4.8	5.8	5.1													

   $X2$ = petal length (c.m.) iris setosa

1.5	1.7	1.4	1.5	1.5	1.6	1.4	1.1	1.2	1.4	1.7	1.0	1.7	1.9	1.6	1.4
1.5	1.4	1.2	1.3	1.5	1.3	1.6	1.9	1.4	1.6	1.5	1.4	1.6	1.2	1.9	1.5
1.6	1.4	1.3	1.7	1.5	1.7										

   File names   Excel: Tvis02.xls
                       Minitab: Tvis02.mtp
                       SPSS: Tvis02.sav
                       TI-83 Plus and TI-84 Plus/ASCII:
                                   X1 data is stored in Tvis02L1.txt
                                   X2 data is stored in Tvis02L2.txt

**03. Sepal Width Of Iris Versicolor Versus Iris Virginica
(Two variable independent larage samples)**
The following data represent sepal width (cm.) for a random sample of 40 iris versicolor and a random sample of 42 iris virginica
Reference: Anderson, E., *Bull. Amer. Iris Soc.*

X1 = sepal width (c.m.) iris versicolor
3.2 3.2 3.1 2.3 2.8 2.8 3.3 2.4 2.9 2.7 2.0 3.0 2.2 2.9 2.9 3.1
3.0 2.7 2.2 2.5 3.2 2.8 2.5 2.8 2.9 3.0 2.8 3.0 2.9 2.6 2.4 2.4
2.7 2.7 3.0 3.4 3.1 2.3 3.0 2.5

X2 = sepal width (c.m.) iris virginica
3.3 2.7 3.0 2.9 3.0 3.0 2.5 2.9 2.5 3.6 3.2 2.7 3.0 2.5 2.8 3.2
3.0 3.8 2.6 2.2 3.2 2.8 2.8 2.7 3.3 3.2 2.8 3.0 2.8 3.0 2.8 3.8
2.8 2.8 2.6 3.0 3.4 3.1 3.0 3.1 3.1 3.1

File names	Excel: Tvis03.xls
	Minitab: Tvis03.mtp
	SPSS: Tvis03.sav
	TI-83 Plus and TI-84 Plus/ASCII:
	X1 data is stored in Tvis03L1.txt
	X2 data is stored in Tvis03L2.txt

04. Archaeology, Ceramics (Two variable independent large samples)
The following data represent independent random samples of shard counts of painted ceramics found at the Wind Mountain archaeological site.
Reference: Woosley and McIntyre, *Mimbres Mogollon Archaeology*, Univ. New Mexico Press

X1 = count Mogollon red on brown

52	10	8	71	7	31	24	20	17	5
16	75	25	17	14	33	13	17	12	19
67	13	35	14	3	7	9	19	16	22
7	10	9	49	6	13	24	45	14	20
3	6	30	41	26	32	14	33	1	48
44	14	16	15	13	8	61	11	12	16
20	39								

X2 = count Mimbres black on white

61	21	78	9	14	12	34	54	10	15
43	9	7	67	18	18	24	54	8	10
16	6	17	14	25	22	25	13	23	12
36	10	56	35	79	69	41	36	18	25
27	27	11	13						

File names	Excel: Tvis04.xls
	Minitab: Tvis04.mtp
	SPSS: Tvis04.sav
	TI-83 Plus and TI-84 Plus/ASCII:
	X1 data is stored in Tvis04L1.txt
	X2 data is stored in Tvis04L2.txt

05. Agriculture, Water Content of Soil (Two variable independent large samples)
The following data represent soil water content (% water by volume) for independent random samples of
soil from two experimental fields growing bell peppers.
Reference: *Journal of Agricultural, Biological, and Environmental Statistics*, Vol. 2, No. 2,
            p 149-155

X1 = soil water content from field I
```
 15.1  11.2  10.3  10.8  16.6   8.3   9.1  12.3   9.1  14.3
 10.7  16.1  10.2  15.2   8.9   9.5   9.6  11.3  14.0  11.3
 15.6  11.2  13.8   9.0   8.4   8.2  12.0  13.9  11.6  16.0
  9.6  11.4   8.4   8.0  14.1  10.9  13.2  13.8  14.6  10.2
 11.5  13.1  14.7  12.5  10.2  11.8  11.0  12.7  10.3  10.8
 11.0  12.6  10.8   9.6  11.5  10.6  11.7  10.1   9.7   9.7
 11.2   9.8  10.3  11.9   9.7  11.3  10.4  12.0  11.0  10.7
  8.8  11.1
```

X2 = soil water content from field II
```
 12.1  10.2  13.6   8.1  13.5   7.8  11.8   7.7   8.1   9.2
 14.1   8.9  13.9   7.5  12.6   7.3  14.9  12.2   7.6   8.9
 13.9   8.4  13.4   7.1  12.4   7.6   9.9  26.0   7.3   7.4
 14.3   8.4  13.2   7.3  11.3   7.5   9.7  12.3   6.9   7.6
 13.8   7.5  13.3   8.0  11.3   6.8   7.4  11.7  11.8   7.7
 12.6   7.7  13.2  13.9  10.4  12.8   7.6  10.7  10.7  10.9
 12.5  11.3  10.7  13.2   8.9  12.9   7.7   9.7   9.7  11.4
 11.9  13.4   9.2  13.4   8.8  11.9   7.1   8.5  14.0  14.2
```

File names   Excel: Tvis05.xls
                    Minitab: Tvis05.mtp
                    SPSS: Tvis05.sav
                    TI-83 Plus and TI-84 Plus/ASCII:
                             X1 data is stored in Tvis05L1.txt
                             X2 data is stored in Tvis05L2.txt

06. Rabies (Two variable independent small samples)
The following data represent number of cases of red fox rabies for a random sample of 16 areas in each of
two different regions of southern Germany.
Reference: Sayers, B., *Medical Informatics*, Vol. 2, 11-34

X1 = number cases in region 1
```
 10  2  2  5  3  4  3  3  4  0  2  6  4  8  7  4
```

X2 = number cases in region 2
```
  1  1  2  1  3  9  2  2  4  5  4  2  2  0  0  2
```

File names   Excel: Tvis06.xls
                    Minitab: Tvis06.mtp
                    SPSS: Tvis06.sav
                    TI-83 Plus and TI-84 Plus/ASCII:
                             X1 data is stored in Tvis06L1.txt
                             X2 data is stored in Tvis06L2.txt

07.   Weight of Football Players Versus Weight of Basketball Players
(Two variable independent small samples)
The following data represent weights in pounds of 21 randomly selected pro football players, and 19 randomly selected pro basketball players.
Reference: *Sports Encyclopedia of Pro Football*, and *Official NBA Basketball Encyclopedia*

X1 = weights (lb) of pro football players

245	262	255	251	244	276	240	265	257	252	282
256	250	264	270	275	245	275	253	265	270	

X2 = weights (lb) of pro basketball

205	200	220	210	191	215	221	216	228	207
225	208	195	191	207	196	181	193	201	

File names	Excel: Tvis07.xls
	Minitab: Tvis07.mtp
	SPSS: Tvis07.sav
	TI-83 Plus and TI-84 Plus/ASCII:
	X1 data is stored in Tvis07L1.txt
	X2 data is stored in Tvis07L2.txt

08.   Birth Rate (Two variable independent small samples)
The following data represent birth rate (per 1000 residential population) for independent random samples of counties in California and Maine.
Reference: *County and City Data Book* 12th edition, U.S. Dept. of Commerce

X1 = birth rate in California counties

14.1	18.7	20.4	20.7	16.0	12.5	12.9	9.6	17.6
18.1	14.1	16.6	15.1	18.5	23.6	19.9	19.6	14.9
17.7	17.8	19.1	22.1	15.6				

X2 = birth rate in Maine counties

15.1	14.0	13.3	13.8	13.5	14.2	14.7	11.8	13.5	13.8
16.5	13.8	13.2	12.5	14.8	14.1	13.6	13.9	15.8	

File names	Excel: Tvis08.xls
	Minitab: Tvis08.mtp
	SPSS: Tvis08.sav
	TI-83 Plus and TI-84 Plus/ASCII:
	X1 data is stored in Tvis08L1.txt
	X2 data is stored in Tvis08L2.txt

09.   Death Rate (Two variable independent small samples)
The following data represents death rate (per 1000 resident population) for independent random samples of counties in Alaska and Texas.
Reference: *County and City Data Book* 12th edition, U.S. Dept. of Commerce

X1 = death rate in Alaska counties

1.4	4.2	7.3	4.8	3.2	3.4	5.1	5.4
6.7	3.3	1.9	8.3	3.1	6.0	4.5	2.5

X2 = death rate in Texas counties

7.2	5.8	10.5	6.6	6.9	9.5	8.6	5.9	9.1
5.4	8.8	6.1	9.5	9.6	7.8	10.2	5.6	8.6

File names	Excel: Tvis09.xls

                          Minitab: Tvis09.mtp
                          SPSS: Tvis09.sav
                          TI-83 Plus and TI-84 Plus/ASCII:
                                          X1 data is stored in Tvis09L1.txt
                                          X2 data is stored in Tvis09L2.txt

10.   **Pickup Trucks (Two variable independent small samples)**
      The following data represents retail price (in thousands of dollars) for independent random samples of
      models of pickup trucks.
      Reference: *Consumer Guide* Vol.681

      X1 = prices for different GMC Sierra 1500 models
       17.4   23.3   29.2   19.2   17.6   19.2   23.6   19.5   22.2
       24.0   26.4   23.7   29.4   23.7   26.7   24.0   24.9

      X2 = prices for different Chevrolet Silverado 1500 models
       17.5   23.7   20.8   22.5   24.3   26.7   24.5   17.8
       29.4   29.7   20.1   21.1   22.1   24.2   27.4   28.1

      File names   Excel: Tvis10.xls
                            Minitab: Tvis10.mtp
                            SPSS: Tvis10.sav
                            TI-83 Plus and TI-84 Plus/ASCII:
                                            X1 data is stored in Tvis10L1.txt
                                            X2 data is stored in Tvis10L2.txt

Two Variable Dependent Samples
File name prefix: Tvds followed by the number of the data file

01. **Average Faculty Salary, Males vs Female (Two variable dependent samples)**
    In following data pairs
    　A = average salaries for males ($1000/yr)
    　B = average salaries for females ($1000/yr)
    for assistant professors at the same college or university. A random sample of 22 US. colleges and
    universities was used.
    Reference: *Academe, Bulletin of the American Association of University Professors*

    A: 34.5　30.5　35.1　35.7　31.5　34.4　32.1　30.7　33.7　35.3
    B: 33.9　31.2　35.0　34.2　32.4　34.1　32.7　29.9　31.2　35.5

    A: 30.7　34.2　39.6　30.5　33.8　31.7　32.8　38.5　40.5　25.3
    B: 30.2　34.8　38.7　30.0　33.8　32.4　31.7　38.9　41.2　25.5

    A: 28.6　35.8
    B: 28.0　35.1

File names	Excel: Tvds01.xls
	Minitab: Tvds01.mtp
	SPSS: Tvds01.sav
	TI-83 Plus and TI-84 Plus/ASCII:
	X1 data is stored in Tvds01L1.txt
	X2 data is stored in Tvds01L2.txt

02. **Unemployment for College Graduates Versus High School Only**
    (Two variable dependent samples)
    In the following data pairs
    　A = Percent unemployment for college graduates
    　B = Percent unemployment for high school only graduates
    The data are paired by year
    Reference: *Statistical Abstract of the United States*

    A: 2.8　2.2　2.2　1.7　2.3　2.3　2.4　2.7　3.5　3.0　1.9　2.5
    B: 5.9　4.9　4.8　5.4　6.3　6.9　6.9　7.2　10.0　8.5　5.1　6.9

File names	Excel: Tvds02.xls
	Minitab: Tvds02.mtp
	SPSS: Tvds02.sav
	TI-83 Plus and TI-84 Plus/ASCII:
	X1 data is stored in Tvds02L1.txt
	X2 data is stored in Tvds02L2.txt

03. **Number of Navajo Hogans versus Modern Houses (Two variable dependent samples)**
    In the following data pairs
    　A = Number of traditional Navajo hogans in a given district
    　B = Number of modern houses in a given district
    The data are paired by districts on the Navajo reservation. A random sample of 8 districts were used.
    Reference: *Navajo Architecture, Forms, History, Distributions* by S.C. Jett and V.E. Spencer,
    　　　　Univ. of Arizona Press

    A: 13　14　46　32　15　47　17　18

```
B:  18    16    68    9    11    28    50    50
```

File names   Excel: Tvds03.xls
                              Minitab: Tvds03.mtp
                              SPSS: Tvds03.sav
                              TI-83 Plus and TI-84 Plus/ASCII:
                                            X1 data is stored in Tvds03L1.txt
                                            X2 data is stored in Tvds03L2.txt

04.   Temperatures in Miami versus Honolulu (Two variable dependent samples)

In the following data pairs
      A = Average monthly temperature in Miami
      B = Average monthly temperature in Honolulu
The data are paired by month.
Reference: U.S. Department of Commerce Environmental Data Service

```
A:  67.5  68.0  71.3  74.9  78.0  80.9  82.2  82.7  81.6  77.8  72.3  68.5
B:  74.4  72.6  73.3  74.7  76.2  78.0  79.1  79.8  79.5  78.4  76.1  73.7
```

File names   Excel: Tvds04.xls
                              Minitab: Tvds04.mtp
                              SPSS: Tvds04.sav
                              TI-83 Plus and TI-84 Plus/ASCII:
                                            X1 data is stored in Tvds04L1.txt
                                            X2 data is stored in Tvds04L2.txt

05.   January/February Ozone Column (Two variable dependent samples)
In the following pairs the data represents the thickness of the ozone column in Dobson units: one milli-
centimeter ozone at standard temperature and pressure.
      A = monthly mean thickness in January
      B = monthly mean thickness in February
The data are paired by year for a random sample of 15 years.
Reference: Laboratorium für Atmospharensphysic, Switzerland

```
A:  360    324    377    336    383    361    369    349
B:  365    325    359    352    397    351    367    397
```

```
A:  301    354    344    329    337    387    378
B:  335    338    349    393    370    400    411
```

File names   Excel: Tvds05.xls
                              Minitab: Tvds05.mtp
                              SPSS: Tvds05.sav
                              TI-83 Plus and TI-84 Plus/ASCII:
                                            X1 data is stored in Tvds05L1.txt
                                            X2 data is stored in Tvds05L2.txt

06.   Birth Rate/Death Rate (Two variable dependent samples)
In the following data pairs,
      A = birth rate (per 1000 resident population)
      B = death rate (per 1000 resident population)
The data are paired by county in Iowa
Reference: *County and City Data Book* 12th edition, U.S. Dept. of Commerce

```
A:  12.7   13.4   12.8   12.1   11.6   11.1   14.2
```

B: 9.8 14.5 10.7 14.2 13.0 12.9 10.9

A: 12.5 12.3 13.1 15.8 10.3 12.7 11.1
B: 14.1 13.6 9.1 10.2 17.9 11.8 7.0

File names   Excel: Tvds06.xls
                       Minitab: Tvds06.mtp
                       SPSS: Tvds06.sav
                       TI-83 Plus and TI-84 Plus/ASCII:
                                       X1 data is stored in Tvds06L1.txt
                                       X2 data is stored in Tvds06L2.txt

07. Democrat/Republican (Two variable dependent samples)
In the following data pairs
     A = percentage of voters who voted Democrat
     B = percentage of voters who voted Republican
in a recent national election. The data are paired by county in Indiana
Reference: *County and City Data Book* 12th edition, U.S. Dept. of Commerce

A: 42.2 34.5 44.0 34.1 41.8 40.7 36.4 43.3 39.5
B: 35.4 45.8 39.4 40.0 39.2 40.2 44.7 37.3 40.8

A: 35.4 44.1 41.0 42.8 40.8 36.4 40.6 37.4
B: 39.3 36.8 35.5 33.2 38.3 47.7 41.1 38.5

File names   Excel: Tvds07.xls
                       Minitab: Tvds07.mtp
                       SPSS: Tvds07.sav
                       TI-83 Plus and TI-84 Plus/ASCII:
                                       X1 data is stored in Tvds07L1.txt
                                       X2 data is stored in Tvds07L2.txt

08. Santiago Pueblo Pottery (Two variable dependent samples)
In the following data
     A = percentage of utility pottery
     B = percentage of ceremonial pottery
found at the Santiago Pueblo archaeological site. The data are paired by location of discovery.
Reference: Laboratory of Anthropology, Notes 475, Santa Fe, New Mexico

A: 41.4 49.6 55.6 49.5 43.0 54.6 46.8 51.1 43.2 41.4
B: 58.6 50.4 44.4 59.5 57.0 45.4 53.2 48.9 56.8 58.6

File names   Excel: Tvds08.xls
                       Minitab: Tvds08.mtp
                       SPSS: Tvds08.sav
                       TI-83 Plus and TI-84 Plus/ASCII:
                                       X1 data is stored in Tvds08L1.txt
                                       X2 data is stored in Tvds08L2.txt

09. Poverty Level (Two variable dependent samples)
In the following data pairs
     A = percentage of population below poverty level in 1998
     B = percentage of population below poverty level in 1990
The data are grouped by state and District of Columbia
Reference: *Statistical Abstract of the United States*, 120th edition

```
A: 14.5    9.4   16.6   14.8   15.4    9.2    9.5   10.3   22.3   13.1
B: 19.2   11.4   13.7   19.6   13.9   13.7    6.0    6.9   21.1   14.4


A: 13.6   10.9   13.0   10.1    9.4    9.1    9.6   13.5   19.1   10.4
B: 15.8   11.0   14.9   13.7   13.0   10.4   10.3   17.3   23.6   13.1


A:  7.2    8.7   11.0   10.4   17.6    9.8   16.6   12.3   10.6    9.8
B:  9.9   10.7   14.3   12.0   25.7   13.4   16.3   10.3    9.8    6.3


A:  8.6   20.4   16.7   14.0   15.1   11.2   14.1   15.0   11.2   11.6
B:  9.2   20.9   14.3   13.0   13.7   11.5   15.6    9.2   11.0    7.5


A: 13.7   10.8   13.4   15.1    9.0    9.9    8.8    8.9   17.8    8.8   10.6
B: 16.2   13.3   16.9   15.9    8.2   10.9   11.1    8.9   18.1    9.3   11.0
```

File names   Excel: Tvds09.xls
                    Minitab: Tvds09.mtp
                    SPSS: Tvds09.sav
                    TI-83 Plus and TI-84 Plus/ASCII:
                        X1 data is stored in Tvds09L1.txt
                        X2 data is stored in Tvds09L2.txt

10. **Cost of Living Index (Two variable dependent samples)**
     The following data pairs represent cost of living index for
         A = grocery items
         B = health care
     The data are grouped by metropolitan areas.
     Reference: *Statistical Abstract of the United States*, 120th edition

Grocery
```
A: 96.6    97.5   113.9    88.9   108.3    99.0    97.3    87.5    96.8
B: 91.6    95.9   114.5    93.6   112.7    93.6    99.2    93.2   105.9


A: 102.1   114.5  100.9   100.0   100.7    99.4   117.1   111.3   102.2
B: 110.8   127.0   91.5   100.5   104.9   104.8   124.1   124.6   109.1


A:  95.3    91.1   95.7    87.5    91.8    97.9    97.4   102.1    94.0
B:  98.7    95.8   99.7    93.2   100.7    96.0    99.6    98.4    94.0


A: 115.7   118.3  101.9    88.9   100.7    99.8   101.3   104.8   100.9
B: 121.2   122.4  110.8    81.2   104.8   109.9   103.5   113.6    94.6


A: 102.7    98.1  105.3    97.2   105.2   108.1   110.5    99.3    99.7
B: 109.8    97.6  109.8   107.4    97.7   124.2   110.9   106.8    94.8
```

File names   Excel: Tvds10.xls
                    Minitab: Tvds10.mtp
                    SPSS: Tvds10.sav
                    TI-83 Plus and TI-84 Plus/ASCII:
                        X1 data is stored in Tvds10L1.txt
                        X2 data is stored in Tvds10L2.txt

Simple Linear Regression
File name prefix: Slr followed by the number of the data file

01. **List Price versus Best Price for a New GMC Pickup Truck (Simple Linear Regression)**
    In the following data
    X = List price (in $1000) for a GMC pickup truck
    Y = Best price (in $1000) for a GMC pickup truck
    Reference: *Consumer's Digest*

    X: 12.4 14.3 14.5 14.9 16.1 16.9 16.5 15.4 17.0 17.9
    Y: 11.2 12.5 12.7 13.1 14.1 14.8 14.4 13.4 14.9 15.6

    X: 18.8 20.3 22.4 19.4 15.5 16.7 17.3 18.4 19.2 17.4
    Y: 16.4 17.7 19.6 16.9 14.0 14.6 15.1 16.1 16.8 15.2

    X: 19.5 19.7 21.2
    Y: 17.0 17.2 18.6

    File names   Excel: Slr01.xls
                        Minitab: Slr01.mtp
                        SPSS: Slr01.sav
                        TI-83 Plus and TI-84 Plus/ASCII:
                                    X1 data is stored in Slr01L1.txt
                                    X2 data is stored in Slr01L2.txt

02. **Cricket Chirps versus Temperature (Simple Linear Regression)**
    In the following data
    X = chirps/sec for the striped ground cricket
    Y = temperature in degrees Fahrenheit
    Reference: *The Song of Insects* by Dr.G.W. Pierce, Harvard College Press

    X: 20.0 16.0 19.8 18.4 17.1 15.5 14.7 17.1
    Y: 88.6 71.6 93.3 84.3 80.6 75.2 69.7 82.0

    X: 15.4 16.2 15.0 17.2 16.0 17.0 14.4
    Y: 69.4 83.3 79.6 82.6 80.6 83.5 76.3

    File names   Excel: Slr02.xls
                        Minitab: Slr02.mtp
                        SPSS: Slr02.sav
                        TI-83 Plus and TI-84 Plus/ASCII:
                                    X1 data is stored in Slr02L1.txt
                                    X2 data is stored in Slr02L2.txt

03. **Diameter of Sand Granules versus Slope on Beach (Simple Linear Regression)**
    In the following data pairs
    X = median diameter (mm) of granules of sand
    Y = gradient of beach slope in degrees
    The data is for naturally occurring ocean beaches
    Reference: *Physical geography* by A.M King, Oxford Press, England

    X: 0.170 0.190 0.220 0.235 0.235 0.300 0.350 0.420 0.850
    Y: 0.630 0.700 0.820 0.880 1.150 1.500 4.400 7.300 11.300

File names	Excel: Slr03.xls
	Minitab: Slr03.mtp
	SPSS: Slr03.sav
	TI-83 Plus and TI-84 Plus/ASCII:
	X1 data is stored in Slr03L1.txt
	X2 data is stored in Slr03L2.txt

04. National Unemployment Male versus Female (Simple Linear Regression)
In the following data pairs
    X = national unemployment rate for adult males
    Y = national unemployment rate for adult females
Reference: *Statistical Abstract of the United States*

X:	2.9	6.7	4.9	7.9	9.8	6.9	6.1	6.2	6.0	5.1	4.7	4.4	5.8
Y:	4.0	7.4	5.0	7.2	7.9	6.1	6.0	5.8	5.2	4.2	4.0	4.4	5.2

File names	Excel: Slr04.xls
	Minitab: Slr04.mtp
	SPSS: Slr04.sav
	TI-83 Plus and TI-84 Plus/ASCII:
	X1 data is stored in Slr04L1.txt
	X2 data is stored in Slr04L2.txt

05. Fire and Theft in Chicago (Simple Linear Regression)
In the following data pairs
    X = fires per 1000 housing units
    Y = thefts per 1000 population
within the same Zip code in the Chicago metro area
Reference: U.S. Commission on Civil Rights

X:	6.2	9.5	10.5	7.7	8.6	34.1	11.0	6.9	7.3	15.1
Y:	29	44	36	37	53	68	75	18	31	25

X:	29.1	2.2	5.7	2.0	2.5	4.0	5.4	2.2	7.2	15.1
Y:	34	14	11	11	22	16	27	9	29	30

X:	16.5	18.4	36.2	39.7	18.5	23.3	12.2	5.6	21.8	21.6
Y:	40	32	41	147	22	29	46	23	4	31

X:	9.0	3.6	5.0	28.6	17.4	11.3	3.4	11.9	10.5	10.7
Y:	39	15	32	27	32	34	17	46	42	43

X:	10.8	4.8
Y:	34	19

File names	Excel: Slr05.xls
	Minitab: Slr05.mtp
	SPSS: Slr05.sav
	TI-83 Plus and TI-84 Plus/ASCII:
	X1 data is stored in Slr05L1.txt
	X2 data is stored in Slr05L2.txt

06. Auto Insurance in Sweden (Simple Linear Regression)
In the following data
    X = number of claims
    Y = total payment for all the claims in thousands of Swedish Kronor

for geographical zones in Sweden
Reference: Swedish Committee on Analysis of Risk Premium in Motor Insurance

X:	108	19	13	124	40	57	23	14	45	10
Y:	392.5	46.2	15.7	422.2	119.4	170.9	56.9	77.5	214.0	65.3

X:	5	48	11	23	7	2	24	6	3	23
Y:	20.9	248.1	23.5	39.6	48.8	6.6	134.9	50.9	4.4	113.0

X:	6	9	9	3	29	7	4	20	7	4
Y:	14.8	48.7	52.1	13.2	103.9	77.5	11.8	98.1	27.9	38.1

X:	0	25	6	5	22	11	61	12	4	16
Y:	0.0	69.2	14.6	40.3	161.5	57.2	217.6	58.1	12.6	59.6

X:	13	60	41	37	55	41	11	27	8	3
Y:	89.9	202.4	181.3	152.8	162.8	73.4	21.3	92.6	76.1	39.9

X:	17	13	13	15	8	29	30	24	9	31
Y:	142.1	93.0	31.9	32.1	55.6	133.3	194.5	137.9	87.4	209.8

X:	14	53	26
Y:	95.5	244.6	187.5

File names   Excel: Slr06.xls
                          Minitab: Slr06.mtp
                          SPSS: Slr06.sav
                          TI-83 Plus and TI-84 Plus/ASCII:
                                      X1 data is stored in Slr06L1.txt
                                      X2 data is stored in Slr06L2.txt

07.   Gray Kangaroos (Simple Linear Regression)

In the following data pairs
      X = nasal length (mm ×10)
      Y = nasal width (mm × 10)
for a male gray kangaroo from a random sample of such animals.
Reference: *Australian Journal of Zoology*, Vol. 28, p607-613

X:	609	629	620	564	645	493	606	660	630	672
Y:	241	222	233	207	247	189	226	240	215	231

X:	778	616	727	810	778	823	755	710	701	803
Y:	263	220	271	284	279	272	268	278	238	255

X:	855	838	830	864	635	565	562	580	596	597
Y:	308	281	288	306	236	204	216	225	220	219

X:	636	559	615	740	677	675	629	692	710	730
Y:	201	213	228	234	237	217	211	238	221	281

X:	763	686	717	737	816
Y:	292	251	231	275	275

File names   Excel: Slr07.xls
                          Minitab: Slr07.mtp
                          SPSS: Slr07.sav

TI-83 Plus and TI-84 Plus/ASCII:
X1 data is stored in Slr07L1.txt
X2 data is stored in Slr07L2.txt

08. Pressure and Weight in Cryogenic Flow Meters (Simple Linear Regression)
In the following data pairs
X = pressure (lb/sq in) of liquid nitrogen
Y = weight in pounds of liquid nitrogen passing through flow meter each second
Reference: *Technometrics*, Vol. 19, p353-379

X:	75.1	74.3	88.7	114.6	98.5	112.0	114.8	62.2	107.0
Y:	577.8	577.0	570.9	578.6	572.4	411.2	531.7	563.9	406.7

X:	90.5	73.8	115.8	99.4	93.0	73.9	65.7	66.2	77.9
Y:	507.1	496.4	505.2	506.4	510.2	503.9	506.2	506.3	510.2

X:	109.8	105.4	88.6	89.6	73.8	101.3	120.0	75.9	76.2
Y:	508.6	510.9	505.4	512.8	502.8	493.0	510.8	512.8	513.4

X:	81.9	84.3	98.0
Y:	510.0	504.3	522.0

File names   Excel: Slr08.xls
Minitab: Slr08.mtp
SPSS: Slr08.sav
TI-83 Plus and TI-84 Plus/ASCII:
X1 data is stored in Slr08L1.txt
X2 data is stored in Slr08L2.txt

09. Ground Water Survey (Simple Linear Regression)
In the following data
X = pH of well water
Y = Bicarbonate (parts per million) of well water
The data is by water well from a random sample of wells in Northwest Texas.
Reference: Union Carbide Technical Report K/UR-1

X:	7.6	7.1	8.2	7.5	7.4	7.8	7.3	8.0	7.1	7.5
Y:	157	174	175	188	171	143	217	190	142	190

X:	8.1	7.0	7.3	7.8	7.3	8.0	8.5	7.1	8.2	7.9
Y:	215	199	262	105	121	81	82	210	202	155

X:	7.6	8.8	7.2	7.9	8.1	7.7	8.4	7.4	7.3	8.5
Y:	157	147	133	53	56	113	35	125	76	48

X:	7.8	6.7	7.1	7.3
Y:	147	117	182	87

File names   Excel: Slr09.xls
Minitab: Slr09.mtp
SPSS: Slr09.sav
TI-83 Plus and TI-84 Plus/ASCII:
X1 data is stored in Slr09L1.txt
X2 data is stored in Slr09L2.txt

10. **Iris Setosa (Simple Linear Regression)**
    In the following data
        X = sepal width (cm)
        Y = sepal length (cm)
    The data is for a random sample of the wild flower iris setosa.
    Reference: Fisher, R.A., *Ann. Eugenics,* Vol. 7 Part II, p 179-188

X:	3.5	3.0	3.2	3.1	3.6	3.9	3.4	3.4	2.9	3.1
Y:	5.1	4.9	4.7	4.6	5.0	5.4	4.6	5.0	4.4	4.9

X:	3.7	3.4	3.0	4.0	4.4	3.9	3.5	3.8	3.8	3.4
Y:	5.4	4.8	4.3	5.8	5.7	5.4	5.1	5.7	5.1	5.4

X:	3.7	3.6	3.3	3.4	3.0	3.4	3.5	3.4	3.2	3.1
Y:	5.1	4.6	5.1	4.8	5.0	5.0	5.2	5.2	4.7	4.8

X:	3.4	4.1	4.2	3.1	3.2	3.5	3.6	3.0	3.4	3.5
Y:	5.4	5.2	5.5	4.9	5.0	5.5	4.9	4.4	5.1	5.0

X:	2.3	3.2	3.5	3.8	3.0	3.8	3.7	3.3
Y:	4.5	4.4	5.0	5.1	4.8	4.6	5.3	5.0

    File names   Excel: Slr10.xls
                           Minitab: Slr10.mtp
                           SPSS: Slr10.sav
                           TI-83 Plus and TI-84 Plus/ASCII:
                                     X1 data is stored in Slr10L1.txt
                                   X2 data is stored in Slr10L2.txt

11. **Pizza Franchise (Simple Linear Regression)**
    In the following data
        X = annual franchise fee ($1000)
        Y = start up cost ($1000)
    for a pizza franchise
    Reference: *Business Opportunity Handbook*

X:	25.0	8.5	35.0	15.0	10.0	30.0	10.0	50.0	17.5	16.0
Y:	125	80	330	58	110	338	30	175	120	135

X:	18.5	7.0	8.0	15.0	5.0	15.0	12.0	15.0	28.0	20.0
Y:	97	50	55	40	35	45	75	33	55	90

X:	20.0	15.0	20.0	25.0	20.0	3.5	35.0	25.0	8.5	10.0
Y:	85	125	150	120	95	30	400	148	135	45

X:	10.0	25.0
Y:	87	150

    File names   Excel: Slr11.xls
                           Minitab: Slr11.mtp
                           SPSS: Slr11.sav
                           TI-83 Plus and TI-84 Plus/ASCII:
                                   X1 data is stored in Slr11L1.txt
                                   X2 data is stored in Slr11L2.txt

12. **Prehistoric Pueblos (Simple Linear Regression)**
   In the following data
      X = estimated year of initial occupation
      Y = estimated year of end of occupation
   The data are for each prehistoric pueblo in a random sample of such pueblos in Utah, Arizona, and Nevada.
   Reference *Prehistoric Pueblo World* by A. Adler, Univ. of Arizona Press

X:	1000	1125	1087	1070	1100	1150	1250	1150	1100
Y:	1050	1150	1213	1275	1300	1300	1400	1400	1250

X:	1350	1275	1375	1175	1200	1175	1300	1260	1330
Y:	1830	1350	1450	1300	1300	1275	1375	1285	1400

X:	1325	1200	1225	1090	1075	1080	1080	1180	1225
Y:	1400	1285	1275	1135	1250	1275	1150	1250	1275

X:	1175	1250	1250	750	1125	700	900	900	850
Y:	1225	1280	1300	1250	1175	1300	1250	1300	1200

   File names   Excel: Slr12.xls
                       Minitab: Slr12.mtp
                       SPSS: Slr12.sav
                       TI-83 Plus and TI-84 Plus/ASCII:
                                X1 data is stored in Slr12L1.txt
                                X2 data is stored in Slr12L2.txt

Multiple Linear Regression
File name prefix: Mlr followed by the number of the data file

01. Thunder Basin Antelope Study (Multiple Linear Regression)
   The data (X1, X2, X3, X4) are for each year.
   X1 = spring fawn count/100
   X2 = size of adult antelope population/100
   X3 = annual precipitation (inches)
   X4 = winter severity index (1=mild , 5=severe)

X1	X2	X3	X4
2.90	9.20	13.20	2.00
2.40	8.70	11.50	3.00
2.00	7.20	10.80	4.00
2.30	8.50	12.30	2.00
3.20	9.60	12.60	3.00
1.90	6.80	10.60	5.00
3.40	9.70	14.10	1.00
2.10	7.90	11.20	3.00

File names   Excel: Mlr01.xls
Minitab: Mlr01.mtp
SPSS: Mlr01.sav
TI-83 Plus and TI-84 Plus/ASCII:
        X1 data is stored in Mlr01L1.txt
        X2 data is stored in Mlr01L2.txt
        X3 data is stored in Mlr01L3.txt
        X4 data is stored in Mlr01L4.txt

02. Section 10.5, problem #3 Systolic Blood Pressure Data (Multiple Linear Regression)
   The data (X1, X2, X3) are for each patient.
   X1 = systolic blood pressure
   X2 = age in years
   X3 = weight in pounds

X1	X2	X3
132.00	52.00	173.00
143.00	59.00	184.00
153.00	67.00	194.00
162.00	73.00	211.00
154.00	64.00	196.00
168.00	74.00	220.00
137.00	54.00	188.00
149.00	61.00	188.00
159.00	65.00	207.00
128.00	46.00	167.00
166.00	72.00	217.00

File names   Excel: Mlr02.xls
Minitab: Mlr02.mtp
SPSS: Mlr02.sav
TI-83 Plus and TI-84 Plus/ASCII:
        X1 data is stored in Mlr02L1.txt

X2 data is stored in Mlr02L2.txt
X3 data is stored in Mlr02L3.txt

03. Section 10.5, Problem #4 Test Scores for General Psychology (Multiple Linear Regression)
The data (X1, X2, X3, X4) are for each student.
X1 = score on exam #1
X2 = score on exam #2
X3 = score on exam #3
X4 = score on final exam

X1	X2	X3	X4
73	80	75	152
93	88	93	185
89	91	90	180
96	98	100	196
73	66	70	142
53	46	55	101
69	74	77	149
47	56	60	115
87	79	90	175
79	70	88	164
69	70	73	141
70	65	74	141
93	95	91	184
79	80	73	152
70	73	78	148
93	89	96	192
78	75	68	147
81	90	93	183
88	92	86	177
78	83	77	159
82	86	90	177
86	82	89	175
78	83	85	175
76	83	71	149
96	93	95	192

File names   Excel: Mlr03.xls
                    Minitab: Mlr03.mtp
                    SPSS: Mlr03.sav
                    TI-83 Plus and TI-84 Plus/ASCII:
                        X1 data is stored in Mlr03L1.txt
                        X2 data is stored in Mlr03L2.txt
                        X3 data is stored in Mlr03L3.txt
                        X4 data is stored in Mlr03L4.txt

04. Section 10.5, Problem #5 Hollywood Movies (Multiple Linear Regression)
The data (X1, X2, X3, X4) are for each movie
X1 = first year box office receipts/millions
X2 = total production costs/millions
X3 = total promotional costs/millions
X4 = total book sales/millions

X1	X2	X3	X4
85.10	8.50	5.10	4.70
106.30	12.90	5.80	8.80

50.20	5.20	2.10	15.10
130.60	10.70	8.40	12.20
54.80	3.10	2.90	10.60
30.30	3.50	1.20	3.50
79.40	9.20	3.70	9.70
91.00	9.00	7.60	5.90
135.40	15.10	7.70	20.80
89.30	10.20	4.50	7.90

File names   Excel: Mlr04.xls
                             Minitab: Mlr04.mtp
                             SPSS: Mlr04.sav
                             TI-83 Plus and TI-84 Plus/ASCII:

                                      X1 data is stored in Mlr04L1.txt
                                      X2 data is stored in Mlr04L2.txt
                                      X3 data is stored in Mlr04L3.txt
                                      X4 data is stored in Mlr04L4.txt

05. Section 10.5, Problem #6 All Greens Franchise (Multiple Linear Regression)
The data (X1, X2, X3, X4, X5, X6) are for each franchise store.
X1 = annual net sales/$1000
X2 = number sq. ft./1000
X3 = inventory/$1000
X4 = amount spent on advertizing/$1000
X5 = size of sales district/1000 families
X6 = number of competing stores in district

X1	X2	X3	X4	X5	X6
231.00	3.00	294.00	8.20	8.20	11.00
156.00	2.20	232.00	6.90	4.10	12.00
10.00	0.50	149.00	3.00	4.30	15.00
519.00	5.50	600.00	12.00	16.10	1.00
437.00	4.40	567.00	10.60	14.10	5.00
487.00	4.80	571.00	11.80	12.70	4.00
299.00	3.10	512.00	8.10	10.10	10.00
195.00	2.50	347.00	7.70	8.40	12.00
20.00	1.20	212.00	3.30	2.10	15.00
68.00	0.60	102.00	4.90	4.70	8.00
570.00	5.40	788.00	17.40	12.30	1.00
428.00	4.20	577.00	10.50	14.00	7.00
464.00	4.70	535.00	11.30	15.00	3.00
15.00	0.60	163.00	2.50	2.50	14.00
65.00	1.20	168.00	4.70	3.30	11.00
98.00	1.60	151.00	4.60	2.70	10.00
398.00	4.30	342.00	5.50	16.00	4.00
161.00	2.60	196.00	7.20	6.30	13.00
397.00	3.80	453.00	10.40	13.90	7.00
497.00	5.30	518.00	11.50	16.30	1.00
528.00	5.60	615.00	12.30	16.00	0.00
99.00	0.80	278.00	2.80	6.50	14.00
0.50	1.10	142.00	3.10	1.60	12.00
347.00	3.60	461.00	9.60	11.30	6.00
341.00	3.50	382.00	9.80	11.50	5.00
507.00	5.10	590.00	12.00	15.70	0.00
400.00	8.60	517.00	7.00	12.00	8.00

File names   Excel: Mlr05.xls
                              Minitab: Mlr05.mtp
                              SPSS: Mlr05.sav
                              TI-83 Plus and TI-84 Plus/ASCII:
                                          X1 data is stored in Mlr05L1.txt
                                          X2 data is stored in Mlr05L2.txt
                                          X3 data is stored in Mlr05L3.txt
                                          X4 data is stored in Mlr05L4.txt
                                          X5 data is stored in Mlr05L5.txt
                                          X6 data is stored in Mlr05L6.txt

06. Crime (Multiple Linear Regression)

This is a case study of education, crime, and police funding for small cities in ten eastern and south eastern states. The states are New Hampshire, Connecticut, Rhode Island, Maine, New York, Virginia, North Carolina, South Carolina, Georgia, and Florida.

The data (X1, X2, X3, X4, X5, X6, X7) are for each city.
X1 = total overall reported crime rate per 1million residents
X2 = reported violent crime rate per 100,000 residents
X3 = annual police funding in dollars per resident
X4 = percent of people 25 years and older that have had 4 years of high school
X5 = percent of 16 to 19 year-olds not in highschool and not highschool graduates.
X6 = percent of 18 to 24 year-olds enrolled in college
X7 = percent of people 25 years and older with at least 4 years of college

Reference: *Life In America's Small Cities,* By G.S. Thomas

X1	X2	X3	X4	X5	X6	X7
478	184	40	74	11	31	20
494	213	32	72	11	43	18
643	347	57	70	18	16	16
341	565	31	71	11	25	19
773	327	67	72	9	29	24
603	260	25	68	8	32	15
484	325	34	68	12	24	14
546	102	33	62	13	28	11
424	38	36	69	7	25	12
548	226	31	66	9	58	15
506	137	35	60	13	21	9
819	369	30	81	4	77	36
541	109	44	66	9	37	12
491	809	32	67	11	37	16
514	29	30	65	12	35	11
371	245	16	64	10	42	14
457	118	29	64	12	21	10
437	148	36	62	7	81	27
570	387	30	59	15	31	16
432	98	23	56	15	50	15
619	608	33	46	22	24	8
357	218	35	54	14	27	13
623	254	38	54	20	22	11
547	697	44	45	26	18	8
792	827	28	57	12	23	11
799	693	35	57	9	60	18
439	448	31	61	19	14	12
867	942	39	52	17	31	10

Data continued

X1	X2	X3	X4	X5	X6	X7
912	1017	27	44	21	24	9
462	216	36	43	18	23	8
859	673	38	48	19	22	10
805	989	46	57	14	25	12
652	630	29	47	19	25	9
776	404	32	50	19	21	9
919	692	39	48	16	32	11
732	1517	44	49	13	31	14
657	879	33	72	13	13	22
1419	631	43	59	14	21	13
989	1375	22	49	9	46	13
821	1139	30	54	13	27	12
1740	3545	86	62	22	18	15
815	706	30	47	17	39	11
760	451	32	45	34	15	10
936	433	43	48	26	23	12
863	601	20	69	23	7	12
783	1024	55	42	23	23	11
715	457	44	49	18	30	12
1504	1441	37	57	15	35	13
1324	1022	82	72	22	15	16
940	1244	66	67	26	18	16

File names   Excel: Mlr06.xls
                  Minitab: Mlr06.mtp
                  SPSS: Mlr06.sav
                  TI-83 Plus and TI-84 Plus/ASCII:
                          X1 data is stored in Mlr06L1.txt
                          X2 data is stored in Mlr06L2.txt
                          X3 data is stored in Mlr06L3.txt
                          X4 data is stored in Mlr06L4.txt
                          X5 data is stored in Mlr06L5.txt
                          X6 data is stored in Mlr06L6.txt
                          X7 data is stored in Mlr06L7.txt

07. **Health (Multiple Linear Regression)**
This is a case study of public health, income, and population density for small cities in eight Midwestern states: Ohio, Indiana, Illinois, Iowa, Missouri, Nebraska, Kansas, and Oklahoma.

The data (X1, X2, X3, X4, X5) are by city.
X1 = death rate per 1000 residents
X2 = doctor availability per 100,000 residents
X3 = hospital availability per 100,000 residents
X4 = annual per capita income in thousands of dollars
X5 = population density people per square mile

Reference: *Life In America's Small Cities*, by G.S. Thomas

X1	X2	X3	X4	X5
8.0	78	284	9.1	109
9.3	68	433	8.7	144
7.5	70	739	7.2	113

8.9	96	1792	8.9	97
10.2	74	477	8.3	206
8.3	111	362	10.9	124
8.8	77	671	10.0	152
8.8	168	636	9.1	162
10.7	82	329	8.7	150
11.7	89	634	7.6	134
8.5	149	631	10.8	292
8.3	60	257	9.5	108
8.2	96	284	8.8	111
7.9	83	603	9.5	182
10.3	130	686	8.7	129
7.4	145	345	11.2	158
9.6	112	1357	9.7	186
9.3	131	544	9.6	177
10.6	80	205	9.1	127
9.7	130	1264	9.2	179
11.6	140	688	8.3	80
8.1	154	354	8.4	103
9.8	118	1632	9.4	101
7.4	94	348	9.8	117
9.4	119	370	10.4	88
11.2	153	648	9.9	78
9.1	116	366	9.2	102
10.5	97	540	10.3	95
11.9	1 76	680	8.9	80
8.4	75	345	9.6	92
5.0	134	525	10.3	126
9.8	161	870	10.4	108
9.8	111	669	9.7	77
10.8	114	452	9.6	60
10.1	142	430	10.7	71
10.9	238	822	10.3	86
9.2	78	190	10.7	93
8.3	196	867	9.6	106
7.3	125	969	10.5	162
9.4	82	499	7.7	95
9.4	125	925	10.2	91
9.8	129	353	9.9	52
3.6	84	288	8.4	110
8.4	183	718	10.4	69
10.8	119	540	9.2	57
10.1	180	668	13.0	106
9.0	82	347	8.8	40
10.0	71	345	9.2	50
11.3	118	463	7.8	35
11.3	121	728	8.2	86
12.8	68	383	7.4	57
10.0	112	316	10.4	57
6.7	109	388	8.9	94

File names   Excel: Mlr07.xls

Minitab: Mlr07.mtp

SPSS: Mlr07.sav

TI-83 Plus and TI-84 Plus/ASCII:

         X1 data is stored in Mlr07L1.txt

X2 data is stored in Mlr07L2.txt
X3 data is stored in Mlr07L3.txt
X4 data is stored in Mlr07L4.txt
X5 data is stored in Mlr07L5.txt

08. Baseball (Multiple Linear Regression)
A random sample of major league baseball players was obtained.

The following data (X1, X2, X3, X4, X5, X6) are by player.
X1 = batting average
X2 = runs scored/times at bat
X3 = doubles/times at bat
X4 = triples/times at bat
X5 = home runs/times at bat
X6 = strike outs/times at bat
Reference: *The Baseball Encyclopedia* 9th edition, Macmillan

X1	X2	X3	X4	X5	X6
0.283	0.144	0.049	0.012	0.013	0.086
0.276	0.125	0.039	0.013	0.002	0.062
0.281	0.141	0.045	0.021	0.013	0.074
0.328	0.189	0.043	0.001	0.030	0.032
0.290	0.161	0.044	0.011	0.070	0.076
0.296	0.186	0.047	0.018	0.050	0.007
0.248	0.106	0.036	0.008	0.012	0.095
0.228	0.117	0.030	0.006	0.003	0.145
0.305	0.174	0.050	0.008	0.061	0.112
0.254	0.094	0.041	0.005	0.014	0.124
0.269	0.147	0.047	0.012	0.009	0.111
0.300	0.141	0.058	0.010	0.011	0.070
0.307	0.135	0.041	0.009	0.005	0.065
0.214	0.100	0.037	0.003	0.004	0.138
0.329	0.189	0.058	0.014	0.011	0.032
0.310	0.149	0.050	0.012	0.050	0.060
0.252	0.119	0.040	0.008	0.049	0.233
0.308	0.158	0.038	0.013	0.003	0.068
0.342	0.259	0.060	0.016	0.085	0.158
0.358	0.193	0.066	0.021	0.037	0.083
0.340	0.155	0.051	0.020	0.012	0.040
0.304	0.197	0.052	0.008	0.054	0.095
0.248	0.133	0.037	0.003	0.043	0.135
0.367	0.196	0.063	0.026	0.010	0.031
0.325	0.206	0.054	0.027	0.010	0.048
0.244	0.110	0.025	0.006	0.000	0.061
0.245	0.096	0.044	0.003	0.022	0.151
0.318	0.193	0.063	0.020	0.037	0.081
0.207	0.154	0.045	0.008	0.000	0.252
0.320	0.204	0.053	0.017	0.013	0.070
0.243	0.141	0.041	0.007	0.051	0.264
0.317	0.209	0.057	0.030	0.017	0.058
0.199	0.100	0.029	0.007	0.011	0.188
0.294	0.158	0.034	0.019	0.005	0.014
0.221	0.087	0.038	0.006	0.015	0.142
0.301	0.163	0.068	0.016	0.022	0.092
0.298	0.207	0.042	0.009	0.066	0.211

0.304	0.197	0.052	0.008	0.054	0.095
0.297	0.160	0.049	0.007	0.038	0.101
0.188	0.064	0.044	0.007	0.002	0.205
0.214	0.100	0.037	0.003	0.004	0.138
0.218	0.082	0.061	0.002	0.012	0.147
0.284	0.131	0.049	0.012	0.021	0.130
0.270	0.170	0.026	0.011	0.002	0.000
0.277	0.150	0.053	0.005	0.039	0.115

File names   Excel: Mlr08.xls
                     Minitab: Mlr08.mtp
                     SPSS: Mlr08.sav
                     TI-83 Plus and TI-84 Plus/ASCII:
                               X1 data is stored in Mlr08L1.txt
                               X2 data is stored in Mlr08L2.txt
                               X3 data is stored in Mlr08L3.txt
                               X4 data is stored in Mlr08L4.txt
                               X5 data is stored in Mlr08L5.txt
                               X6 data is stored in Mlr08L6.txt

09.   Basketball (Multiple Linear Regression)

A random sample of professional basketball players was obtained.

The following data (X1, X2, X3, X4, X5) are for each player.
X1 = height in feet
X2 = weight in pounds
X3 = percent of successful field goals (out of 100 attempted)
X4 = percent of successful free throws (out of 100 attempted)
X5 = average points scored per game
Reference: *The official NBA basketball Encyclopedia*, Villard Books

X1	X2	X3	X4	X5
6.8	225	0.442	0.672	9.2
6.3	180	0.435	0.797	11.7
6.4	190	0.456	0.761	15.8
6.2	180	0.416	0.651	8.6
6.9	205	0.449	0.900	23.2
6.4	225	0.431	0.780	27.4
6.3	185	0.487	0.771	9.3
6.8	235	0.469	0.750	16.0
6.9	235	0.435	0.818	4.7
6.7	210	0.480	0.825	12.5
6.9	245	0.516	0.632	20.1
6.9	245	0.493	0.757	9.1
6.3	185	0.374	0.709	8.1
6.1	185	0.424	0.782	8.6
6.2	180	0.441	0.775	20.3
6.8	220	0.503	0.880	25.0
6.5	194	0.503	0.833	19.2
7.6	225	0.425	0.571	3.3
6.3	210	0.371	0.816	11.2
7.1	240	0.504	0.714	10.5
6.8	225	0.400	0.765	10.1
7.3	263	0.482	0.655	7.2
6.4	210	0.475	0.244	13.6
6.8	235	0.428	0.728	9.0

7.2	230	0.559	0.721	24.6
6.4	190	0.441	0.757	12.6
6.6	220	0.492	0.747	5.6
6.8	210	0.402	0.739	8.7
6.1	180	0.415	0.713	7.7
6.5	235	0.492	0.742	24.1
6.4	185	0.484	0.861	11.7
6.0	175	0.387	0.721	7.7
6.0	192	0.436	0.785	9.6
7.3	263	0.482	0.655	7.2
6.1	180	0.340	0.821	12.3
6.7	240	0.516	0.728	8.9
6.4	210	0.475	0.846	13.6
5.8	160	0.412	0.813	11.2
6.9	230	0.411	0.595	2.8
7.0	245	0.407	0.573	3.2
7.3	228	0.445	0.726	9.4
5.9	155	0.291	0.707	11.9
6.2	200	0.449	0.804	15.4
6.8	235	0.546	0.784	7.4
7.0	235	0.480	0.744	18.9
5.9	105	0.359	0.839	7.9
6.1	180	0.528	0.790	12.2
5.7	185	0.352	0.701	11.0
7.1	245	0.414	0.778	2.8
5.8	180	0.425	0.872	11.8
7.4	240	0.599	0.713	17.1
6.8	225	0.482	0.701	11.6
6.8	215	0.457	0.734	5.8
7.0	230	0.435	0.764	8.3

File names

Excel: Mlr09.xls
Minitab: Mlr09.mtp
SPSS: Mlr09.sav
TI-83 Plus and TI-84 Plus/ASCII:

X1 data is stored in Mlr09L1.txt
X2 data is stored in Mlr09L2.txt
X3 data is stored in Mlr09L3.txt
X4 data is stored in Mlr09L4.txt
X5 data is stored in Mlr09L5.txt

10. **Denver Neighborhoods (Multiple Linear Regression)**
A random sample of Denver neighborhoods was obtained.

The data (X1, X2, X3, X4, X5, X6, X7) are for each neighborhood
X1 = total population (in thousands)
X2 = percentage change in population over past several years
X3 = percentage of children (under 18) in population
X4 = percentage free school lunch participation
X5 = percentage change in household income over past several years
X6 = crime rate (per 1000 population)
X7 = percentage change in crime rate over past several years
Reference: The Piton Foundation, Denver, Colorado

X1	X2	X3	X4	X5	X6	X7
6.9	1.8	30.2	58.3	27.3	84.9	-14.2

8.4	28.5	38.8	87.5	39.8	172.6	-34.1
5.7	7.8	31.7	83.5	26.0	154.2	-15.8
7.4	2.3	24.2	14.2	29.4	35.2	-13.9
8.5	-0.7	28.1	46.7	26.6	69.2	-13.9
13.8	7.2	10.4	57.9	26.2	111.0	-22.6
1.7	32.2	7.5	73.8	50.5	704.1	-40.9
3.6	7.4	30.0	61.3	26.4	69.9	4.0
8.2	10.2	12.1	41.0	11.7	65.4	-32.5
5.0	10.5	13.6	17.4	14.7	132.1	-8.1
2.1	0.3	18.3	34.4	24.2	179.9	12.3
4.2	8.1	21.3	64.9	21.7	139.9	-35.0
3.9	2.0	33.1	82.0	26.3	108.7	-2.0
4.1	10.8	38.3	83.3	32.6	123.2	-2.2
4.2	1.9	36.9	61.8	21.6	104.7	-14.2
9.4	-1.5	22.4	22.2	33.5	61.5	-32.7
3.6	-0.3	19.6	8.6	27.0	68.2	-13.4
7.6	5.5	29.1	62.8	32.2	96.9	-8.7
8.5	4.8	32.8	86.2	16.0	258.0	0.5
7.5	2.3	26.5	18.7	23.7	32.0	-0.6
4.1	17.3	41.5	78.6	23.5	127.0	-12.5
4.6	68.6	39.0	14.6	38.2	27.1	45.4
7.2	3.0	20.2	41.4	27.6	70.7	-38.2
13.4	7.1	20.4	13.9	22.5	38.3	-33.6
10.3	1.4	29.8	43.7	29.4	54.0	-10.0
9.4	4.6	36.0	78.2	29.9	101.5	-14.6
2.5	-3.3	37.6	88.5	27.5	185.9	-7.6
10.3	-0.5	31.8	57.2	27.2	61.2	-17.6
7.5	22.3	28.6	5.7	31.3	38.6	27.2
18.7	6.2	39.7	55.8	28.7	52.6	-2.9
5.1	-2.0	23.8	29.0	29.3	62.6	-10.3
3.7	19.6	12.3	77.3	32.0	207.7	-45.6
10.3	3.0	31.1	51.7	26.2	42.4	-31.9
7.3	19.2	32.9	68.1	25.2	105.2	-35.7
4.2	7.0	22.1	41.2	21.4	68.6	-8.8
2.1	5.4	27.1	60.0	23.5	157.3	6.2
2.5	2.8	20.3	29.8	24.1	58.5	-27.5
8.1	8.5	30.0	66.4	26.0	63.1	-37.4
10.3	-1.9	15.9	39.9	38.5	86.4	-13.5
10.5	2.8	36.4	72.3	26.0	77.5	-21.6
5.8	2.0	24.2	19.5	28.3	63.5	2.2
6.9	2.9	20.7	6.6	25.8	68.9	-2.4
9.3	4.9	34.9	82.4	18.4	102.8	-12.0
11.4	2.6	38.7	78.2	18.4	86.6	-12.8

File names   Excel: Mlr10.xls
                   Minitab: Mlr10.mtp
                   SPSS: Mlr10.sav
                   TI-83 Plus and TI-84 Plus/ASCII:
                        X1 data is stored in Mlr10L1.txt
                        X2 data is stored in Mlr10L2.txt
                        X3 data is stored in Mlr10L3.txt
                        X4 data is stored in Mlr10L4.txt
                        X5 data is stored in Mlr10L5.txt
                        X6 data is stored in Mlr10L6.txt
                        X7 data is stored in Mlr10L7.txt

11. **Chapter 10 Using Technology: U.S. Economy Case Study (Multiple Linear Regression)**
    U.S. economic data 1976 to 1987
    X1 = dollars/barrel crude oil
    X2 = % interest on ten yr. U.S. treasury notes
    X3 = foreign investments/billions of dollars
    X4 = Dow Jones industrial average
    X5 = gross national product/billions of dollars
    X6 = purchasing power u.s. dollar (1983 base)
    X7 = consumer debt/billions of dollars
    Reference: *Statistical Abstract of the United States* 103rd and 109th edition

X1	X2	X3	X4	X5	X6	X7
10.90	7.61	31.00	974.90	1718.00	1.76	234.40
12.00	7.42	35.00	894.60	1918.00	1.65	263.80
12.50	8.41	42.00	820.20	2164.00	1.53	308.30
17.70	9.44	54.00	844.40	2418.00	1.38	347.50
28.10	11.46	83.00	891.40	2732.00	1.22	349.40
35.60	13.91	109.00	932.90	3053.00	1.10	366.60
31.80	13.00	125.00	884.40	3166.00	1.03	381.10
29.00	11.11	137.00	1190.30	3406.00	1.00	430.40
28.60	12.44	165.00	1178.50	3772.00	0.96	511.80
26.80	10.62	185.00	1328.20	4015.00	0.93	592.40
14.60	7.68	209.00	1792.80	4240.00	0.91	646.10
17.90	8.38	244.00	2276.00	4527.00	0.88	685.50

File names   Excel: Mlr11.xls
         Minitab: Mlr11.mtp
         SPSS: Mlr11.sav
         TI-83 Plus and TI-84 Plus/ASCII:
              X1 data is stored in Mlr11L1.txt
              X2 data is stored in Mlr11L2.txt
              X3 data is stored in Mlr113.txt
              X4 data is stored in Mlr114.txt
              X5 data is stored in Mlr115.txt
              X6 data is stored in Mlr116.txt
              X7 data is stored in Mlr117.txt

One-Way ANOVA
File name prefix: Owan followed by the number of the data file

01. Excavation Depth and Archaeology (One-Way ANOVA)
Four different excavation sites at an archeological area in New Mexico gave the following depths (cm) for significant archaeological discoveries.
$X1$ = depths at Site I
$X2$ = depths at Site II
$X3$ = depths at Site III
$X4$ = depths at Site IV
Reference: *Mimbres Mogollon Archaeology* by Woosley and McIntyre, Univ. of New Mexico
        Press

X1	X2	X3	X4
93	85	100	96
120	45	75	58
65	80	65	95
105	28	40	90
115	75	73	65
82	70	65	80
99	65	50	85
87	55	30	95
100	50	45	82
90	40	50	
78	45		
95	55		
93			
88			
110			

File names   Excel: Owan01.xls
                          Minitab: Owan01.mtp
                          SPSS: Owan01.sav
                          TI-83 Plus and TI-84 Plus/ASCII:
                                    X1 data is stored in Owan01L1.txt
                                    X2 data is stored in Owan01L2.txt
                                    X3 data is stored in Owan01L3.txt
                                    X4 data is stored in Owan01L4.txt

02. Apple Orchard Experiment (One-Way ANOVA)

Five types of root-stock were used in an apple orchard grafting experiment. The following data represent the extension growth (cm) after four years.

X1 = extension growth for type I
X2 = extension growth for type II
X3 = extension growth for type III
X4 = extension growth for type IV
X5 = extension growth for type V
Reference: S.C. Pearce, University of Kent at Canterbury, England

X1	X2	X3	X4	X5
2569	2074	2505	2838	1532
2928	2885	2315	2351	2552
2865	3378	2667	3001	3083
3844	3906	2390	2439	2330
3027	2782	3021	2199	2079
2336	3018	3085	3318	3366
3211	3383	3308	3601	2416
3037	3447	3231	3291	3100

File names   Excel: Owan02.xls
Minitab: Owan02.mtp
SPSS: Owan02.sav
TI-83 Plus and TI-84 Plus/ASCII:

        X1 data is stored in Owan02L1.txt
        X2 data is stored in Owan02L2.txt
        X3 data is stored in Owan02L3.txt
        X4 data is stored in Owan02L4.txt
        X5 data is stored in Owan02L5.txt

03. Red Dye Number 40 (One-Way ANOVA)
S.W. Laagakos and F. Mosteller of Harvard University fed mice different doses of red dye number 40 and recorded the time of death in weeks. Results for female mice, dosage and time of death are shown in the data
$X1$ = time of death for control group
$X2$ = time of death for group with low dosage
$X3$ = time of death for group with medium dosage
$X4$ = time of death for group with high dosage
Reference: *Journal Natl. Cancer Inst.*, Vol. 66, p 197-212

X1	X2	X3	X4
70	49	30	34
77	60	37	36
83	63	56	48
87	67	65	48
92	70	76	65
93	74	83	91
100	77	87	98
102	80	90	102
102	89	94	
103		97	
96			

File names

Excel: Owan03.xls
Minitab: Owan03.mtp
SPSS: Owan03.sav
TI-83 Plus and TI-84 Plus/ASCII:
    X1 data is stored in Owan03L1.txt
    X2 data is stored in Owan03L2.txt
    X3 data is stored in Owan03L3.txt
    X4 data is stored in Owan03L4.txt

04. **Business Startup Costs (One-Way ANOVA)**
The following data represent business startup costs (thousands of dollars) for shops.
X1 = startup costs for pizza
X2 = startup costs for baker/donuts
X3 = startup costs for shoe stores
X4 = startup costs for gift shops
X5 = startup costs for pet stores
Reference: *Business Opportunities Handbook*

X1	X2	X3	X4	X5
80	150	48	100	25
125	40	35	96	80
35	120	95	35	30
58	75	45	99	35
110	160	75	75	30
140	60	115	150	28
97	45	42	45	20
50	100	78	100	75
65	86	65	120	48
79	87	125	50	20
35	90			50
85				75
120				55
				60
				85
				110

File names   Excel: Owan04.xls
Minitab: Owan04.mtp
SPSS: Owan04.sav
TI-83 Plus and TI-84 Plus/ASCII:
      X1 data is stored in Owan04L1.txt
      X2 data is stored in Owan04L2.txt
      X3 data is stored in Owan04L3.txt
      X4 data is stored in Owan04L4.txt
      X5 data is stored in Owan04L5.txt

05.   Weights of Football Players (One-Way ANOVA)

The following data represent weights (pounds) of a random sample of professional football players on the following teams.

X1 = weights of players for the Dallas Cowboys
X2 = weights of players for the Green Bay Packers
X3 = weights of players for the Denver Broncos
X4 = weights of players for the Miami Dolphins
X5 = weights of players for the San Francisco Forty Niners
Reference: *The Sports Encyclopedia Pro Football*

X1	X2	X3	X4	X5
250	260	270	260	247
255	271	250	255	249
255	258	281	265	255
264	263	273	257	247
250	267	257	268	244
265	254	264	263	245
245	255	233	247	249
252	250	254	253	260
266	248	268	251	217
246	240	252	252	208
251	254	256	266	228
263	275	265	264	253
248	270	252	210	249
228	225	256	236	223
221	222	235	225	221
223	230	216	230	228
220	225	241	232	271

File names   Excel: Owan05.xls
                    Minitab: Owan05.mtp
                    SPSS: Owan05.sav
                    TI-83 Plus and TI-84 Plus/ASCII:
                              X1 data is stored in Owan05L1.txt
                              X2 data is stored in Owan05L2.txt
                              X3 data is stored in Owan05L3.txt
                              X4 data is stored in Owan05L4.txt
                              X5 data is stored in Owan05L5.txt

Two-Way ANOVA
File name prefix: Twan followed by the number of the data file

01. Political Affiliation (Two-Way ANOVA)
Response: Percent of voters for a recent National Election
Factor 1: counties in Montana
Factor 2: political affiliation
Reference: *County and City Data Book* U.S. Dept. of Commerce

County	Democrat	Republican
Jefferson	33.5	36.5
Lewis/Clark	42.5	35.7
Powder River	22.3	47.3
Stillwataer	32.4	38.2
Sweet Grass	21.9	48.8
Yellowstone	35.7	40.4

File names

Excel: Twan01.xls
Minitab: Twan01.mtp
SPSS: Twan01.sav
TI-83 Plus and TI-84 Plus/ASCII: Twan01.txt

02. Density of Artifacts (Two-Way ANOVA)
Response: Average density of artifacts, number of artifacts per cubic meter
Factor 1: archeological excavation site
Factor 2: depth (cm) at which artifacts are found
Reference: Museum of New Mexico, Laboratory of Antrhopology

Site	50-100	101-150	151-200
I	3.8	4.9	3.4
II	4.1	4.1	2.7
III	2.9	3.8	4.4
IV	3.5	3.3	3
V	5.2	5.1	5.3
VI	3.6	4.6	4.5
VII	4.5	3.7	2.8

File names

Excel: Twan02.xls
Minitab: Twan02.mtp
SPSS: Twan02.sav
TI-83 Plus and TI-84 Plus/ASCII: Twan02.txt

03. Spruce Moth Traps (Two-Way ANOVA)

Response: number of spruce moths found in trap after 48 hours
Factor 1: Location of trap in tree (top branches, middle branches, lower branches, ground)
Factor 2: Type of lure in trap (scent, sugar, chemical)

Location	Scent	Sugar	Chemical
Top	28	35	32
	19	22	29
	32	33	16
	15	21	18
	13	17	20
Middle	39	36	37
	12	38	40
	42	44	18
	25	27	28
	21	22	36
Lower	44	42	35
	21	17	39
	38	31	41
	32	29	31
	29	37	34
Ground	17	18	22
	12	27	25
	23	15	14
	19	29	16
	14	16	19

File names   Excel: Twan03.xls
                             Minitab: Twan03.mtp
                             SPSS: Twan03.sav
                             TI-83 Plus and TI-84 Plus/ASCII: Twan03.txt

04. **Advertising in Local Newspapers (Two-Way ANOVA)**

Response: Number of inquiries resulting from advertisement
Factor 1: day of week (Monday through Friday)
Factor 2: section of newspaper (news, business, sports)

Day	News	Business	Sports
Monday	11	10	4
	8	12	3
	6	13	5
	8	11	6
Tuesday	9	7	5
	10	8	8
	10	11	6
	12	9	7
Wednesday	8	7	5
	9	8	9
	9	10	7
	11	9	6
Thrusday	4	9	7
	5	6	6
	3	8	6
	5	8	5
Friday	13	10	12
	12	9	10
	11	9	11
	14	8	12

File names

Excel: Twan04.xls
Minitab: Twan04.mtp
SPSS: Twan04.sav
TI-83 Plus and TI-84 Plus/ASCII: Twan04.txt

05. Prehistoric Ceramic Sherds (Two-Way ANOVA)
      Response: number of sherds
      Factor 1: region of archaeological excavation
      Factor 2: type of ceramic sherd (three circle red on white, Mogollon red on brown, Mimbres corrugated,
            bold face black on white)
      Reference: *Mimbres Mogollon Archaeology* by Woosley and McIntyre, University of New
            Mexico Press

Region	Red on White	Mogollon	Mimbres	Bold Face
I	68	49	78	95
	33	61	53	122
	45	52	35	133
II	59	71	54	78
	43	41	51	98
	37	63	69	89
III	54	67	44	41
	91	46	76	29
	81	51	55	63
IV	55	45	78	56
	53	58	49	81
	42	72	46	35
V	27	47	41	46
	31	39	36	22
	38	53	25	26

File names   Excel: Twan05.xls
                         Minitab: Twan05.mtp
                         SPSS: Twan05.sav
                         TI-83 Plus and TI-84 Plus/ASCII: Twan05.txt